THE NEURO-CONSUMER

Neuroscientific research shows that the great majority of purchase decisions are irrational and driven by subconscious mechanisms in our brains. This is hugely disruptive to the rational, logical arguments of traditional communication and marketing practices and we are just starting to understand how organizations must adapt their strategies. This book explains the subconscious behavior of the "neuro-consumer" and shows how major international companies are using these findings to cast light on their own consumers' behavior.

Written in plain English for business and management readers with no scientific background, it focuses on:

- how to adapt marketing and communication to the subconscious and irrational behaviors of consumers;
- the direct influence of the primary senses (sight, hearing, smell, taste, touch) on purchasing decisions and the perception of communications by customers' brains;
- implications for innovation, packaging, price, retail environments and advertising;
- the use of "nudges" and artifices to increase marketing and communication efficiency by making them neuro-compatible with the brain's subconscious expectations;
- the influence of social media and communities on consumers' decisions – when collective conscience is gradually replacing individual conscience and recommendation becomes more important than communication; and
- the ethical limits and considerations that organizations must heed when following these principles.

Authored by two globally recognized leaders in business and neuroscience, this book is an essential companion to marketers and brand strategists interested in neuroscience and vital reading for any advanced student or researcher in this area.

Anne-Sophie Bayle-Tourtoulou is an Associate Professor of Marketing at HEC Paris where she holds academic responsibilities at the graduate and executive levels. Her interests deal with marketing strategy, retail and e-tail marketing, neuroscience and artificial intelligence. She is the co-author of *Neuromarketing in Action*.

Michel Badoc is an Emeritus Professor at HEC Paris. He gives courses on neuroscience, marketing and sensory marketing. He develops consultancy activities for companies in Europe and North America. He is the co-author of *Neuromarketing in Action* and many books on banking and insurance marketing.

The Neuro-Consumer

Adapting Marketing and Communication Strategies for the Subconscious, Instinctive and Irrational Consumer's Brain

ANNE-SOPHIE BAYLE-TOURTOULOU
AND
MICHEL BADOC

LONDON AND NEW YORK

First published 2020
by Routledge
2 Park Square, Milton Park, Abingdon, Oxon OX14 4RN

and by Routledge
52 Vanderbilt Avenue, New York, NY 10017

Routledge is an imprint of the Taylor & Francis Group, an informa business

British Library Cataloguing-in-Publication Data
A catalogue record for this book is available from the British Library

Library of Congress Cataloging-in-Publication Data
A catalog record has been requested for this book

ISBN: 978-0-367-89590-7 (hbk)
ISBN: 978-1-003-01997-8 (ebk)

Typeset in Sabon
by Swales & Willis, Exeter, Devon, UK

CONTENTS

Acknowledgments ix

Introduction 1

PART I THE ADVENT OF THE NEURO-CONSUMER 7

Introduction 7

1 **From philosophical theory to neuroscience 9**

2 **Consumer behavior studied by marketing research 14**

3 **Limits of marketing studies and the advent of the "neuro-consumer" 18**

4 **The brain: underestimated by science for centuries 23**

5 **Neuromarketing and the neuro-consumer: two 21st-century inventions 29**

6 **Neuroscientific techniques used to understand the neuro-consumer's behavior 39**

Points to remember 53

PART II UNDERSTANDING THE NEURO-CONSUMER'S BRAIN 55

Introduction 55

7 The neuro-consumer's brain: a complex organ 57

8 Is the brain free or programmed? 64

9 How age and gender condition the brain 78

10 How memory conditions the brain 92

11 The influence of emotions and desires 99

Points to remember 112

PART III THE NEURO-CONSUMER'S BRAIN INFLUENCED BY THE SENSES 113

Introduction 113

12 The neuro-consumer and the sense of sight 115

13 The neuro-consumer and the sense of hearing 125

14 The neuro-consumer and the sense of smell 133

15 The neuro-consumer and the sense of touch 144

16 The neuro-consumer and the sense of taste 156

Points to remember 164

PART IV THE NEURO-CONSUMER'S BRAIN INFLUENCED BY INNOVATION 167

Introduction 167

17 The influence of innovation, design and packaging 169

18 The influence of the price of products and services 180

19 The influence of experiential and sensory marketing 188

20 The salesman faced with the neuro-consumer 205

Points to remember 211

PART V THE NEURO-CONSUMER'S BRAIN INFLUENCED BY COMMUNICATION 213

Introduction 213

21 Advertising language 215

22 Subliminal influences of communication on the brain and the nudge concept 231

23 The subliminal influence of brands 250

Points to remember 269

PART VI THE NEURO-CONSUMER'S BRAIN INFLUENCED BY THE DIGITAL REVOLUTION 271

Introduction 271

24 When the digital revolution changes the brain's perception 272

25 The brain and the emergence of social networks 288

Points to remember 302

Conclusion: a vision of the future 304

Selected bibliography 307

Index 310

ACKNOWLEDGMENTS

The authors would like to thank HEC Paris and the Ile de France Chamber of Commerce and Industry for their encouragement and support in writing this book.

They would also like to thank all the professionals who answered their questions and shared their knowledge and expertise.

Finally, they would like to thank their students and executive participants for their genuine interest in the field and their constructive interactions during class sessions.

INTRODUCTION

"Understanding the human brain in biological terms has emerged as science's main challenge for the 21st century."[1] This quote from Austrian neuropsychiatrist Eric Richard Kandel, awarded the Nobel Prize in Medicine in 2000, struck us not only on a personal level, but also in our professional capacity as marketing and communication professors. We asked ourselves whether it applies to the two subjects that we teach and research. To try and answer this question, over the last few years we have devoted some of our time to meeting – during our lectures and at conferences, seminars and in individual meetings all over the world – a large number of company directors, marketing and communication managers, neuroscientists and neuromarketing experts, etc. These many encounters have showed us how new ways of thinking and changes in practical applications are appearing.

In English-speaking countries – particularly the United States and Britain – and in Asia, the directors of big international groups are showing a real interest in applying neuroscientific research to the fields of marketing, sales and communication. On the other hand, however, those in southern countries, notably in France, are showing less interest.

On most continents, academic research in this field has seen quasi-exponential growth over the last few years. Using techniques drawn from neuroscience, consultants specializing in neuromarketing are becoming more numerous and meeting with real success. Most American companies use their services as are the advisers to certain American politicians and presidents.

These observations have encouraged us to try and better understand how the human brain functions and how it reacts when solicited by marketing, commerce and communication. In this, we have received much help from neurosurgery professor Patrick M. Georges. Our collaboration with him led to the joint publication of a first book, *Neuromarketing in Action*,[2] which has met with international success.

In this book, we look at a different aspect of the question.

We have been surprised to discover that most marketing and communication are still based on the hypothesis that consumers are rational, both in their purchasing decisions and in their reactions to advertising. Much of the research and many of the surveys carried out today are limited to asking them what they think.

Research carried out using neuroscientific techniques shows that in fact the opposite is true. Neuroscientists, as we will see throughout this book, have shown that a great majority of our actions are directly governed by our brains, outside the conscious mind. When we get up in the morning, brush our teeth, drive to work, etc., we perform most of these actions without thinking. Nevertheless, millions of neurons, controlled by the brain, work to carry them out correctly without us being aware of it.

The same is true when we buy something. North American neuromarketing experts, such as A.K. Pradeep and Martin Lindstrom, go as far as to say that 70% to 80% of purchases are irrational. Their observation has led them to wonder about the incongruity of marketing and communication, where large sums of money continue to be spent, based on the supposition that the customer is rational, when in fact this is only true to an extremely limited extent.

This striking observation has led us to become interested in the brain's modes of perception and subconscious behavior when solicited by marketing and communication.

After talking to many professionals and neuroscience experts and consulting numerous pieces of research, we decided to write a book on the subject that studies the perceptions and decisions, subconscious or unconscious, of the consumer's brain when solicited by marketing, commerce and communication. To simplify matters, in this situation we call them "neuro-consumers."

Attempts to understand the subconscious reactions of individuals presented with commercial and communication propositions go back well before the emergence of neuroscience.

In ancient Greece, after the Sophists, philosophers, at the instigation of Aristotle, were quick to understand the importance of appealing to the audience's subconscious to make the most of a speech. They invented rhetoric, a method designed to seduce the audience and that makes great use of irrational artifices aimed at giving the speaker's words credibility. During the following centuries, first in Rome then elsewhere in Europe, many thinkers highlighted the importance of the subconscious in human behavior and decision-making. Among them, Sigmund Freud had a major influence that still exists today through different schools of psychoanalysis.

In Part I of this book, we refer to those thinkers, whose ideas have been rediscovered by today's neuroscientists. The main works by the famous neuroscience professor Antonio Damasio refer to Spinoza and Descartes. We look at marketing's progress in understanding the consumer's behavior and describe its contributions. We underline the limitations of the main techniques, for the most part interview based, often used in this field. A discussion of their relevance, which too often neglects customers' subconscious perceptions, contributes to our thinking.

This leads us to seek neuroscience's contribution to knowledge of customers' irrational attitudes and decisions. From the beginning of the 21st century, marketing researchers in the United States, such as Read Montague and his colleagues, started using magnetic resonance imaging (MRI) to better understand the direct role played by the brain in the perception and influence of brands. The era of neuromarketing was born. Research was quickly followed by applications in companies to improve the effectiveness of marketing, sales and communication. These applications did not aim to replace traditional marketing, but to complement it by widening its methods of investigation.

To better understand the interest of this approach, we had to acquire deeper knowledge of the functioning of the brain. The way in which it works, considerably underestimated for centuries, remains relatively little known today. We look at the main techniques in use. They range from observation methods such as neuro-linguistic programming (NLP) and transactional analysis (TA) to more complex neuroscientific tools using MRI, electroencephalography (EEG) and other less elaborate procedures such as eye tracking, neuro-endocrinology, neuroscientific diagnosis, etc. Knowledge of these methods appears to be essential to a scientific understanding of the subconscious perceptions of the human brain. To consolidate the first part of this book, we present the results of neuroscientific research.

Part II deals with understanding the unconscious behavior of the neuro-consumer's brain. After studying the principles of how it works, our research deals with the endogenous and exogenous factors conditioning its behavior. Neuroscientists have found that certain factors, such as the subject's sex and age, can deeply modify the brain's perceptions and attitudes. Conditioned by genetics and structure, the male brain has very different behavioral predispositions to that of the female brain. Age also produces significant changes in the individual's perception of his environment and in his expectations.

The brain is also influenced by the "somatic markers" discovered by Antonio Damasio and his colleagues and the "mirror neurons" brought to light by Giacomo Rizzolatti, Corrado Sinigaglia and their teams.

At the beginning of this century, scientists and practitioners began to accord importance to emotional intelligence, as opposed to intellectual intelligence, measured by the well-known intelligence quotient (IQ).

The role played by emotions and desires, already revealed by psychoanalysts, has been confirmed by neuroscientific research. Some professionals go so far as to recommend that marketing and communication be improved by developing "emotional marketing" or "desire-based marketing." Having explained how the brain reacts unconsciously and what modifies its perceptions and its behavior, we go on to deal with the applications resulting from these observations.

Part III deals with how the senses can subconsciously influence the behavior of the neuro-consumer's brain. The senses have the characteristic of being able to short-circuit the conscious mind and directly access the internal zones of the brain. Sensory perception can alter reasoning. Numerous scientific experiments show that it is possible to create false impressions and inaccurate memories in the brain, to make it feel unreal emotions and to lead it to make spontaneous decisions devoid of any reason. In Part III, we look at the way in which the five main senses – sight, hearing, smell, taste and touch – reveal themselves to be capable of orientating behavior. "Sensory marketing" makes great use of the influence of the senses to increase sales, make products or services more attractive, help consumers to better remember a brand, etc. Part III includes many examples of concrete applications from retail chains.

Part IV analyzes the way in which the neuro-consumer's brain can be influenced by the cognitive ergonomics of products, pricing, distribution and sales. Good knowledge of the brain's perception plays an important role in making different aspects of marketing – innovation, new packaging, the pricing of a product or service – more effective. The creation of a sensory ambiance at a point of sale influences the number of people entering, the length of time they stay and, consequently, the volume of sales. Major producers of products and services frequently call on neuromarketing professionals and cognitive specialists to improve their creations and the attractiveness of the goods aimed at their customers. Big retail chains and independent shops often consult experts in sounds, smells, etc. to create elements corresponding to the sensory experience that they want their customers to have. Many companies now have a sensory strategy to complement their communication policy and Part IV includes some concrete examples.

Part V deals with the subliminal relationships that exist between the brain, communication and brands. The amount of money that companies allocate to these relationships demonstrates the extent to which they want to ensure their effectiveness. This subject is not new: As we have seen, orators have long sought to make their words more credible by using subliminal techniques, i.e., by reaching their listeners' unconscious brains. Some of these techniques – such as using messages that cannot be detected by the sense of sight but are directly registered by the brain – are forbidden in many countries. Subliminal effects in perfectly legitimate forms make messages more effective. In a book that has become a best-seller, Vance Packard denounces the growth of the "hidden persuaders" in Western civilization. In Part V, we reveal most of the artifices used in communication. There are various forms ranging from classic tricks – advertainment, art, rules of harmony such as the famous "golden number" and the proportions of the Vitruvian Man, eroticism – to more elaborate methods such as "nudging." Neuroscience can shed new light on this field. Recent research helps us to better understand the influence of certain communication tricks from writing, images, videos and pages on various screens. Neuro-compatible, they please the brain and make the messages aimed at it more agreeable and convincing. Reference to current scientific research illustrated by cases involving real-life applications provides concrete, operational content.

Part VI concerns changes to the brain resulting from the scale of the digital revolution, the emergence of communities and social networks and the interconnection of objects with each other and with human beings. Thanks to the Internet and mobile phones, it is now possible for the neuro-consumer to obtain a multitude of information and contacts, in real time, anywhere in the world and often at extremely low tariffs. The search for interaction with virtual friends, as well as with brands, has been found to profoundly change the behavior of an individual's brain, to the point of creating veritable addictions. Some observers go as far as claiming that the mobile phone is becoming an inseparable "comforter" for individuals connected by social networks and online communities. Their conscience becomes collective. They are gradually becoming "neuro-conso-actors" influenced by the communities that they belong to. Companies' marketing and communication need to adapt quickly to changes in the brain's perceptions.

Part VI tries to understand and explain the changes induced in the brain by the profound transformations caused by the advent of a digital and community-based environment. It looks at how brands have to deal with a radical change in both the individual and collective conscience. It reflects on how the brains of "neuro-Internet-users" who belong to social networks and communities feel the need for relational interaction.

To deal with this, the brands use new techniques arising from "big data" and "social customer relationship management" (CRM). They employ people from new professions, such as web master, community manager, social network officer, etc. Part VI explores a vision of what the experts in this field recommend and offers an analysis of concrete experience. Changes in marketing and communication have become essential. These two disciplines can only benefit from making use of the new opportunities arising from neuroscientific research.

This book in no way aims to indulge in proselytism about the use of neuroscience in marketing, sales and communication and even less to provide an apology for neuromarketing. It confines itself to analyzing the many pieces of research and concrete applications that have already come to fruition and to questioning the interest that they have for companies and consumers. It does not neglect the ethical aspects that are essential to consider when these techniques are used and that are also the subject of much debate.

Nor does the book aim to question the interest of traditional marketing and communication practices. It simply tries to improve them by suggesting new approaches, now possible, which call on neuroscientific knowledge. Through better knowledge of the behavior of the consumer's brain, particularly its subconscious and irrational perceptions – for a long time underestimated by research and studies in these disciplines – we hope to contribute to perfecting these new approaches in the combined interest of consumers and brands.

The book is aimed at companies that want to make themselves more attractive to consumers. It is also aimed at customers who no longer wish to be caught out by communication or sales tricks, by improving their knowledge of their brain's subliminal perception and subconscious reactions, and those who are concerned about the exponential growth of the digital world and social networks, particularly when these suggest the possibility of a gradual transformation of people's individual conscience into a collective conscience.

It is equally aimed at professionals in research, sales and distribution, publicity, design and communication in all its forms making use of the Internet and mobile devices and to those who are setting up digital marketing projects. It is likely to interest senior managers and board members by providing them with a new vision of the behavior of their company's greatest asset: its customers. It will contribute to informing lecturers and students, including the in-service training of professionals in the above-mentioned fields. We hope that it will contribute to widening the perspectives for research and course content.

Notes

1. Eric Richard Kandel, *In Search of Memory: The Emergence of New Science of Mind*, W.W. Norton & Company, 2000.
2. Patrick Georges, Anne-Sophie Bayle-Tourtoulou and Michel Badoc, *Neuromarketing in Action: How to Talk and Sell to the Brain*, Kogan Page, 2013.

PART I

The advent of the neuro-consumer

Man's "subconscious, instinctive, irrational" behavior has long remained an enigma for philosophers, psychologists and sociologists, as it has for those involved in marketing. By using neuroscientific techniques that enable us to visualize and observe how the brain functions, it has become possible to shed light on the mystery. Studying the "neuro-consumer" makes an essential contribution to understanding consumer behavior.

Introduction

A significant number of works have been devoted to studying consumer behavior. Their contributions allow companies to improve their marketing efforts thanks to a better understanding of their clients. While their quality is not in question, these studies have multiple limitations, which often result from their research and analysis methodology. New techniques in the field of neuroscience are now available to enrich consumer behavior knowledge. These techniques are not intended to replace the research that has been conducted through other methods but rather to complement them. The use of new technologies such as magnetic resonance imaging (MRI) or electroencephalography (EEG), which allow researchers to observe the brain's reaction to marketing and communication stimuli, offers greater depth to the information that is collected through traditional studies. Thanks to neuroscience, it is possible to confirm or invalidate observations made by philosophers or marketing researchers but also to explain and complement them. Direct observation of the consumer's brain reveals a lot of surprises.

In these early days of the 21st century, scientists from a variety of different fields are showing growing interest in how the brain works. Having long been underestimated over the last centuries, knowledge of the human brain has now become an increasingly popular field of research.

After psychologists and sociologists, marketing researchers now feel the need to understand the instinctive and most often subconscious behavior of the neuro-consumer. This trend is a relatively recent one. The first studies led by Read Montague and his American colleagues date back to the turn of this century. By looking into the evolution of neuroscientific knowledge, using techniques stemming from neuroscience, researchers are discovering new horizons enabling a better insight into consumer behavior, when buying processes and perceptions are under the direct influence of the brain's automatisms.

The efficiency of these new techniques raises questions related to the possible intrusion into individual freedom as well as manipulation. It is essential to build a set of ethical and deontological rules as safeguards against potential abuse.

CHAPTER 1

From philosophical theory to neuroscience

Neuroscientists express genuine interest in the various theories of thinkers and philosophers on human behavior. Some of these theories enable neuroscientists to build hypotheses linked to the behavior of the human brain when submitted to environmental stimuli. They are then verified with the aid of neuroscientific techniques.

World-renowned neurologist Antonio Damasio draws inspiration from philosophers such as Descartes or Spinoza to devise hypotheses relative to his neuroscientific experiences. American philosopher Patricia Churchland highlights the strong links between philosophy and neuroscience.

The pursuit of happiness and consumer behavior

Neurologists often consider that one of our brain's main functions is to contribute to our happiness by ensuring that body and mind work in a balanced and harmonized manner thanks to the hormonal system. This process, called homeostasis, affects the behavior of individuals in consumer situations.

The pursuit of happiness is one of the fundamental concerns of many philosophical schools of thought and we start this book by referring to some of their ideas that could potentially orient consumer behavior. The book by Frédéric Lenoir, *Happiness: A Philosopher's Guide*,[1] as well as other works by various philosophers, helped us summarize the ideas that are interesting for 21st-century neuroscientists.

A first dichotomy separates thinkers who believe that happiness cannot be achieved in the world we live in and those who believe the opposite. This divergence of beliefs triggers different – sometimes opposite – buying behaviors.

The former school of thought is strongly sustained by Judeo-Christian religions. It is supported by a large number of philosophers such as Socrates,

Pascal, Descartes and Kant. As happiness cannot be achieved in this world, it is essential to comply with moral rules that will enable to achieve it in another. Socrates, according to Plato, would talk about a "good life" on Earth, based on virtue and the values of the city, rather than a "happy life." Like Jesus, Socrates did not hesitate to sacrifice his own life in the name of a greater truth and high values, aspiring in this way to achieve genuine happiness after death. In his work, philosopher Immanuel Kant (1724–1804) summarizes this idea. For Kant, "full and complete happiness does not exist on earth: it is an ideal of imagination."[2] One can only hope to achieve ideal happiness and eternal bliss after death. It is a reward bestowed by God on those who have lived a just and moral life.

This philosophical approach is attached to achieving sainthood rather than wisdom. It can trigger behavioral responses among some categories of consumers whose brain is culturally influenced by these beliefs.

> *Those who support these ideas might sometimes embrace a somewhat ascetic behavior when they are facing enticements from marketing and communication. They exhibit a preference for what is natural and reject what they consider "overrated." This might lead certain people to reject consumer society and brands. Those who abide by this creed are attracted to unbranded products, "hard discount" or low-cost products. They choose the offers they deem compatible with their principles or morals (fair-trade or sustainably manufactured goods or goods offered by socially responsible corporations), products that are in harmony with nature and crafts that are closely related to it (natural or organic, strongly marked by local customs and traditions).*

On the contrary, many philosophical schools of thought believe that happiness can and should be found in our world. This, however, does not necessarily mean that there cannot also be a certain form of happiness after death, except in the case of nonbelievers.

Aristotle and Epicurus are renowned for advocating a lifestyle full of pleasure.

Alexander the Great's teacher, Aristotle (384–322 BC) left the academy of his master, Plato, to found his own school, the Lyceum, in Athens in 335 BC. In his work *Nicomachean Ethics*,[3] which extensively deals with the idea of happiness, he wrote that "there is no happiness without pleasure." For this philosopher, pleasure is an enjoyable feeling linked to the satisfaction of a need or desire of the body, but also of the mind. Pleasure is the main driver of our actions. He also advocates adopting a behavior that leads to "seeking the highest level of pleasure with the highest level of reason."

Epicurus (341–270 BC), whose name remains to this day associated with the notion of the pursuit of pleasure, also founded his own school in Greece, the Garden. In his teachings and main writings, *Letter to Menoeceus* and

Letters and Sayings,[4] he makes a distinction between three types of desires. The first type includes the natural, necessary desires (food, drink, clothes, accommodation). The second are natural, nonessential desires (luxury clothes, fine cuisine, comfortable dwelling). The third are desires he considers neither natural nor fundamental (power, honour, pomp). To achieve happiness in our world, he recommends behaving so as to satisfy the first type, to seek to satisfy the second but to avoid the third. His pursuit of happiness was moderate and does not correspond to the image of Epicureanism we have today, which is sometimes associated with debauchery, luxury and the quest for immoderate pleasure.

In the 16th century, Michel de Montaigne (1533–1592), whom Friedrich Nietzsche would later refer to as the first modern thinker, suggested in his *Essays*[5] to find happiness by creating "a joyful path of happiness, consistent with one's nature." He advocates loving life and enjoying the pleasures it offers in a balanced and flexible manner according to the needs of one's own nature. The behavior he recommends, which he also adopts, consists of being as happy as possible according to his own aspirations, by enjoying the day-to-day pleasures that life offers.

In *West–East Divan*,[6] Johann Wolfgang Von Goethe (1749–1832) suggests that "happiness consists of living according to one's nature, developing one's personality to be able to enjoy life and the world with heightened levels of sensitivity."

> *Consumers abiding by this type of philosophy are more likely to adopt a hedonistic, even "epicurean" lifestyle. They strive to enjoy the present moment and show an interest in the acquisition of consumer goods or pleasure, of luxury items.*

From philosophical theory to neuroscience

Neuroscientists often refer to philosophical concepts to test some hypotheses in order to better understand the fundamentals of brain behavior, through specific techniques.

Such is the case when trying, for instance, to explain the relationship between body and mind, the role of emotions, of memory, of desire, etc.

Neurologist Antonio Damasio, in two famous works,[7] questions Descartes' theories, which state that body and mind are separate and, on the contrary, supports Spinoza's theories, in which the philosopher highlights a deep interaction between these two components of human nature.

The writings of Arthur Schopenhauer (1788–1860),[8] which show the fundamental role of health on the aptitude of happiness or unhappiness ("a healthy beggar is happier than a sick king") interest the neuroscientists, who study the link between the body's well-being and individual behavior.

Antonio Damasio is also very attentive to *Ethics*,[9] the book of Baruch Spinoza, later known as Benedict de Spinoza (1632–1677). He recognizes the philosopher's contribution in highlighting the key role played by what Spinoza named "affects" – i.e., emotions – in the conditioning of human behavior. He strives to test these theories through neuroscientific analysis within the neurology department of the Iowa Institute in the United States.

The relations between behavior and memory have been the basis of many studies by neurologists.

In the *Philebus*,[10] Plato (428–348 BC) insists on the memory's role on happiness, as well as on the influence of the recall of bodily pleasures on behavior: "It is because I have kept in memory the intense pleasure I felt when drinking good wine that I am happy not only to remember it but also to taste it again." More recently, Proust (1871–1922) in *In Search of Lost Times*,[11] describes the interrelation between memory and some senses, such as the sense of smell. Through smell, an individual is able to relive in the present happy moments he experienced in the past. This phenomenon is the basis of neuroscientific and neuromarketing research in influencing consumers' senses, especially on the point of sale. Regarding memory, Antonio Damasio elaborated his "somatic marker" theory, which we will develop in a later chapter.

Desire, as a source of pleasure and driver of consumption, is also currently being studied by several neuroscientific laboratories in Europe as well as in the United States. Spinoza offered them something important to ponder when he said, "It is not because things are beautiful or good that I desire them; it is because I desire them that they are beautiful or good."[12]

Joseph Breuer (1842–1925) and Sigmund Freud (1856–1939), along with Carl Gustav Jung (1856–1939)[13] are considered as the precursors of the recognition of the role of the unconscious in explaining human attitudes and behaviors. By creating psychoanalysis, Sigmund Freud, followed by his disciples, showed how important it was to understand how the brain functions beyond consciousness. Like many other neuroscientists, David Eagleman – director of the Perception and Action laboratory at Baylor College of Medicine in Houston, Texas – pays tribute to them in his book *Incognito: The Secret Lives of the Brain*.[14]

Even from the second half of the 20th century, various thinkers endeavoured to establish close ties between philosophy and neuroscience. A dominant movement, combining different sciences and schools of thought, took root in North America under the name of "cognitive revolution."

Patricia Churchland, whose many books include *Braintrust: What Neuroscience Tells Us About Morality*,[15] became a precursor by creating the concept of "neurophilosophy." She recommends replacing the "armchair" philosophy with one that bears a closer connection to the progress offered by neuroscience.

In the mid-1980s, French neurobiologist Jean-Pierre Changeux, author of the famous *Neuronal Man*,[16] offered a series of recommendations to restructure the field of research in cognitive sciences, "sciences that aim to study the thought mechanisms – language, psychology, memory, thought, reasoning, behavior organisation." He recommends a necessary interdisciplinarity in the field. In 1990, the CNRS (France's national scientific research center) created a programme named "Cognisciences," headed by André Holley. Later, in 1998, the Institute of Cognitive Sciences was set up in Bron, on the outskirts of Lyon. In this institute, Marc Jeannerod (1935–2011), whose many books include *Motor Cognition: What Actions Tell to the Self*,[17] sought to understand the links between the mind and the brain through a tight relationship between biology and philosophy, based on the history of this relationship.

Notes

1. Frédéric Lenoir, *Happiness: A Philosopher's Guide*, Melville House, 2015.
2. Immanuel Kant, *Groundwork of the Metaphysic of Morals* [1785], Cambridge University Press, 2012.
3. Aristotle, *Nicomachean Ethics*, Hackett, 1999.
4. Epicurus, *Letter to Menoeceus*, CreateSpace Independent Publishing Platform, 2016 and *Letters and Sayings of Epicurus*, Barnes & Noble Books, 1990.
5. Michel de Montaigne, *The Complete Essays*, Penguin Classics, 1993.
6. Johann Wolfgang von Goethe, *West–East Divan*, Global Academic Publishing, 2010.
7. Antonio Damasio, *Descartes' Error: Emotion, Reason and the Human Brain*, Vintage, 2006 and *Looking For Spinoza: Joy, Sorrow and the Feeling Brain*, Vintage, 2004.
8. Arthur Schopenhauer, *On the Basis of Morality* [1840], Hackett, 1999.
9. Benedict de Spinoza, *Ethics* [1675], Penguin Classics, 1996.
10. Plato, *Philebus*, Oxford University Press, 1975.
11. Marcel Proust, *In Search of Lost Time* [1913–1926], Everyman, 2001.
12. Spinoza, *Ethics*, Penguin Classics Series, 2004.
13. Sigmund Freud and Joseph Breuer, *Studies on Hysteria* [1895], Read Books, 2013; Sigmund Freud, *Civilization and Its Discontents* [1930], W. W. Norton & Company, 2010; Carl Gustav Jung, *The Archetypes and the Collective Unconscious*, Princeton University Press, 1981 and *Psychological Types*, Routledge Classics, 2016.
14. David Eagleman, *Incognito: The Secret Lives of the Brain*, Vintage, 2012.
15. Patricia Churchland, *Braintrust: What Neuroscience Tells Us About Morality*, Princeton University Press, 2012.
16. Jean-Pierre Changeux, *Neuronal Man*, Princeton University Press, 1997.
17. Marc Jeannerod, *Motor Cognition: What Actions Tell to the Self*, Oxford University Press, 2006.

CHAPTER 2

Consumer behavior studied by marketing research

Finding the button that triggers consumers' purchases is the dream of marketing researchers. Since the creation of the discipline, the analysis of customer tastes, needs and expectations through various types of consumer behavior studies provides useful information to companies. There are several companies that specialize in conducting these studies in most countries. Some of them have gained international reputation, such as Nielsen, Kantar, Quintiles IMS, Ipsos, GFK and IRI. Professors in the marketing departments of leading universities and management schools are actively involved in trying to improve their knowledge of consumer behavior and write multiple papers on the subject in scientific journals. By improving their knowledge in this field, companies can better adapt what they offer to customer expectations. Many different techniques are used to do so. Beyond marketing, they tap into other disciplines such as psychology, psychoanalysis, sociology, sociometry and semiology.

There are several books[1] and publications focusing on consumer behavior research, which interested readers can consult for deeper knowledge of the topics covered in this chapter.

Understanding consumer behavior through interview-based studies

The first type of study is designed to answer the many questions marketing departments have regarding consumers, including: Why do they buy? When do they buy? Where do they buy? What do they buy? In their family, who buys? Who prescribes? Who influences? Who consumes? Outside the family, who influences them in their purchases? How did they find out about the offer in question? What do they like about the point of sale or the salesperson's visit? These studies want to know more about what forms the consumers' current or future needs, their tastes, interests, motivations or intentions. They try to answer the various concerns of the companies that try to meet the requirements of their customers through a deeper knowledge of their expectations. To help them come up with answers, researchers have developed a broad range of methodologies described in great detail in reference books.[2]

The main techniques that are used include:

- face-to-face interviews;
- mail, telephone or Internet-based interviews;
- "focus groups" or group interviews usually carried out by a psychologist;
- customer feedback, consisting of a free interview carried out at home or in the workplace.

Understanding consumer behavior through observation-based research

The interview-based research techniques relating to consumer behavior have certain limitations. This is the case when the interviewees are embarrassed by the interviewer's presence or when they cannot answer, for example when we would like to know the tastes of animals or babies. It is also the case when the interviewees have difficulty in answering some questions or when the answers fail to reflect their purchasing behavior.

> *At an experiment, wine connoisseurs were asked what, in their opinion, constitutes the quality of the product. Not surprisingly, one of the most popular answers in the interview was "the taste." However, when the same people were subject to other test types, particularly blind tests (wine tasting in unlabeled bottles of different shapes or in identical bottles with different labels), it became clear that the shape of the bottle and the presence of the label had greater influence on quality perception than taste.*

In order to partly tackle these issues, researchers complement traditional interview-based studies with observation-based studies. These include: the use of projective tests, semiology, sign, symbol or icon analysis, ethnology-based research, studies of brand preference and attractiveness, the study of consumer behavior observed in experimental or real points of sale through the collection of cash register data.

Over the last several years, the ability to stock, manage and analyze massive databases in large customer relationship management (CRM) systems has made observation techniques relevant again. The advent of what experts are calling "big data" has granted a new dimension to the evolution of observation-based research and their concrete and effective use in marketing, e-marketing (Internet-based marketing) and m-marketing (mobile-based marketing).

Understanding consumer behavior through client segmentation

Marketing research has long been used to identify market segments that consist of consumers with similar behaviors.

Segmenting a market means dividing the total potential market of a product or a service into a certain number of homogeneous subdivisions, to help a company better adapt its sales policy to each, or some, of these subdivisions.

> *In the case of food products, the baby market has a specific behavior, very different from that of the teenager, adult or senior markets. It is a homogeneous market segment.*

Segmentation consists of identifying present and future markets according to specific client groups who to some extent have similar tastes, expectations and attitudes. This knowledge is essential to detect the emergence of a new client base or one that differs from the "average client" as presented by statistical analysis and to draw conclusions in terms of the marketing, sales and communication policies.

The most frequently used segmentation criteria are divided into four main categories: demographic, geographic, social and economic; personality and lifestyle; behavioral; psychological attitude towards the product or service to be sold.

On top of traditional segmentation criteria (for individuals: age, gender, income or assets, habitat, marital status; for companies: turnover, number of employees, sector of activities, company status, age and training of manager), there are some constantly evolving behaviors that lead to more elaborate behavioral typologies.

In the United States, James F. Engel, Roger D. Blackwell and Paul W. Miniard in their book *Consumer Behavior*,[3] a classic in the field, look into ethnic, cultural, social and family segmentations that are likely to influence consumer behavior. In Europe, there are temporal segmentations, which consist of looking for moments in time when populations have similar behaviors whatever their socio-professional categories (inheritance, receiving a large amount of money, birth of a child or grandchild).

Marketing researchers are trying to understand the evolution of consumer behaviors overtime. To do so, they look at studies of lifestyles and socio-cultural trends, the first of which were attributed to sociologist Max Weber (1864–1920). These studies have been the subject of countless publications in both the United States and Europe.

Traditional research in consumer behavior is an ever-evolving discipline. It sheds better light on the many questions marketing departments may have about their customers. However, the current methods have certain limitations that can be overcome by the use of techniques offered by neuroscience.

Notes

1. Michael R. Solomon, *Consumer Behavior: Buying, Having, and Being*, Pearson Education, 2014; Robert East, Malcom Wright and Marc Vanhuele, *Consumer Behaviour: Applications in Marketing*, Sage, 2008; James F. Engel, Roger D. Blackwell and Paul W. Miniard, *Consumer Behavior*, Dryden Press-Harcourt, Brace, 1994.
2. Ibid.
3. Engel et al., *Consumer Behaviour*.

CHAPTER 3

Limits of marketing studies and the advent of the "neuro-consumer"

The advent of neuroscience from the second half of the 20th century and its use in marketing research roughly starting from the 21st century have made it possible to improve our understanding of consumer behavior. Researchers are now able to directly study the brain's reaction when confronted with environmental stimuli, by avoiding the interviewer's interference and offering a different vision of what conditions the way humans act. The integration of this new knowledge and neuroscientific techniques makes it possible to overcome certain bias. The use of neuroscience will lead to the evolution of marketing research in the behavior of consumers who become neuro-consumers, when it comes to investigating the unconscious modes of the brain functioning.

Traditional studies and their limits

The criticism concerning the limitations of traditional marketing studies is nothing new. David Ogilvy (1911–1999), whose books are considered essential reading in the advertising world, used to say in his conferences that marketing studies tend to be used "as a drunkard uses a lamppost: for support, not for illumination."[1]

More recently, Martin Lindstrom in his book *Buy.Ology: How Everything We Believe About What We Buy Is Wrong*,[2] which reflects the results of his experiments and has become a best-seller despite some criticism, insists on the insufficiencies of traditional consumer studies. He reflects on the fact that despite the billions of dollars spent each year in the United States on research into the discipline, the failure rate for new products remains close to 80% after the first three months following product launch. He wonders why despite the hundreds of billions of dollars invested every year in advertising and all the related research – these budgets have been increasing exponentially over the last 40 years – memorization rates for adverts have dropped steadily. Despite sky-high costs, memorization rates on television can be below 5%, even a few dozen of minutes right after watching the advert. This leads Lindstrom to a pretty stern conclusion: "What people say does not reflect their behavior."

Also driven by the desire to overcome the limitations of traditional marketing studies, Dr. A.K. Pradeep, a researcher and author of the book *The Buying Brain: Secrets for Selling to the Subconscious Mind*,[3] founded the company NeuroFocus, which was later acquired by Nielsen (world leader in the field of market studies).

Both of these researchers are trying to complement our knowledge of consumer behavior that is studied through traditional techniques by using new tools that stem from neuroscience.

> *Many international corporations help them to fund their research or use the services they provide. The companies mentioned in their books include many different names such as: Glaxo Smith Kline, Fremantle and Bertelsmann (Lindstrom); CBS, Microsoft, Google, PayPal and Citibank (Pradeep).*

There are various criticisms regarding the methodologies used in traditional studies. The main one concerns the presence of either an interviewer who could disrupt the interviewee's answers, or of observers who may project their own feelings in the interpretations of the results. This remains the case in spite of the many precautions taken to avoid these biases.

One type of cognitive bias stems from the context in which the research is generally conducted. They can come from answers that get distorted by a variety of factors including: social conformism (the desire to give answers that seem socially acceptable, fear of being assessed or judged by the interviewer); verbalization (difficulty in finding the right words or sentences to express a sensation, an emotion, a feeling); and context (refusal to answer or discomfort in front of the interviewer because the question was not understood correctly). These kinds of difficulties are particularly common when the consumer has to address delicate issues, such as sex, money, death, old age, fear, relationships or family life.

Consumers may also have trouble translating into speech or writing the cognitive process that leads them to a specific choice or to what they sense. In their book,[4] Agnès Giboreau and Laurence Body insist on this difficulty. The authors are convinced that it is difficult to transform the consumers' perceptions into words. Consumers have poor control of sensory terms. The same word might encompass different notions depending on who uses it. It is therefore difficult to find the relevant word for all the different consumers in a single sample. Perception may also vary between consumers and experts, who are familiar with expert words.

Other notions such as those of the beautiful and the ugly, the appreciation of innovative forms, are difficult to apprehend with the help of traditional marketing studies. Their knowledge can, however, be very useful to decision makers in a number of areas, such as choosing a product design or packaging.

> *The automotive industry is keen to uncover consumer perceptions about the design of a future vehicle that can only be put on the market a few years after its design. It's no coincidence that the world's largest automaker Toyota has announced its association with the Riken Brain Science Institute in Japan. The purpose of this association is to better understand the physiology of the brain as well as the processing of data. The Japanese manufacturer plans to use neuroscience to supplement the information obtained from traditional studies so far.*

What makes the study of the brain so attractive is that its reactions are highly independent from the observer. The brain does not lie, nor does it cheat. Some countries resort to direct answers from the brain observed through magnetic resonance imaging (MRI) as lie detector tests.

Traditional studies on consumer behavior can offer enough information to enlighten marketing departments and resolve a large number of their concerns. The use of tools emanating from neuroscience, now applied by neuromarketing, can offer complementary knowledge when it is deemed necessary. The goal is not to replace traditional studies but rather to complement them

The advent of the neuro-consumer

Neuro-consumers have been around since men and women started consuming. The existence of the neuro-consumer, however, was only revealed by the emergence of neuroscience in marketing research. The study of their behavior strives to understand what pushes them to make potentially unconscious purchase decisions. Philosopher Carl Gutav Jung wrote: "In each and every one of us there is another being we do not know."[5]

Neuroscientists insist on the fact that the vast majority of our actions and decisions are taken without the intervention of our conscience. When we brush our teeth, wash, shave and in most of our day-to-day activities, millions of neurons are automatically activated by our brain without us even noticing.

Martin Lindstrom laments that the traditional research into consumer behavior only looks at 15% of the purchase decisions that are conscious, while the remaining 85% are unconscious. For this expert – who carried out 2,081 MRI-based studies on volunteers from five countries: the United States, United Kingdom, Germany, Japan and China – most of our purchase decisions, which are often linked to emotions, are irrational: "A brand is nothing but emotion; you cannot see it, smell it or feel it."[6] It is only in retrospect that they are rationalized.

Martin Lindstrom's studies, whose methodological rigor is sometimes criticized, have caught the attention of many large corporation managers in the United States. Any study that questions the results of traditional research in consumer behavior is scrutinized by marketing and communication departments.

The conclusions of some of Lindstrom's experiments have led to some big surprises among professionals.

Such was the case when he claimed to have demonstrated the ineffectiveness of "negative messages," often used on cigarette packets.[7] While a large proportion of smokers will declare during an interview that these kinds of messages could encourage them to quit smoking, studies carried out under MRI seem to point the other way. Direct observation of what is going on inside a patient's brain when submitted to these messages shows that they trigger a positive reaction in the parts of the brain that are generally favorable to addiction. In other words, the use of these messages in anti-smoking campaigns has the opposite effect to what was anticipated: they make the viewer want to smoke.

Several of Lindstrom's findings are published in his books or in various American newspapers and journals. One of his articles, published in the *New York Times*,[8] deals with the high addiction risk that mobile phones present for consumers, who might even "fall in love" with their devices. The article led to a heated debate in the United States.

The number of players offering improved knowledge of consumer behavior through neuroscience has soared around the world. They are catching the eye of many marketing and communication professionals, who are increasingly inclined to call on their services. The study of the neuro-consumer's behavior is becoming reality and is beginning to offer operational applications for companies, beyond its research field.

The label of neuro-consumer is new in the marketing literature, while that of neuromarketing is increasingly being vulgarized. As far as we are concerned, we define neuro-consumers as "consumers who are greatly subject to the unconscious or subconscious automatisms of their brain, whose decisions and buying processes are mainly driven by their emotions and desires."

Notes

1. David Ogilvy, *Ogilvy on Advertising*, Crown Publishing, 1983.
2. Martin Lindstrom, *Buy.Ology: How Everything We Believe About What We Buy Is Wrong*, Random House Business, 2009.
3. A.K. Pradeep, *The Buying Brain: Secrets for Selling to the Subconscious Mind*, Wiley, 2010.

4. Agnès Giboreau and Laurence Body, *Le marketing sensoriel: De la Stratégie à la Mise en Œuvre*, Vuibert, 2007.
5. Carl Gustav Jung, *The Archetypes and the Collective Unconscious*, Princeton University Press, 1981.
6. Lindstrom, *Buy.Ology*.
7. Ibid.
8. Martin Lindstrom, "You Love Your Iphone Literally," *New-York Times*, September 30, 2011.

CHAPTER 4

The brain

Underestimated by science for centuries

While the interest in knowledge of the human brain has become extremely popular in the past few years, this was far from true in the millennia leading up to the 20th century. Our history has long minimized the importance of the brain as a center for thought, emotion and reflection.

For many of our ancestors, thought and feelings came from the heart, or even the liver, hence some expressions that are still used today: "He's got a good heart," when it would be more accurate to say, "He's got a good brain." In his book *A History of the Brain: From Stone Age Surgery to Modern Neuroscience*,[1] Andrew P. Wickens helps the reader gain deeper insight into the evolution of our knowledge of this organ over the centuries.

The brain: from antiquity to the Middle Ages

Although there have been cases of trepanning dating back to the earliest Western antiquity (3000 BC), in Africa or in Peru (2000 BC), the importance of the brain as a center for thought and decision-making has not been made clear.

One of the most ancient civilizations in the world, the Egyptian civilization, considered the heart as being the center of thought and of life. During the embalming process, priests would remove the mummy's brain without any form of consideration while other organs would be kept religiously in funerary urns.

In Greece, Alcmaeon of Croton (6th century BC) was the first Westerner to suggest that the brain might govern the human body, drawing an essential behavioral distinction between men and animals.

Later on, still in Greece, Aristotle (384–322 BC), one of the most renowned philosophers of his time, maintained the belief that the heart was the center for thought. He believed that one of the rare functions of the brain was to cool down the blood that had been overheated by the heart's emotional agitation. While modern knowledge goes against this theory, vocabulary has kept a trace of this with the expression "cold-blooded." Hippocrates

(460–379 BC), the great physician, probably influenced by Alcmaeon of Croton's theories, contradicted Aristotle by saying that the brain was the true center for feelings and intelligence. Greek philosophers Plato (428–347 BC) and Democritus (460–370 BC) both believed that the brain was the organ in charge of thought and senses. Unfortunately for science, Aristotle's vision would remain the most popular for the following centuries.

Another leading medical figure from antiquity, Claudius Galen (AD 130–201), a Greek physician living in Rome, carried out many dissections on monkeys (human dissections were forbidden in those days). By transposing his animal experiments onto humans, Galen studied the influence of nerves on muscular movements. His research into spinal cord trauma was used up until the 19th century. Galen's ideas would remain a strong base in the medical field throughout the Middle Ages, just as Aristotle's thoughts would endure for several centuries.

The Romans' legacy in the medical field, as far as knowledge of the brain is concerned, is barely worth mentioning.

The brain from the Middle Ages to modern times

Galen's research marked the beginning of a dark age for Western medical research. Under growing influence from Christian churches, the dissection of the human body as well as the study of the anatomy remained banned. It was believed that thoughts can only come from the soul, a direct emanation from God, and whose location inside the body remains somewhat a blur. The beliefs regarding its origin are for a large part dictated by the religious dogma that governed citizens' behaviors. In that period, people seemed more interested in what would become of their soul after death than trying to locate it within the human body.

In the 17th century, René Descartes (1596–1650) wrote in his *Treatise of Man* (1662, in Latin) about a duality between the body and the mind. Physical functions are ruled by "humors," while mental functions are governed by God. The pineal gland acts as a link between the two entities. This theory, which also proved erroneous, was widely accepted for decades after. Roughly half a century later, physician and philosopher Julien Onfray de La Métrie (1709–1751), eliminated the need to resort, like Descartes, to a link (the pineal gland), by replacing the soul with the brain, which was designed as a machine.

In those times, a handful of scientists started examining the physiology of the brain, with utmost discretion. In the 16th century, anatomist André Vésale (1515–1564) defied taboos by digging up corpses to dissect them. While he helped improve the knowledge of the brain, he was forced to remain discreet to avoid being condemned. Leonardo da Vinci (1452–1519) drew many anatomical sketches. He even went as far as to imagine that the soul might be located in the brain, and that this organ might play a central part in controlling the body.

In the United Kingdom, in the 17th century, physician anatomist Thomas Willis (1621–1675) also studied corpses, despite the prohibition. He discovered a part of the brain's vascular system, since named "circle of Willis." He is considered as the father of neurology, although some people believe this title should be awarded to Jean-Martin Charcot, two centuries later.

By making it legal to dissect corpses, the French Revolution helped the medical field make giant strides. Studies of the human brain benefitted from this opportunity.

Research was rife in the 19th century, some of which would only be concluded in the following century. Paul Broca (1824–1880) showed that the brain was compartmented, and that one of the areas, still called the "Broca area," is responsible for language. A few other conclusions this founder of the French anthropological society came to were perhaps not as pertinent. He weighed 432 male and female brains, and made the following conclusion in the 1861 *Anthropological Society Bulletin*: "The relative smallness of the female brain depends on both her physical and intellectual inferiority,"[2] which was refuted a century later. After his death, the brain of Albert Einstein, who was considered the most intelligent man in the world, was donated for medical research and weighed. To everyone's surprise, its weight was closer to that of a female brain than a male one.

Around the turn of the 20th century, research on the human brain really took off. The murderous wars – first the Civil War in the United States, then the First World War – with the staggering number of wounded and traumatized left in their wake, offered researchers a hugely fertile ground for experimentation in a variety of areas related to medicine, anatomy and psychology. Many of them left their mark on the field they specialized in, be it medicine, anthropology, biology, physics, philosophy, psychology or psychoanalysis. Experts will remember the most famous among them, which include: Franz Joseph Gall (1758–1828), Pierre Flourens (1794–1867), Théodore Schwann (1810–1882), Jean-Baptiste Lamarck (1744–1829), Charles Darwin (1809–1882), Ramon y Cajal (1852–1934), Jean-Martin Charcot (1824–1893), Franz Nissi (1860–1919), Camilo Golgi (1843–1926), Pierre Janet (1859–1947) and William James (1842–1912). Thanks to their work, knowledge of this often underestimated organ improved considerably.

By studying the cases of two patients (Phineas Gage and Elliot) with trauma in the same area of the brain, Professor Antonio Damasio demonstrates the essential role played by the brain in managing emotions.[3] He observed that the brain conditions a large part of our psychological and relational balance.

Modern times and the advent of neuroscience

The First and Second World Wars, which shed death and destruction during the first half of the 20th century, enabled us to further our knowledge of how the human brain works and of the important role it plays in human physical and psychological behavior.

This understanding led to a shift in mentalities. As stated by Serge Ginger: "In the years around 1914, it was not rare for a woman to see a doctor if she experienced pleasure during sexual intercourse."[4] Female orgasm was often considered in those days as a sign of hysteria, and even perversion.

In another field, during the Second World War, the US Army was the first to consider a soldier who "lost it" due to excessive pressure in combat as possibly suffering from extreme depression and that he should therefore be seen as a patient and offered treatment, rather than executed for cowardice.

The shift in mentalities gradually transformed views on "senility," "autism" and "madness."

As early as 1906, famous German neurologist and psychologist Dr. Alois Alzheimer (1864–1915) defended the idea that senility was not a natural degeneration but rather due to a disease affecting the brain, which since has carried his own name.

Austrian pediatrician Hans Asperger (1906–1980) demonstrated that autism was a disease. He demanded that patients be treated as such, and not labeled "mad." Let us not forget that in those days, methods used to treat the "mad" or "insane" were both traumatic and inappropriate. Autistic patients were given the "packing" treatment, which consisted of wrapping the patient in wet sheets and heating them. For other patients deemed "mad," treatment ranged from cold showers to electrical shocks, even lobotomy or leucotomy. Film director Milos Forman showed these methods in his film *One Flew Over the Cuckoo's Nest*, based on the famous novel by Ben Kesey. Portuguese neurologist Egas Moniz (1864–1955) was awarded the 1949 Nobel Prize in Medicine for his contribution to the invention of "frontal leucotomy," a partial lobotomy technique.

Thanks to these major strides in terms of knowledge, our understanding of the brain has improved faster in the Western world over the last 100 years than it had in the previous 6,000 years.

During the second half of the 20th century, knowledge of the human brain made significant progress. This was largely due to a large number of inventions in a variety of fields – pharmacy, chemistry, medicine, physics, radiology, information technology (IT), psychology – and above all to a close collaboration between these fields at the research level.

In pharmacology, as early as the 1950s, Professor Henri Laborit created Largactil, a psychotropic product aimed at the psyche. It offers spectacular improvements for the treatment of psychiatric ailments considered as madness and treated through traumatic methods. Other researchers such as Professor Gowlinsky improved the effectiveness of these drugs, leading to the creation of Temesta and Prozac.

The evolution of our knowledge of the human mind took giant steps forward with the development of schools of psychology, communication, psychiatry, psychoanalysis and, more recently, neuropsychology. Beyond Freud and his disciples, many authors such as Antonio Damasio, Jean Pierre Changeux, Read Montague, Martin Lindstrom, Robert Ornstein, Michael Gazzaniga, Arthur and Mitchell Bard, Serge Ginger, David Servan-Schreiber, Giacomo Rizzolati, among others, have contributed significantly to neuroscience and neuromarketing.

The middle of the 20th century saw the birth of two important landmarks in the representation of the human brain: cybernetics and functional imagery.

Cybernetics

Cybernetics considers the brain as equivalent to a computer and that the brain makes comparable logical operations to those performed by a machine. Theories strive to represent supposedly underlying logical circuits. Images of connected neurons look more like logical or electronic diagrams than histological drawings. Proponents of this discipline favor functional representations over physiognomic presentations. They would rather know what the system does rather than understand what it is.

Physiognomic representations are no longer fashionable, as they were all too often used by fascist and nationalist regimes in order to prove the superiority of some races over others.

The cybernetics approach was developed by authors such as John von Neumann (1903–1957) in his book *The Computer and the Brain*,[5] as well as by conferences of the Macy Foundation instigated by neurologist Warren McCulloch (1898–1969).

The Palo Alto school

As of the 1950s, an important school of thought and research was developed in the town of Palo Alto, California, under the name of the "Palo Alto School." It was founded by Gregory Bateson (1904–1980) and a group of colleagues from a variety of different backgrounds: psychologists, psycho-sociologists, IT engineers, communication specialists, anthropologists, sociologists and physicians. Gregory Bateson detailed his views in several books, including *Steps to an Ecology of Mind*.[6] Influenced by the work carried out in the field of cybernetics, the school aimed to construct a general science of the mechanics of the mind. The authors consider mental illness as a method of adaptation to a pathological structure of family relations. This theory led to an upheaval of psychiatric conceptions at the time and played a part in the emergence of family therapy. The practitioner no longer considers patients as an isolated being, but rather pays attention to their interactions with the environment. He seeks to understand how the system as a whole maintains the homeostasis, or the harmony balance of the whole group. From individual and linear, the explanation becomes systemic and circular. Beyond cybernetics, researchers from that school developed new psycho-sociological approaches such as the notion of interaction, anti-psychiatry and constructivism.[7]

Functional imagery

Since the 1990s, image has made a strong comeback in the field of research into the brain.

The most significant contribution was offered by magnetic resonance imagery (MRI), invented by several American researchers at the end of the 1970s, or more simply by electroencephalography (EEG), which was invented in the 1950s. Thanks to these new procedures, it is now possible to study the brain without opening the cranium. It is also possible to better understand its interrelations with the human body as well as the manner in which it responds to external stimuli.

Notes

1. Andrew P. Wickens, *A History of the Brain: From Stone Age Surgery to Modern Neuroscience*, Psychology Press, 2014.
2. Paul Broca, "Sur le volume et la forme du cerveau suivant les individus et suivant les races," *Bulletin de la Société d'anthropologie*, vol. 2 (1861), pp. 200–202.
3. Antonio Damasio, *Descartes' Error: Emotion, Reason and the Human Brain*, Vintage, 2006.
4. Serge Ginger, *Gestalt Therapy: The Art of Contact*, Routledge, 2007.
5. John von Neumann, *The Computer and the Brain* [1958] (3rd ed.), Yale University Press, 2012.
6. Gregory Bateson, *Steps to an Ecology of Mind*, University of Chicago Press, 2000.
7. Dominique Picard and Edmond Marc, *L'Ecole de Palo Alto: Un nouveau regard sur les relations humaines*, Retz, 1984.

CHAPTER 5

Neuromarketing and the neuro-consumer

Two 21st-century inventions

After applications in psychology, sociology and sexology, it appears logical to look at the use of neuroscience in marketing. "Neuromarketing" appeared at the beginning of the 2000s. Its name comes from the contraction of the terms neuroscience and marketing, as is the case in other fields such as "neuropsychology" or "neuropsychiatry." The origin of the name is often attributed to a team of American researchers from Baylor College of Medicine, in Texas, including Samuel McClure and Read Montague. It was coined following a scientific study of the brands Pepsi and Coca-Cola, using functional magnetic resonance imaging (fMRI). The experiment was published in 2004 by Read Montague and his colleagues in the scientific journal *Neuron*.

According to Chistophe Morin,[1] experiments linked to the application of cognitive science to marketing were already carried out a few years beforehand, by companies such as BrightHouse and Salesbrain. The term "neuromarketing" was allegedly already used by the company BrigthHouse.

By introducing neuroscience techniques into marketing, and initiating the concept of the neuro-consumer, new research and application perspectives emerged for the discipline. In 2015, Hilke Plassmann, Vinod Venkatraman, Scott Huettel and Carolyn Yoon offered a comprehensive and challenging overview of the first decade of consumer neuroscience research. They suggested concrete ways in which neuroscience could more effectively affect marketing theory and practice.[2]

The emergence of neuromarketing

In 2002, a team of American researchers were intrigued by the fact that in many blind tests, where brands were concealed, consumers tended to prefer the taste of Pepsi while they favored Coca-Cola when the brand was apparent. They decided to work on these two chemically very similar brands to investigate, through the use of fMRI, whether and how brand knowledge could influence consumers' sensory preferences. They studied 67 volunteers for whom they first determined their preference for one brand or the other, both by asking them and by subjecting them to blind taste tests. Then, they scanned their brains when they were giving them sips of one drink or the other preceded by either anonymous cues of flashes of light or images of a Coca-Cola or Pepsi can. When the brands were concealed, the same part of the brain was activated regardless of the product tested: it was the ventromedial prefrontal cortex, an "area of the brain strongly implicated in signaling basic appetitive aspects of reward"[3] and its relative activity depending on the beverage tested could predict people's preference. When the brand Pepsi was signaled, no change was noticed. On the contrary, the brand knowledge of Coca-Cola influenced preferences and activated other brain areas including the dorsolateral prefrontal cortex and the hippocampus, "both implicated in modifying behavior based on emotion and affect."[4] Memorized images and messages conveyed in Coca-Cola's communication appeared to be triggered and took precedence over the consumers' own judgment in the brain, altering their preferences. Researchers have brought neural evidence that the knowledge of very strong brands such as Coca-Cola affects sensory preferences and activates memory-related brain regions that recall cultural influences. This study based on fMRI has highlighted the effect of a very strong brand on the consumer's brain.

After the conclusions of this study were published, the use of neuroscience as a marketing aid gained popularity, first of all in the United States and then in the rest of the world.

More recently, an article published by Justin R. Garcia and Gad Saad[5] looked at the influence of price on neuron activity in the field of wine consumption. The experiment carried out on wine consumers under magnetic resonance imaging (MRI) revealed a dramatic increase in neuronal activity in the area of the brain dedicated to pleasure when they tasted wine presented as more expensive, while all wines were actually identical. The study shows that price recognition has a direct impact on the neuro-consumer's brain, at the expense of taste, as the high price creates an image of better quality for the product.

As of 2002, academic research started showing increasing interest in neuromarketing. Christophe Morin[6] pointed out that at the time, their numbers were relatively low. According to him, there are various reasons for that, including:

- personal ethics that create a barrier when searching the secrets of the brain;
- fear of being criticized by the scientific community or the media for the same reasons; and
- lack of training of marketing researchers in the field of neuroscience.

On top of these, we can mention the low level of cooperation between consumer behavior research developed in marketing and that carried out in different fields. Mentalities change rapidly. Today there are thousands of university research projects as well as publications in scientific journals across the world. The study of the neuro-consumer, even though the word might not appear explicitly, has become a subject of interest for marketing departments in leading international universities and management schools.

In the United States, the first book on the subject, *Selling to the Old Brain*, was written by Patrick Renvoisé and Christophe Morin in 2003.[7] It focuses mainly on how the "reptilian" brain works and the influence its understanding might have on improving sales.

The first colloquia were held in 2005 in the United States. A large number of research projects were undertaken, with financial backing from large corporations.

The largest research project was carried out by Martin Lindstrom between 2004 and 2007, and the results were published in his book *Buy.Ology*.[8] Some 2,081 volunteers were studied under MRI in five countries, with a budget of around US$7 million. The information gathered contradicted several recommendations emanating from traditional marketing studies. Apart from questioning the effectiveness of messages printed on cigarette packets dramatizing the effects of smoking and showing that sex in advertising does not necessarily sell, Martin Lindstrom's research, following Read Montague's, sought to highlight the strong influence that brands have on the brain.

Other researchers looked at its role in choices and decisions made by the neuro-consumer. Among others, we can mention the experiments carried out by Brian Knutson at Stanford University in California, which focused on purchase predictions, by analyzing the brain's reactions observed by means of MRI. The work carried out by his team and published in 2007[9] is extremely popular with marketing departments in many companies in the United States.

On the other side of the Atlantic, researchers were also keen to use tools and knowledge emanating from neuroscience to better understand neuro-consumers' reactions when subjected to marketing or communication messages. In the United Kingdom, Gemma Calvert at Oxford then at Warwick University, was considered a precursor. She founded NeuroSense, the first company specialized in studies in this field, in 1999. Other European researchers have also carried out extensive research in this field, including Fabio Fabiloni in Rome, Arnaud Pètre and Patrick Georges in Belgium, Olivier Oullier and Olivier Droulers in France, all of whom have carried out neuromarketing studies in relation to companies. In Asia, Zhejiang University, located in Hangzhou (China), has undertaken research on brand perception. In 2010 in France, filmmaker Laurence Serfati presented a film showing various experiments of this type with the evocative title *Citizens Under influence* (*Citoyens sous influence*).

Several seminal books were published in the 2010s both in the United States and in Europe.[10]

To respond to the growing interest expressed by corporations in the knowledge of neuro-consumers, companies specializing in this field started appearing across the world. The oldest was NeuroSense, founded by Gemma Calvert in Oxford, UK. One of the most famous is NeuroFocus, founded by A.K. Pradeep in Berkeley, USA, later taken over by the leader in market studies, Nielsen. In most developed countries, there are subsidiaries or independent companies offering neuromarketing studies: Brighthouse, Neuroinsight, Neurofocus, Salesbrain in the United States; PhdMedia in Canada; NeuroInsight in Australia; Brain Impact, NeuroSense, Shop Consult, Map Brain Communication, Neuroco, Comao, Delphi in Europe, etc. On top of this nonexhaustive list, there are scores of companies offering services in the field of sensory marketing.

Many shops across a whole variety of different fields (consumer goods, luxury, retail, hotel, pharmacy, banking, communication) resort to neuromarketing or sensory marketing techniques. In the United States and the English-speaking world, experts are keen to communicate their experiments. In Europe, particularly in Latin countries, communication in these fields is of a much more confidential nature.

Neuromarketing is now an integral part of the most advanced companies' broader marketing strategies. It even plays a part in politics. The latest American president's communications consultants were quick to use these services, and didn't conceal it.

The study of neuro-consumer behavior appears essential to improve marketing efficiency and reduce the considerable amount of money spent on communication, distribution, design and development. Its place is bound to play an even greater role over the coming years, and the notion of the neuro-consumer should gradually complete that of the consumer.

The Technology, Media and Telecommunications (TMT) Predictions 2012 of the Deloitte's Center for Technology, Media and Telecommunications, which managers of leading corporations worldwide pay close attention to, mentions the emergence of neuromarketing for the first time.[11] This is a landmark in the role that this discipline should play in companies' marketing strategies.

The techniques used to understand and "woo" the neuro-consumer are increasingly efficient. They should go hand in hand – internally and externally – with a strong set of ethical and deontological rules to avoid any possible fraud or manipulation of the consumer. Indeed, neurological research carried out on the brain shows that it can easily be tricked or exploited by false information.

Evolution of scientific knowledge about brain behavior

Investments in marketing and communication in multinational corporations aimed at the wider public are increasing steadily. Managers are constantly on the lookout for ways of optimizing them. Particular attention is being paid to research aimed at improving our knowledge of what could trip the "purchase button" with neuro-consumers.

Concurrently, there has been increasing interest in neuroscientific experiments within research centers worldwide, with high levels of investment poured into them. In Lausanne, Switzerland, the Human Brain Project (HBP) is attempting to create a computerized copy of the human brain. In France, as part of NeuroSpin, a project belonging to the National Centre for Scientific Research (CNRS), Professor Denis Le Bihan and a team of researchers are using an ultra-powerful diffusion MRI to try to highlight a neural code governing the spatial layout of neurons. In the United States, there is also acute interest in the brain. According to an official report, President Obama was in favor of a project aimed at creating a brain activity map (BAM).[12] This study would make it possible to understand how nerve messages are exchanged, and how the connections between cells enable memory to be formed. This ambitious, costly program can only come to fruition if the American government invests massively.

Outside the United States and Europe, research is constantly being developed, most notably in Asia, particularly in China, South Korea and Japan. Some of this research is oriented directly towards marketing and communication, more specifically studies focused on memory, implication and manipulation.

Memory

At the Santa Lucia Foundation, located in Rome, Italy, neuroscience specialist Professor Fabio Babiloni is working on decrypting how memory works. He is studying memorization and omission phenomena in the context of advertising communication. Some of his experiments lean towards the idea that the memorization of an advertisement is not necessarily linked to the desire to buy the product. He is even convinced that people can remember bad adverts or products they do not wish to buy.

At the University of Geneva, Switzerland, neuropsychologist Professor Martial van der Linden has shown through his work that it is easy for our memory to confuse reality and the imaginary. It is possible to create a fake memory in the mind of a patient. Memory is formed through a reconstruction process, in which memories are remodeled by values, beliefs and life strategies. They can be hidden if they are too embarrassing or reorganized according to events that never actually happened to make them more enjoyable or gratifying.

In the United States, Professor Elizabeth Loftus, a psychologist at Irvine University, California, has managed to create a fake memory. Over the course of several experiments, she deliberately implanted fake memories of events that never actually took place in the minds of patients, who then declared they were certain they had experienced them.

In the United Kingdom, Gemma Calvert and her firm NeuroSense carry out MRI experiments for music channel MTV. They focus on viewer implication when watching an advert on TV. The results highlight the importance of the context in which the message should be placed to achieve maximum memorization.

Research into the memorization of communication, brands and logos is being carried out at an increasing rate at universities in the United States and around the world.

Mind, implication and brain manipulation

Other neuroscientific experiments are grabbing the attention of marketing and communications experts, particularly those aimed at reading subjects' minds or studying the influence of implication or manipulation.

In the United States, Professor Marcel Just and his teams at the Carnegie Mellon University Center for Cognitive Brain Imaging, in Pittsburgh, are working on reading subjects' minds. They have developed a program able to recognize words subjects are thinking under MRI. Psychologist and researcher at Stanford University, Professor Brian Knutson, is working on the role of emotions in the decision-making process. As shown in the above-mentioned French film by Laurence Serfati *Citizens Under Influence* (*Citoyens sous influence*), he can even predict a purchase by simply studying activity in the brain before the client has expressed an opinion.

In Switzerland, Professor Daria Knoch[13] at the Social Psychology Department of Basel University has succeeded in altering a subject's behavior through the stimulation of an area of the prefrontal cortex that plays a part in the decision-making process. Behavioral changes following this stimulation can include becoming more impulsive, less polite or other types of alterations.

The perspectives offered by neuroscientific studies will become broader over the course of the 21st century. They will enable us to achieve a great understanding of the neuro-consumer's behavior, leading to legitimate concerns over the risks this knowledge might entail linked to the concept of freedom of choice and decision. These fears, widely echoed by scientists and journalists, make it necessary to organize and implement a set of deontological and ethical rules.

The need to draw up deontological and ethical rules to approach the neuro-consumer

Studies in the field of neuroscience have generated such passion throughout the world that their progression seems inevitable. This growth can only go hand in hand with heightened knowledge of the neuro-consumer's behavior. As popular lore goes, "When an idea reaches maturity, nothing can stop it." This idea was often stated by Victor Hugo in his conversations with scientists. Those who would purely and simply ban these experiments are already considered by most as having joined the old guard.

Experiments concerning marketing studies carried out in hospitals from MRI are banned in some European countries, such as France. This ban does not prevent companies from these countries from carrying them out in the United Kingdom, Belgium or the United States, offering those countries substantial sources of funding to equip their hospitals with MRI technology.

To date, the use of MRI is considered harmless for patients, with no damaging effects ever having been detected. Thus opponents to the regulations on MRI used for marketing purposes in some countries argue that they are too restrictive and claim there is no solid basis to substantiate them. They even point out that contrary to MRI, using the Internet and social media is a highly risky activity that can lead to illness and even death, especially among young subjects, for example individual suicides due to online bullying or collective suicides within suicide "communities," groups encouraging anorexia, gambling addiction, etc. They wonder why those who wish to censor MRI-based studies aimed at improving the understanding of consumer behavior do not also call for the ban of the Internet and social media, which arguably could pose an even greater threat to the population.

Beyond the controversy, it remains essential to draw up a set of ethical and deontological rules governing both research into neuro-consumers and its application in the field of neuromarketing.

Do neuro-consumers need to be protected from neuromarketing?

As early as 2007 in the United States, Nick Lee, Amanda J. Broderick and Laura Chamberlain, in an article published in the *Journal of Psychology* titled "What Is Neuromarketing? A Discussion and Agenda for Future Research,"[14] discuss the existing controversies that arose in the science world around the use of new tools generating high risks of intrusion into the consumer's freedom.

Also in the United States, the Center for Digital Democracy (CDD) criticizes the use of neuromarketing based on its intrusiveness as well as the issues it raises on an ethical level. They warn of the potential misuses that could stem from the manipulation of consumers' minds in order to influence their choices.

There are countless debates surrounding ethics and deontology. They have led some states to react by banning subliminal advertising (which affects the brain directly without being picked up by the senses) or the use of some techniques such as MRI for marketing studies and also create laws to give the brain time to go back on its purchase decision and for the consumer to change their mind.

Experts' opinions

Managers of companies using neuromarketing also offer their opinions. A vast majority among them agree on the fact that it is essential to erect a strong set of ethical and deontological rules governing the use of neuroscientific techniques in the marketing and communication fields. They point out that these rules have already been implemented internally in most serious firms. They recognize the importance of the roles played by consumer protection organizations, politicians and states regarding these questions. They do, however, question the inclination that these organizations have to often "throw the baby out with the bath water" by suggesting outright bans without trying to understand how neuroscientific studies work, and how they are used by professionals. A clear understanding of how they work, as well as their limitations, could make it possible to come up with intelligent and effective directives offering greater protection to consumers against any deceptive or manipulative practices.

When this subject is brought up, most professionals (e.g., A.K. Pradeep, Gemma Calvert and Brian Knutson) deny any manipulation on their behalf. During the conferences he gives, A.K. Pradeep, founding manager of NeuroFocus, often uses the following image: "A candle can give light. A candle can burn down a building. We must be careful how we use a candle, and not blame the candle."

If we refer to neuroscientific knowledge of the brain, the best protection against any risks of abuse and manipulation resides in trusting the most highly reflective part of the neuro-consumer's brain: the neocortex. As we will see, the brain always ends up making the right decision provided it is given enough time to think and change its mind. The framework of practices and laws granting consumers a few days to rescind their decision without any penalty can prove highly effective in protecting them against the risk of being misled or manipulated.

Finally, it is not in a company's best interest to deceive or manipulate its clients. Good marketing practices show that their aim should rather be to develop customer loyalty and gradually build an attractive brand image. Any deception or manipulation of the consumer goes against this aim. What's more, nowadays any misleading or manipulative communication on behalf of a company is often swiftly picked up on and spread like wildfire on social media, thereby backfiring against the company responsible for it. This can only go against directives given by most managers.

Neuromarketing: manipulation or conviction of the neuro-consumer?

The debate on the border between manipulation and conviction is an ongoing one. Contrary to what is often claimed, neuroscientific tecniques can't be considered in themselves as being manipulative. They are by essence neutral just like any other techniques. Manipulation can only stem from the manner and purpose in wish firms wish to use them.

If the goal of the company is to maximize profit by misleading consumers through the use of more effective tools than those traditionally offered by marketing and advertising, this constitutes a clear case of manipulation.

If, on the other hand, the main aim of the company is to anticipate and respond through innovation to genuine needs that could prove useful to their clients, to communicate more creatively and more in tune with their expectations, to make their points of sales more user-friendly, then does this really amount to manipulation? Resorting to effective techniques that make it possible to gain greater awareness of what consumers feel and what their deep-seated beliefs might be, in order to better meet their expectations, can be seen as an acceptable approach. This situation reflects the clear ambition of a great number of firms, even though it also constitutes an intelligent way of creating value. In this case, we believe such an approach resembles more an attempt to convince than to manipulate.

Firms resorting to neurosciences to study and approach their clients must surround themselves with a set of ethical and deontological precautions, as their image is likely to be at risk. They should carry out a deontological and ethical audit before taking any steps in that direction, however commendable their intentions may be. It is also advisable for them to seek the assistance of internal or external specialists, and to offer proper training to their managers to anticipate and prepare the adequate response to any questions that might arise internally or externally.

Notes

1. Christophe Morin, *Neuromarketing: The New Science of Consumer Behavior*, Springer Science + Business Media LLC, 2011.
2. Hilke Plassmann, Vinod Venkatraman, Scott Huettel and Carolyn Yoon, "Consumer Neuroscience: Applications, Challenges, and Possible Solutions," *Journal of Marketing Research*, vol. 52 (August 2015), pp. 427–435.
3. Samuel M. McClure, Jian Li, Damon Tomlin, Kim S. Cypert, Latané M. Montague and P. Read Montague, "Neural Correlates of Behavioral Preference for Culturally Familiar Drinks," *Neuron*, vol. 44 (2004), pp. 379–387.

4. Ibid.
5. Justin R. Garcia and Gad Saad, "Evolutionary Neuromarketing: Darwinizing the Neuroimaging Paradigm for Consumer behavior," *Journal of Consumer Behavior*, vol. 7 (2007), pp. 397–414.
6. Morin, *Neuromarketing*.
7. Patrick Renvoisé and Christophe Morin, *Selling to the Old Brain*, SalesBrain, 2003.
8. Martin Lindstrom, *Buy.Ology: How Everything We Believe About What We Buy Is Wrong*, Random House Business, 2009.
9. Brian Knutson, Scott Rick, G. Elliott Wimmer, Drazen Prelec and George Loewenstein, "Neural Predictory of Purchases," *Neuron*, vol. 53, no. 1 (2007), pp. 147–156.
10. A.K. Pradeep, *The Buying Brain: Secrets for Selling to the Subconscious Mind*, Wiley, 2010; Leon Zurawicki, *Neuromarketing: Exploring the Brain of the Consumer*, Springer, 2010; Roger Dooley, *Brainfluence: 100 Ways to Persuade and Convince Consumers with Neuromarketing*, John Wiley & Sons, 2012; Patrick Georges, Anne-Sophie Bayle-Tourtoulou and Michel Badoc, *Neuromarketing in Action: How to Talk and Sell to the Brain*, Kogan Page, 2013; Stephen J. Genco and Andrew P. Pohlmann, *Neuromarketing For Dummies*, John Wiley & Sons, 2013; Bernard Roullet and Olivier Droulers, *Neuromarketing: Le marketing revisité par les neurosciences du consommateur*, Dunod, 2010.
11. Deloitte, TMT Predictions 2012, www2.deloitte.com/za/en/pages/technology-media-and-telecommunications/articles/tmt-media-predictions-2012-4k-kicks-off-deloitte-tmt-predictions.html.
12. Paul Alivisatos, Miyoung Chun, George M. Church, Ralph J. Greenspan, Michael L. Roukes and Rafael Yuste, "Brain Activity Map and the Challenge of Functional Connectomic," *Neuron*, vol. 74, no. 6 (2012), pp. 970–974.
13. Daria Knoch, Frédéric Schneider, Daniel Schunk, Martin Hofhmann and Ernst Fehr, "Disrupting the Prefrontal Cortex Diminishes the Human Ability to Build a Good Reputation," *Proceedings of the National Academy of Sciences of the United States*, vol. 106, no. 49 (2009), pp. 20895–20899.
14. Nick Lee, Amanda J. Broderick and Laura Chamberlain, "What Is Neuromarketing? A Discussion and Agenda for Future Research," *International Journal of Psychology*, vol. 63 (2007), pp. 199–204.

CHAPTER 6

Neuroscientific techniques used to understand the neuro-consumer's behavior

There are a wide variety of neuroscientific techniques used to improve our knowledge of neuro-consumers' deep-seated expectations, sensory perceptions and behaviors. These range from costly, sophisticated tools such as magnetic resonance imaging (MRI) or electroencephalography (EEG) to simple neuromarketing diagnostics offered by cognitive scientists specialized in this discipline. Their implementation depends on various factors such as the relevance of their contribution as opposed to traditional marketing studies, as well as the budget to be allotted to them.

The use of neurosciences was preceded by methods aimed at understanding people's behavior using expressions and signs deciphered by experts. These were developed by social psychology theorists, who were interested in the impact of third-party environment on people's behavior. The study of "influence" is a favorite among these researchers.

One such researcher, psychology professor Robert Cialdini of Arizona State University, carried out a three-year study of three different publics in a situation of persuasion, including: used-car salespeople, telemarketing companies and charity organizations. In his book, *Influence: Psychology of Persuasion*,[1] he described the results of his experiments and showed that individuals operate according to preprogrammed mechanisms that condition their reflexes.

Various schools founded on sensory observation were created in the last century. They offer methods or programs seeking to better understand individuals by observing and deciphering expressions emanating from their senses. The most renowned of these programs include neuro-linguistic programming (NLP) and transactional analysis (TA).

NLP and TA: two precursors in the understanding of the neuro-consumer

NLP, just like other approaches such as TA and personal development groups that are based on a sensorial understanding of individuals, seek to help individuals muster their internal resources and use their senses.

Neuro-linguistic programming (NLP)

Neuro-linguistic programming (NLP), a registered trademark, was invented in 1972 by linguistics professor John Grinder and mathematician-psychotherapist Richard Bandler. Their book, *Reframing: Neuro-Linguistic Programming and the Transformation of Meaning*,[2] includes a set of communication and self-help techniques, laying the emphasis on people's reactions rather than where behaviors come from. It offers an observation grid to improve the perception of oneself as well as of others. The method is based on the use of the language as well as the body language produced by each individual from the five senses. It endeavors to program and reproduce its own success models. The designers consider that subjects construct their representation of the world through their senses. Their goal is to establish a link between the sensorial aspects of a person's thoughts and their emotional reactions.

Programming Programming in NLP refers to learning reflexes. NLP experts program through different models designed from a set of predetermined questions and observation. Programming consists in putting into formula the manner in which people think, feel and behave, in various everyday situations. NLP offers success models in a variety of fields, such as psychotherapy, sports, pedagogy, but also creativity, communication, management and sales.

Neuro NLP is based on the analysis of neurological capacity. According to Robert Dilts and Judith Delozier, whose contribution to the evolution of NLP have been immense, "There is a link between NLP and other psychological schools, as NLP is designed from neurology and cognitive science."[3]

Linguistic The two first models of NLP, branded "meta models," are essentially linguistic. They are founded on 12 questions and are aimed at revealing the transformation mechanisms of the sensorial experience in language. On the initiative of researchers but also teachers and consultants such as Robert Dilts,[4] they evolve through the study of the different senses. The "sensorial channel" model, or visual, auditory, kinesthetic, olfactory and gustatory (VAKOG), plays a part in this evolution. The VAKOG model is based on the idea that each individual has a favorite communication mode. It contends that senses place the person in contact with their environment.

Originally centered on psychotherapy, NLP broadened to other fields such as marketing and communication on the initiative of John Grinder who left the University of California Santa Cruz in 1977. In 1979, the first certification training was organized in the United States. The certification made it possible to create several levels of knowledge. NLP training started taking off on the international stage as of the 1980s. This led to the emergence of various institutes or training centers in different European countries, such as France, Germany and the United Kingdom. NLP offers a variety of models, which endeavor, through appropriate techniques, to rethink personal communication, assess management skills, offer sales teams a better understanding of customer psychology and behavior and improve marketing and communication.

Companies can choose from a host of different training courses, taught by a variety of different speakers: managers with specific training, coaches who have been given more in-depth training or master practitioners confirmed by NLP schools.

As is the case with any science aimed at decoding the brain, NLP is on the receiving end of criticism both of a scientific and an ethical nature.[5] The discipline, which favors observation methods, is sometimes considered as a precursor of neuromanagement and neuromarketing approaches. The use of neuroscience can help reach a new milestone in understanding the neuroconsumer.

Transactional analysis (TA)

Transactional analysis (TA) is a theory regarding personality and communication that was invented in the 1950s by American psychiatrist Eric Berne (1910–1970), at Englewood hospital, New Jersey.[6] The theory behind TA offers a method aiming to reach an awareness of "what is at play here and now" in relationships between two people or within a group. This improved understanding of relationships is established based on three "ego states" and the study of intrapsychic phenomena that are observed through relational exchanges named "transactions." TA offers a method, an interpretative framework and intervention methods aimed at shedding light on and solving these problems.

Eric Berne postulates that the broad orientations in our lives are decided as early as childhood, and can take the shape of a "life scenario." The understanding of relationships between people is achieved through observation of their "ego state" during this relationship. He describes three different "ego states" representing "a coherent system of thoughts, of emotions and associated behaviors": the parent, the adult and the child. They are observed from "recognition signs," which Eric Berne defines as "strokes."[7]

Since its inception, TA has given rise to various schools of thought and treatment. In Europe, it is currently headed by an official body named the

European Association for Transactional Analysis (EATA). The structure offers different levels of training leading to certification and applications can be found in various fields ranging from psychotherapy of individuals to psychotherapy aimed at solving interpersonal conflicts or conflicts within a group.

The objective of TA is to improve interactions between people in an environment where clients may act aggressively, such as post offices, banks or hospitals. It prepares the subject's brain to help them deal with conflicts and try to downplay the situation when faced with angry or aggressive interlocutors. It is also used to make some interlocutors' routine contacts with customers or users friendlier (cashier, toll-booth employees). To this end, the EATA offers a variety of training courses tailored to their needs. Just like NLP, TA is a highly debated topic and often criticized.[8]

MRI and EEG: two sophisticated techniques

In 1890, English physiology Professor Sir Charles Scott Sherrington (1857–1952, Nobel Prize in Medicine in 1932) and his friend Charles Smart Roy (1854–1897), established a link between the brain's activity and the blood flow. When an area of the brain is called upon, it receives more hemoglobin loaded with oxygen and glucose. A century later, this discovery would lead to revolutionary inventions used for better knowledge of the brain. Some of these, like MRI or EEG are used to gain greater understanding of the neuro-consumer. A large number of these inventions occurred in the second half of the 20th century and are described by doctors Arthur S. Bard and Mitchell G. Bard in their book, *The Complete Idiot's Guide to Understanding the Brain.*[9]

A first breakthrough was made by German neurologist Hans Berger (1873–1941) who managed to record electric current in the brain through electrodes placed on the scalp in 1929. This technique, which would later become EEG, is capable of detecting brain waves on a conscious patient.

It was not until 1968 that David Cohen, a physician at Illinois University, discovered the emission of magnetic signals, leading to the creation of magnetoencephalography (MEG).

In 1972, British engineer Godfrey Honsfield (1919–2004) invented "computer-assisted axial tomography," better known as CT scan or CAT scan. During this experiment, patients place their head in a cylinder. The data collected thanks to X-rays are computer processed in order to offer a cross section of much higher quality than an ordinary X-ray.

Two years later, a new technique was invented called "positron emission tomography," or PET scan. This more complex technique requires the injection of glucose or water with a radioactive component that gathers in the areas of the brain when they become active. The premise is a simple one: The more the brain cells are active, the more radioactive glucose they consume. Radioactive elements appear in the PET images, the most active generally in red and the least active in blue. They allow the researcher to see which areas of the brain react to certain specific actions.

In 1946, Americans Felix Bloch (1905–1983) and Edward Mills Purcell (1912–1997) discovered the principle of nuclear magnetic resonance, a discovery that would earn them the 1952 Nobel Prize in physics. This paved the way for the invention of MRI a few years later.

Some of these tools are used to know what the neuro-consumer is thinking. As A.K. Pradeep put it, "What our researchers are interested in is discovering why and how the consumer buys."[10]

The way these techniques work is both relatively simple to explain and highly complicated to achieve. When a neuron is active, it produces electricity and requires more fresh blood. Its metabolism changes. Neuroscientific devices receive the signals resulting from these changes, which are then processed by complex software that offer representations, or 3-D color images, of the areas of the brain that are activated when the patient performs an action, has a thought or makes a decision. Researchers know what roles are performed by the activated areas that light up (areas dedicated to pleasure, stress, memorization, attention, emotion and desire).

> *By analyzing the effects produced by the stimulus on some specific areas of the brain, cognitive experts study the efficiency of the tested actions (accepting a price, interest in the presentation of a product, memorizing communication).*

Brain imagery using MRI and its applications in gaining knowledge about the neuro-consumer

The first studies concerning MRI were published in 1973 in the journal *Nature* by American chemical engineer Paul Lauterbur (1929–2007, 2005 Nobel Prize in physics). At first, MRI was called nuclear magnetic resonance (NMR), but the word nuclear was dropped to avoid frightening patients away. Denis Le Bihan, founder of NeuroSpin, a French institution belonging to the CEA (the French Center for Atomic Energy), offers readers in-depth knowledge of this technique in his book, *Le cerveau de cristal.*[11]

Functional magnetic resonance imagery (fMRI) makes it possible to study the activity of a person's brain, placing it in the center of a powerful magnet, with a magnetic field 30,000 times higher than that of the Earth. It is even more powerful in the latest generation MRIs. When an area of the brain receives a stimulus, there is an increase in the flow of fresh blood containing high levels of hemoglobin transported by red blood cells. Hemoglobin has hydrogen and oxygen nuclei. The magnet makes their atoms resonate.

The waves emitted by the nuclei are then recovered by the scanner's central computer and sorted by the computer's "reconstruction" software according to their location inside the brain. The researcher will have previously programmed the map of the magnetic field into the computer, and the information is translated into images according to a set of definitions, 2-D, 3-D, etc. The view obtained thanks to MRI is not a direct representation of what is happening in the brain but rather a reconstitution that can be made into a picture thanks to complicated software.

Thanks to the work by scientists such as Denis le Bihan, MRIs are becoming increasingly powerful and efficient. They make it possible to carry out in-depth research into how the brain works. Functional MRIs allow us to see the brain in action, while it is thinking. Researchers have discovered that the mere action of thinking creates a mental visual image in the brain. They realize that an action can be seen in the brain before it actually happens.[12]

Functional MRI has become a highly appreciated tool in the development of behavioral studies and the largest psychology laboratories in the United States are adding these machines to their equipment.

A host of services companies such as NeuroSense in the United Kingdom or university research centers like Stanford use MRI, often in partnership with hospitals to carry out experiments in neuromarketing.

> *A variety of multinational companies (Coca-Cola, Apple, e-Bay, Google, Facebook, McDonald's) use these methods to improve their marketing and communication. Some companies even own their own MRI. It is most frequently used in the following areas: choosing new products or packaging, testing a brand and communications strategies. MRI establishes a direct reaction between the visualization of an image, an advert, a logo, a slogan and the effect it triggers in the brain. From the information obtained, companies can then improve their offer, or make their communications campaigns more attractive.*

Despite its advantages, MRI does also have some drawbacks that limit its applications. First of all, not every company can afford the cost of such a technique. Costs vary depending on the provider and the type of use required. In Europe, a machine costs around €2,000,000 to purchase, €3,000 to €3,500 to rent for a session and €10,000 for a day's rental, generally excluding consultant fees. It is usually advised to perform at least 15 to 20 sessions.

Another aspect that is often criticized is the artificial aspect of the experiments, which have to be carried out inside a tube, as well as the fact that volunteers are often paid to take part. This is a far cry from the normal environment a purchase would usually take place in. Although painless and nonintrusive, MRI sessions are far from fun. The magnet makes a deafening noise. MRIs are not recommended for people suffering from claustrophobia. They could also be dangerous and are therefore forbidden for people with any metal object inside their body (e.g., pacemaker, pins, etc.).

EEG, a more flexible – albeit less precise – technique for the study of the neuro-consumer

EEG and MEG are also frequently used in neuromarketing. EEG is the better known of the two. It consists of capturing in real time the electric activity of a consumer's neurons while being submitted to marketing or communication stimuli. Electrodes are placed on the subject's head, and they can either remain still or wander around a sales point.

> *Neuromarketing world leader, NeuroFocus-Nielsen, uses this technique in studies they carry out for international corporations such as Pepsi-Cola, Intel, CBS-New, PayPal and Cheetos chips. They designed "Mynd," a portable EEG unit they use for their studies, which sends the data in real time to a computer via Bluetooth. EEG is more practical and user-friendly than MRI, and can be used to study neuro-consumers' behaviors as they move around a sales point as well as study their thoughts in front of products displayed on shelves. It is used in studies to test the impact of communication or e-communication on the brain. Experts seek to find out whether the firm's communication triggers attention, interest, emotion and if it is remembered.*

A.K. Pradeep, founder and chief executive officer (CEO) of NeuroFocus, states in the frequent conferences he gives that thanks to these techniques, it is now possible to predict whether an advertising campaign is liable to fail and explain the causes for this.

EEG is also used to try to apprehend the effect on the client's unconscious of a product's design or a brand. Although less accurate than MRI, EEG offers a huge advantage in terms of ease of use and mobility. Moreover, it is much less costly to use, although the cost of data interpretation and marketing advice remains high. Prices vary depending on the provider and the type of research. In Europe, they can range between €20,000 and €50,000 for a test on a product based on a sample of 20 to 30 consumers.

Some practitioners have highlighted the limitations of EEG in research into the neuro-consumer. These are mainly linked to the lack of spatial accuracy inside the brain and the fact that it is close to impossible to record the deep areas of the brain.

Simpler, cheaper tools

When marketing and communication experts wish to improve their understanding of neuro-consumers' behavior, MRI and EEG remain the preferred techniques for certain big companies, whereas simpler, cheaper tools categorized as "peripheral" by neuroscientists, are used by companies with more limited means. Most of these tools also detect the emotions felt by the neuro-consumer's brain when he is confronted with different demands from his environment. A variety of methods can be used, ranging from measuring hormonal secretions through the various "peripheral" tools listed below to a simple cognitive diagnosis.

Hormonal secretion or neuroendocrinology and its applications

Nerve messages are transmitted in the form of electrical signals by hundreds of billions of neurons. Nerve impulses travel along the axons like wildfire. Communication between the neurons is made by the synapses, which form gaps between them. When it arrives in the terminal part of the axon, the nerve message must find a means of crossing the gap. Its electrical energy is then converted into chemical energy in the form of a hormonal substance known as a "neurotransmitter." When the brain is solicited in certain ways, certain types of hormone are produced in greater quantity. The quantity secreted conditions the individual's behavior. Neurotransmitters can be measured in the blood, the urine and sometimes in the saliva. The quantity of hormone detected indicates what the neuro-consumer is feeling at the moment that he is solicited. For example, according to professor of neurosurgery Patrick Georges,[13] desire and pleasure are correlated with three hormonal neurotransmitters: dopamine (tension related to desire), noradrenalin (excitement, shared pleasure), endorphins (well-being, rest). A heavy secretion of cortisol, on the other hand, indicates a high level of stress.

Measurement of neurotransmitter secretions has the advantage of being fairly precise and relatively cheap. Experiments on blood or urine samples are not very practical for research into marketing or communication but tests on saliva are much easier to carry out.

> *Studies based on hormone samples, particularly those using saliva, provide interesting information to the catering industry. They are used, for example, to test the attractiveness of restaurant menus and provide useful advice on their content and presentation.*
>
> *They are also used to understand the acceptance or resistance behavior of populations who may be subject to significantly stressful situations. A heavy secretion of cortisol can reveal a high level of stress. If an individual has low stress tolerance, it may be*

dangerous to place him in a situation or context where he risks secreting too great a quantity of the hormone. This can be the case for sales staff subject to high levels of uncertainty that generate stress.

The stress tolerance of different categories of customer is also used to optimize sensory marketing at sales points.

Using peripheral tools

There are different peripheral tools that companies can use to carry out neuromarketing studies.[14]

Electrodermal activity (EDA) Electrodermal activity (EDA) measures micro-perspiration using electrodes placed on the fingers. If perspiration levels change while the subject is watching a video or looking at a specific image, it indicates that there is a response from the nervous system and more precisely emotional involvement, be it positive or negative.

Electrocardiogram (ECG) An electrocardiogram (ECG) measures heart rhythm. Generally, it involves an increase in body temperature, respiration rate or palpitations. This technique is similar to EDA in that it provides understanding of the consumer's emotions when presented with a given product or service.

Facial electromyography (fEMG) Facial electromyography (fEMG) measures the tension of the face muscles using electrodes. It aims to indicate the degree of emotion.

Eye tracking Eye tracking is based on how the eyes work, tracking the direction in which they are looking at each instant. Making use of a laser instrument, market research companies use this technique widely to test the impact of packaging and how products are laid out in supermarkets. Mock-ups of supermarket shelves are filled with the products concerned. The subject walks between the shelves while wearing the apparatus. The subject may also be asked to look at videos, Internet sites and publicity images.

Eye tracking enables the direction of the consumer's eyes to be followed and the amount of time he spends at each place he looks at. This technique is very useful for understanding consumer preferences. It is used in different tests that help to determine the best place to put products on a shelf, which aspects of packaging, a poster, a publicity video, an Internet site, a mobile application, a

> *Facebook page, etc. are perceived favorably. They help to find the ideal location for an advertisement.*

Studies using eye tracking are simple and relatively inexpensive. They are used in artificial sales points like those set up by Eric Singler for BVA in Europe, Asia and the United States. Eye tracking is of particular interest in behavioral studies when it is coupled with other telemetric procedures that measure emotion, for example EDA or ECG.[15]

Telemetry On the basis of different characteristics (wetting of the eyes, acceleration of the heart rate measured using the pulse, skin coloration, sweating, facial analysis, etc.), telemetry provides a good indication of what the brain feels in different situations where demands are placed upon it.

Neurologists have shown that the brain experiences a sensation a few microseconds before expressing it. The sensation creates a variation in different factors such as the degree of wetness of the eyes, particular signs on the face, increased blood pressure, etc. that correspond to a type of emotion being felt. It is very difficult to dissimulate them. The use of appropriate techniques employing microsensors or facial surveillance cameras coupled with analytical software enables them to be detected and interpreted.

> *These methods, relatively simple to implement, are sometimes used by "spin doctors" or communication consultants in the culture industry. Some publishers use them to help authors rewrite their books so as to render them more attractive to the brain. Directors such as James Cameron use some of these techniques to make their films more attractive. They were used, for example, on* Titanic, *and more recently,* Avatar.

Producers of advertising, be it on the Internet, mobile devices or social networks, are becoming increasingly interested in these tools, which enable the neuro-consumer's behavior to be analyzed and emotions to be detected. A growing number of marketing and communication departments are making use of them when they do not use more sophisticated techniques such as MRI or EEG.

During a television broadcast,[16] Nicolas Delattre, director of marketing development at Perception Média, presented experiments using a wristband measuring several physiological parameters such as the heartbeat and micro-perspiration coupled with eye tracking. These applications help to improve various fields of marketing and communication. The technique is used to analyze the emotion produced by an advertisement, which helps the audience remember it, the layout of a display unit and the design of a product. Simpler than using MRI or EEG, telemetry can provide good information at a more modest cost. The cost of a study to measure the impact of an advertisement or a product is around €15,000.

Neuroscientific diagnosis and the emergence of new professions: cognitive specialists and "spin doctors"

A new profession of neuromanagement or neuromarketing expert has been born. It includes professionals from medical, psychological or marketing backgrounds trained in cognitive science and neuroscience. Frequently grouped under different names (cognitive specialists, "spin doctors," etc.), they are able to advise companies on how to develop and implement actions compatible with the way in which their customers' brains work. Their recommendations involve improvements to products, pricing, sales, distribution, communication, etc. that take account of recent neuroscientific discoveries. Their advice may be limited to an expert assessment and practical solutions that can be quickly implemented. This is frequently the case when they are interested in evaluating the sensory characteristics of sales points. They may also recommend complementary research using more elaborate neuroscientific tools. The cost of an authorized expert is between €2,500 and €3,000 per day. Depending on the case, the consultancy can take between three days and one week but may sometimes require longer. Figure 6.1 shows an example of the diagnostic methodology and cognitive recommendations for a sensory marketing policy to be used at a sales point for goods or services.

Courses

There are still relatively few structured courses at international higher educational establishments on neuromarketing or neuro-consumers' behavior. Up to now, such courses have rather been offered by private trainers and companies like SalesBrain in the United States. In Europe, Patrick Georges, professor of neurosurgery, and his teams offer complete training courses to companies over several days. One-day master classes are offered in various cities throughout the world by the Neuromarketing Science and Business Association (NMSBA), based in the Netherlands. Courses on the application of neuroscience to marketing, communication and sensory marketing are being developed in several European universities, which are also encouraging PhD theses in the field. Schools like ISCOM, in Paris, are taking an interest in this new management discipline. The training center at HEC Paris

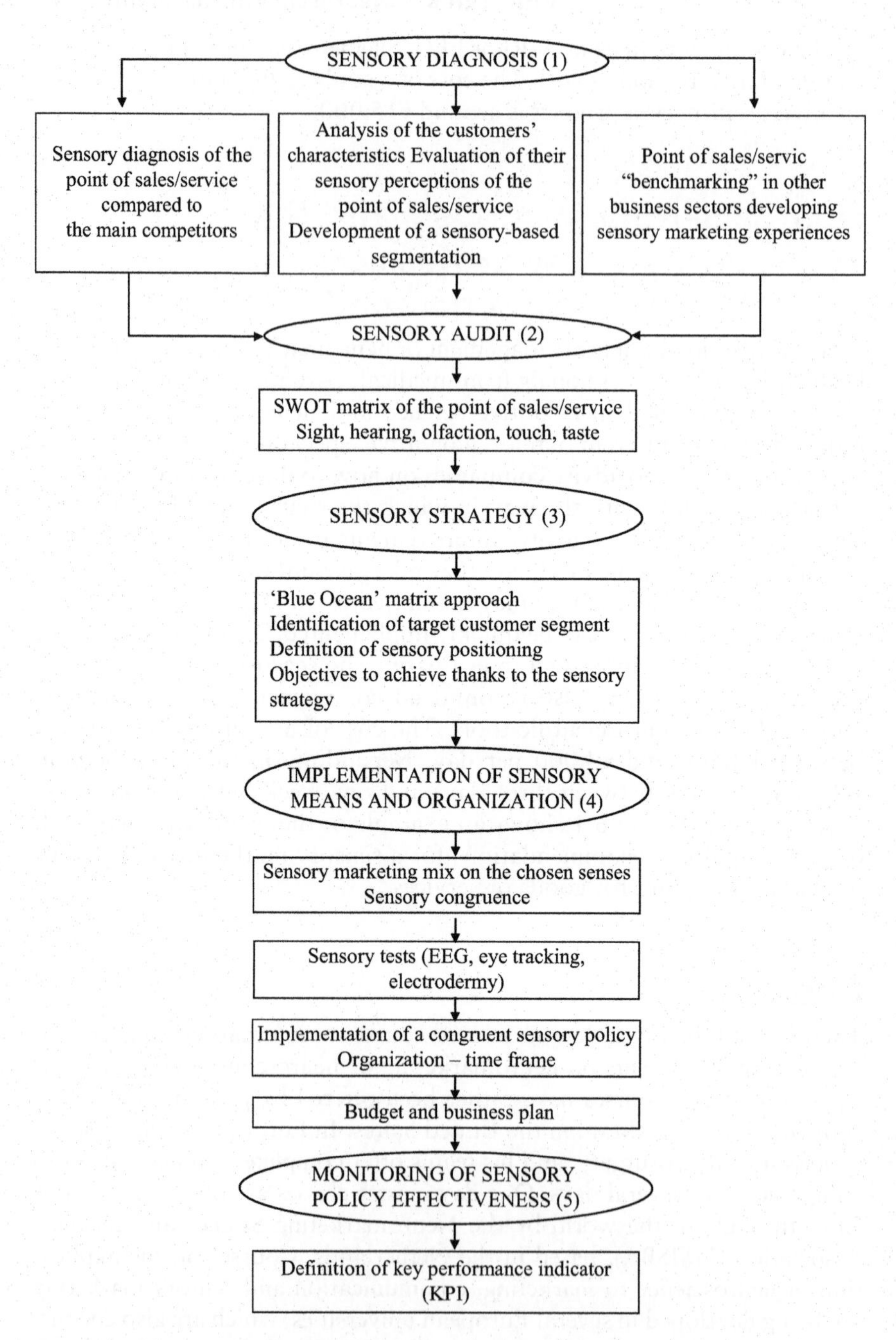

Figure 6.1 Method for producing a sensory policy for a sales or service point

for directors and senior managers, HEC EXED, also offers master classes on the subject in France and has included it in its program on digital marketing and new trends in marketing. The school is giving more space to this theme in its marketing and communication-related courses, be they aimed at students or professionals.

In the future, teaching on this subject could go in different directions:

- The establishment of schools that certify, coach and offer expertise in neuromarketing similar to that occurring in the teaching of NLP and TA.
- The development of modules or complete seminars included in traditional teaching of consumer behavior, marketing, digital marketing, communication, e-communication, etc. for both students and professionals. We consider this second solution to be more probable.

A new profession of "neuromarketer" is emerging, aimed at better understanding and satisfying the subconscious perceptions of the neuro-consumer's brain. Apart from essential training in the principles and techniques of neuromarketing, the neuromarketer must keep abreast of all the new neuroscientific tools that appear regularly and could be of interest to his profession. Inventive engineers are continually perfecting existing tools and creating new ones with the aim of improving our knowledge of the brain's behavior. Apart from biology, the main fields where innovations are already appearing are cerebral imaging and statistical processing of information. They include:

- In the field of imaging:
  - tractography;
  - transcranian magnetic simulation (TMS).
- In the field of information processing:
  - independent components analysis (ICA);
  - multiple discriminant analysis;
  - multivoxel pattern analysis.

Knowledge of the behavior of the human brain is still in its infancy. The rapid progress being made during this century will certainly have major repercussions on our overall knowledge of medicine, psychology and sociology, but it will also improve our knowledge of the neuro-consumer's behavior.

Notes

1. Robert B. Cialdini, *Influence: The Psychology of Persuasion*, Harper Business, 2006.
2. Richard Bandler and John Grinder, *Reframing: Neuro-Linguistic Programming and the Transformation of Meaning*, Real People Press, 1983.

3. Robert Dilts and Judith Delozier, *Encyclopedia of Systemic Programming and NLP New Coding*, NLP University Press, 2000.
4. Robert Dilts, *Roots of NLP*, Metamorphous Press, 1983.
5. Philippe Breton, *La parole manipulée*, La Découverte, coll. Essais, 2004.
6. Eric Berne, *Transactional Analysis in Psychotherapy*, Martino Fine Books, 2015; Eric Berne, *What Do You Say After You Say Hello?* Corgi, 1975.
7. Berne, *What Do you Say After You Say Hello?*
8. Michel Tougne, *Ni prince ni crapaud: l'analyse transactionnelle, savoir ou mystification?* CFP Edition, 1996.
9. Arthur S. Bard and Mitchell G. Bard, *The Complete Idiot's Guide to Understanding the Brain*, Alpha, 2002.
10. A.K. Pradeep, *The Buying Brain: Secrets for Selling to the Subconscious Mind*, Wiley, 2010.
11. Denis le Bihan, *Le cerveau de cristal: Ce que nous révèle la neuro-imagerie*, Odile Jacob, 2012.
12. *Le cerveau et ses automatismes: Le pouvoir de l'inconscient* [film], Arte, Part 1: December 1, 2011, Part 2: April 14, 2014.
13. Patrick Georges, Anne-Sophie Bayle-Tourtoulou and Michel Badoc, *Neuromarketing in Action: How to Talk and Sell to the Brain*, Kogan Page, 2013.
14. Darren Bridger, *Decoding the Irrational Consumer: How to Commission, Run and Generate Insights from Neuroscience Research*, Kogan Page, 2015.
15. Emmanuelle Ménage, *Consommateur pris au piège* [film], France 5, January 12, 2014.
16. "La quotidienne," France 5, March 2, 2016.

Points to remember

- Technologies originating from neuroscience (MRI, EEG, eye tracking, etc.) make information collected in traditional marketing studies more relevant, often complementing it but sometimes calling it into question or contradicting it.
- Thinkers have sought to understand the behavior of individuals and consumers since antiquity. Over the centuries, certain philosophers have developed theories tinged with modernity. They serve as the basis for the hypotheses of many contemporary researchers who then try to verify them using neuroscientific tools.
- The famous neurologist Antonio Damasio and his colleagues often refer to the writings of Descartes and Spinoza. Ancient theories put forward by philosophers on themes such as the relationship between the body and the mind, the role of the emotions, memory, desire and the importance of the subconscious are analyzed on the basis of scientific observations of the brain's reactions.
- Researchers such as Joseph Breuer, Carl Gustav Jung and Sigmund Freud were veritable pioneers, bringing to light the subconscious mind's fundamental role in human behavior. Today's researchers, with effective tools for neuroscientific analysis, can delve more deeply into their theories. Patricia Churchland, author of many internationally renowned books, has created the concept of "neurophilosophy," which aims to associate philosophical thinking with neuroscientific research. Many academics follow her ideas.
- It is important to understand the limits of research into consumer behavior for marketing and to see where and how neuroscientific analysis will bring improvements.
- We define the "neuro-consumer" as "a consumer significantly subject to automatic reflexes in the subconscious or unconscious part of his brain, whose purchasing decisions and processes are strongly influenced by his emotions and desires."
- After neuroscience had been applied to numerous fields ranging from medicine to psychology, it was inevitable that marketing specialists would become interested in these new research techniques. The first experiments in consumer behavior took place in the United States at the end of the 20th and beginning of the 21st century.
- After having been neglected for centuries, the function of the brain became a subject of increasing interest from the 19th century and an increasing number of studies were carried out throughout the world in the following decades. They have been a major source of research in many fields, including marketing, in the first part of the third millennium.

They have benefited from approaches such as those developed at Palo Alto and from tools used in fields such as cybernetics and functional imaging.

- The first applications in marketing began in the 2000s. The American researcher Read Montague and his colleagues had the idea of using MRI in an experiment relating to the Coca-Cola and Pepsi brands.
- The term "neuromarketing" is often attributed to Montague. Experiments aimed at better understanding consumer behavior by applying neuroscientific techniques have become increasingly frequent throughout the world. Numerous specialized companies have come into being in recent years: SalesBrain, NeuroSense, NeuroFocus, etc. and experimentation has developed in the research centers of major universities. Various books describe their applications. The majority of major international companies use neuromarketing to make their marketing and communication more effective.
- Several observation-based approaches have led to the development of techniques such as NLP and TA, etc. Neuroscientific techniques allow researchers to go deeper into the understanding of the neuro-consumer's behavior and range from the more complex, such as MRI and EEG, to cheaper, simpler tools, including hormonal secretion, EDA, ECG, eye tracking and simple neuroscientific diagnosis. Some of these techniques are coupled with other neuroscientific tools or with traditional marketing and communication studies. Their objective is not to call existing research into question, but to improve on it and complement it.

PART II

Understanding the neuro-consumer's brain

Understanding how the neuro-consumer's brain works and its irrationality is essential to deciphering its behavior. Its perception of reality is largely conditioned by elements acting outside its consciousness. Factors such as age, gender, memory, emotion and desire influence its attitude. Somatic markers and mirror neurons, recently discovered by neuroscientists, condition its responses to its environment. Neuroscience is helping to improve knowledge in this field.

Introduction

While neglected for centuries, the understanding of how the functioning of the human brain behaves is attracting increasing interest in this new millennium. Neuropsychologists Bryan Kolb and Ian Q. Whishaw from the University of Lethbridge in Canada have written a major book on the subject.[1] The discovery of the extraordinary, and hitherto largely unknown, capacities of this organ fascinates researchers in a wide range of fields. Medical professors Arthur and Mitchell Bard write at the beginning of their book *The Brain*:

> The brain is a fascinating organ. It determines what we think and how we interpret the world. It generates our dreams and our nightmares. It tells us to be happy or sad. We eat, drink and indulge in sexual activity on the instruction of our brain. By better understanding what the brain is like, how it is assembled and how it works, we hope to improve our lives and those of our children. We hope to find ways of improving our memory, of modifying our children's behavior, and of overcoming the anxieties that govern our actions and affect our happiness.[2]

These two authors consider that curiosity alone is a very good reason for studying the brain. We share their opinion, even though knowledge of how it works remains embryonic and is thus likely, as it develops, to call into question some of the current ideas on the subject. Neuroscience researchers sometimes say: "We know more about the workings of the universe than those of the brain!" Modesty that reveals the extreme degree to which knowledge in this field is limited.

To better understand the neuro-consumer's brain, we will look briefly at its constitution and then at the automatic reflexes that condition its actions.

The neuro-consumer's subconscious behavior is influenced by the way in which his brain is structured, but also by his moods, which modify emotions and desires. Endogenous factors such as age, gender, memory, somatic markers and mirror neurons play an important role in the brain's responses to the demands from its environment. All these factors condition our perceptions, attitudes and purchasing behavior. In a few moments, emotions and desires can annihilate conscious, rational reasoning. Many research projects are under way all over the world to try to understand the phenomena that influence and condition the brain. This research provides new understanding of the neuro-consumer's behavior, which is essential to improving the efficiency of companies whose objective is to better satisfy the needs and expectations that the neuro-consumer hesitates to express. As Roger Dooley[3] pragmatically states, research into this subject is of great interest, in particular helping companies to make their marketing, communication and sales policies more efficient in order to obtain "better results for less money." It is also useful for neuro-consumers. Clear understanding of their brain's subconscious reactions can help them to avoid succumbing too easily to artifices designed to lead them into making irrational or instinctive purchases.

Notes

1. Bryan Kolb and Ian Q. Whishaw, *An Introduction to Brain and Behavior* (2nd ed.), Worth Publishing, 2006.
2. Arthur S. Bard and Mitchell G. Bard, *The Complete Idiot's Guide to Understanding the Brain*, Alpha, 2002.
3. Roger Dooley, *Brainfluence: 100 Ways to Persuade and Convince Consumers with Neuromarketing*, John Wiley & Sons, 2012.

CHAPTER 7

The neuro-consumer's brain

A complex organ

Understanding the neuro-consumer means first understanding that surprising and marvelous organ of nature: the brain. It can process information coming from the five senses, control the movements of the body and guarantee its cognitive functions. Often compared to an extremely powerful computer or an orchestral conductor, it has unexpected qualities. These are being brought to light by major research projects in laboratories throughout the world thanks to the rapid development of neuroscientific techniques.

To explain the brain and its relationship to the human body, we refer to several articles and books published by professors of medicine and neuroscience. Apart from the book by doctors Arthur S. and Mitchell G. Bard,[1] we have found other publications useful, including that by doctors Frédéric Sedel and Olivier Lyon-Caen.[2] These two books, written in nonscientific language, one in English and one in French, will enable interested readers to delve further into the subject.

The neuro-consumer's brain

The neuro-consumer's brain weighs about 340 g at birth, 930 g at 1 year of age and 1.3 to 1.4 kg when adult. It is lighter than that of the sperm whale, which weighs 7.8 kg, or that of the elephant brain, which weighs 6 kg. There does not appear to be a direct relationship between the weight of the brain and the level of intelligence. Einstein's brain, collected by anatomical pathologist Thomas Stoltz Harvey (1912–2007) in 1955, at Princeton hospital, weighed less than the average. However, it had a greater density of neurons in the cortex.

The volume of the brain is 1,400 cm^3 and the perimeter of the skull measures from 55 to 60 cm.

The organ only represents 2% of the body's weight, but consumes 20% of its energy. Approximately one-fifth of the blood pumped by the heart (around 5 liters per minute when resting) is sent to the brain. It is fed by the oxygen and glucose transported by the blood. It produces the same amount of energy as a 10-watt light bulb. It works 24 hours a day, 7 days a week throughout its life. It is 75% water. It has no moving parts and is totally insensitive to pain, which facilitates certain surgical operations.

At birth, a child possesses 100 billion neurons. Each neuron can connect to 100,000 other neurons. The fetus creates 250,000 neurons every minute. A billion signals circulate in the brain every second. The speed of the impulses can exceed 400 km/s. Apart from the neurons, the brain is composed of neuroglia. Unlike neurons, they have neither axons nor synapses. They help the neurons to repair themselves and they fabricate myelin. Studies of them have begun recently. Among them are "astrocytes," which provide the neurons with nutrients and help them eliminate waste products. They also play a role in memory.

At the age of 20, brain cells start dying at the rate of 10,000 a day. After the age of 45, the brain can lose up to 10% of its weight. At the end of the individual's life, millions of neurons disappear. Professor Pierre-Marie Lledo of the Pasteur Institute[3] has shown that, whatever a person's age, the brain can produce new neurons. This potential for creation varies from person to person. According to the researcher, several factors contribute to this regeneration: "Not being subject to stress, not regularly consuming psychotropic substances, indulging in physical activity, having an active social life, being filled with wonder and wanting to learn."

Sophisticated communication with the human body

The brain communicates with the human body via the multitude of "cables" that make up the nervous system. The spinal cord is the essential communication route between the brain and the muscles. The body contains about 50,000 km of connecting nerves linking sensory and motor nerves, enabling decisions to be taken and executed.

The basic unit of the nervous system is the neuron. The brain contains 100 billion neurons. The Milky Way contains the same number of stars. The rest of the human body contains as many neurons again. The center of the wiring system is concentrated in the head. In terms of volume, 30,000 neurons could fit onto a pinhead.

Neurons have two properties: reactivity and conductibility. They can respond to a stimulus and conduct an electrical signal (or nerve impulse) generated by the stimulus. Contact with the environment is carried out by receptors linked to the senses. They communicate information to the brain by using the "wiring" system.

Each neuro-consumer has 135 million visual receptors, 5 million receptors for smell, 700,000 for touch and 30,000 for hearing. Sight is the dominant sense for the neuro-consumer. Communication and marketing are very effective when they favor the visual sense. Neurons communicate by means of synapses. A synapse is a gap between two neurons. To cross the synapse and transmit a message, the electrical energy is converted into chemical substances called neurotransmitters. These are reconverted into electrical signals.

Researchers have identified more than 50 neurotransmitter hormones including acetylcholine, dopamine, serotonin and noradrenalin. A single neuron can possess up to 100,000 synapses, receiving and transmitting information by means of thousands of neurotransmitters produced by 100,000 neighboring neurons. The reception of the neurotransmitters' secretions gives useful information enabling a researcher to understand what the neuro-consumer is experiencing or feeling when confronted with demands from his environment.

The zones of the neuro-consumer's brain and their functions

The core of the brain is the cortex. Most of the functions associated with thinking, creative memory and intelligence are carried out there. The brain consists of two hemispheres, right and left. Each hemisphere is subdivided into four parts, or "lobes," each with specific functions. If the brain were a cube, the frontal lobe would be the front, the parietal lobe the back, the lateral lobes the sides and the limbic lobe the bottom.

These zones contain areas that have even more specific roles. They were generally described at the beginning of the 20th century by German anatomist Korbinian Brodmann (1868–1918), who produced a map of the brain based on differences in the microscopic architecture of regions of the cortex. Numbered from 1 to 47, each area corresponds to different functions (audition, shape recognition, emotion, memory, olfaction). This mapping is still used today. Over the years, the description of the functions has been refined and improved. Neurologist Wilder Penfield (1891–1976) drew even more detailed maps of the human brain.

Knowledge of the actions linked to the zones and areas is very complex, because it is systemic. It has developed as neuroscientific techniques have become more sophisticated. This is the case with the latest generation of functional magnetic resonance imaging (fMRI) created by Denis Le Bihan and his colleagues at NeuroSpin in France. Or, more recently, with the "clarity" method, developed for the "Brain" project launched by the American government with the objective of producing a complete map of the human brain. This technique, developed by chemists and neuroscientists at the University of Stanford, USA, renders the brain transparent.

As we saw in Part I, during an MRI, areas of the brain "light up" according to the tasks or stimuli:[4]

- The *lateral prefrontal cortex* lights up when we are asked to decide, when will prevails over instincts.

- The *medial prefrontal cortex* corresponds to judgmental behavior, when we assess a value.
- The *nucleus accumbens*, the brain's pleasure center, lights up in the basal forebrain when a person is shown something he or she really desires, something precious to the person, alcohol, sex, a game, food or related products that promise these pleasures.
- The *premotor area* is activated when we see someone make a gesture and we prepare to imitate it.
- The *temporal cortex* lights up when we listen to and when we memorize something.
- The *occipital cortex* is activated when we look.
- The *limbic cortex* kicks in when we are moved emotionally and open up our memory.
- The *cerebral amygdalae* are activated when we are anxious and when something frightens us, thus making us aggressive towards it.
- The *ventral putamen* is activated when we experience a feeling of satisfaction.

Knowledge of the zones and areas that become active in response to stimuli from marketing, communication, sales, etc. is particularly important in understanding what interests, pleases or displeases the neuro-consumer and thus working out his behavior. Such knowledge enables companies to avoid making pointless investments by offering products that do not interest his brain, or advertisements that irritate or leave him indifferent. It helps to improving marketing and communication by providing companies with fundamental information on the depth of his expectations and the reality of how he makes purchasing decisions.

The functioning of conscience is more difficult to grasp. According to Stanislas Dehaene,[5] psychologist, cognitive specialist and neuroscientist, the conscience is a system that distributes information. It is not an area but a network.

The brain observed through its hormonal secretions

The human body, managed by the brain, may be compared to a chemical plant. A large proportion of the neuro-consumer's feelings and attitudes are conditioned by the production of neurotransmitter hormones. Some are related to pleasure, others to aggression, stress or serenity. Among the hormones that can be measured in a neuro-consumer are:

- *serotonin*, the "good-mood" hormone, which protects us against depression and impulsiveness;
- *dopamine*, which facilitates impulsiveness and aggression and also signals pleasure;

- *cortisol* in the saliva, which measures the intensity of stress;
- *testosterone*, which is related to sexual desire;
- *progesterone* and *estrogen*, which are very important in women and are related to love and affection;
- *noradrenaline*, which creates excitement and shared pleasure;
- *adrenaline*, which triggers tension or stress; and
- *endorphins*, which condition well-being and auto-anesthesia.

This list is not exhaustive. Hormonal secretion is one of the subjects studied by neuromarketing specialists to improve knowledge of the neuro-consumer's behavior.

The brain: computer or orchestral conductor?

The neuro-consumer's brain is often compared to a computer. This idea goes back to the 1950s. British mathematician Alan Turing[6] (1912–1954) suggested that computers could be programmed to rival the human brain. He developed the "Turing Test," which would enable researchers to identify what objects, created by a machine, could be considered human.

In 1996, IBM's "Deep Blue" computer managed to beat world chess champion Garry Kasparov. More recently, in 2008, under the direction of neuroscience researcher Henry Markram, the "Human Brain Project" started trying to create a computerized copy of the human brain. The project is run by the École Polytechnique Fédérale de Lausanne (EPFL).

Although comparisons with a supercomputer are possible, many researchers remain skeptical. Some, like professors Arthur and Mitchell Bard, have remarked that, unlike the brain, computers "do not feel, do not think and are not aware of their own existence. The human brain is."[7] Professor Andréas Kleinschmidt, a scientist from Geneva, has shown that the brain "anticipates while we sleep, evaluates hypotheses relating to a situation which could arise in the future and is permanently seeking to establish a balance between internal and external worlds."[8] As early as the 1960s, American doctor Paul Bach-y-Rita (1934–2006) and his psychiatrist brother George discovered that the brain is capable of repairing itself. It has many capacities that are very different from, and above all much more complex than, those of computers. That is why many neuroscientists prefer to compare the brain to an orchestral conductor rather than a supercomputer.

The digestive tract: a second brain

Most people are familiar with sayings like "to be sick with fear," "to take decisions viscerally," "to have a lot of guts," "to have the stomach to do something," etc. On the other hand, they are often unaware that the digestive tract acts as a "second brain."[9] Michael Gershon, professor in the Department of Anatomy and Biology at the University of Columbia in New York, has investigated this subject. The walls of the intestines contain the enteric nervous system, a veritable second brain and double of the first. He writes:

> Two hundred million neurons, as many as in a dog's brain, line the intestinal wall. The cells come from the same germ layer as those in the brain, which they leave at an early stage of development to migrate towards the digestive tract, where they form the enteric nervous system.[10]

Hundreds of millions of neurons that form the local networks regulate intestinal function and release neurotransmitter hormones identical to those controlled by the brain. According to recent discoveries, nearly 95% of the serotonin produced in our bodies is fabricated in the intestines. Remember that this neurotransmitter is involved in the management of the emotions. According to Michael Gershon: "With these endogenous psychoactive substances, the digestive tract can generate discouragement or enthusiasm, impotence or pleasure, depression or accomplishment."[11] These two brains, one for thinking and the other for feeling, communicate all the time, in both directions, through the parasympathetic vagal nerve. This connection manifests itself through stress and anxiety. The digestive tract is an open window on the central nervous system. The amount of research into this subject is constantly growing. Two major projects have recently finished: the Human Microbiome Project (HMP) ran in America from 2008 to 2013 under the aegis of the National Institutes of Health while the Metagenomics of the Human Intestinal Tract (MetaHIT) project ran in Europe from 2008 to 2012. Since Michael Gershon's work, several researchers have written on this subject.[12]

The influence of the digestive tract neurons on human behavior is presented in Cécile Denjean's evocatively titled documentary, *The Digestive Tract, Our Second Brain* (*Le ventre, notre deuxième cerveau*).[13] This film also shows, based on research carried out on humans by Dr. Kirsten Tillisch and her colleagues at the University of California at Los Angeles Medical Center, the role played by certain bacteria in the sensitivity to negative images and the ability to resist the stress generated by these images. According to Kirsten Tillisch, the injection of probiotics (living microorganisms added to food) including certain bacteria modifies the brain's responses to the environment when faced with negative images. She concludes that certain bacteria, in the form of probiotics, have the ability to change the brain's perceptions and to influence human behavior.

No fewer than 100 trillion bacteria inhabit our digestive system. The bacteria inside us weigh from 1 to 2 kg and provide 30% of our calories. Recent research has shown that they play a fundamental role in the elimination of waste, the provision of energy and the body's internal equilibrium. They communicate with the brain in liaison with the intestinal neuronal system.

Professor Stephen M. Collins from McMaster University in Canada, referring to research carried out on animals, shows that the action of bacteria has a direct influence on behavior governed by the brain. He suggests that beyond a second brain, we have a third brain in our digestive tract emanating from the intelligence of the bacteria. They live in an internal ecosystem of our organs known as a "microbiota," an independent environment inside the human body made up of self-sufficient bacteria.

Research projects in these fields are generating a lot of hope of improving our health through the treatment of diseases such as diabetes, obesity and neuropsychiatric disorders like autism, schizophrenia and depression.[14]

No application has yet been found to increase our knowledge of the emotions felt by the neuro-consumer, but advances in research will likely enable applications to be found in this area in the future.

Notes

1. Arthur S. Bard and Mitchell G. Bard, *The Complete Idiot's Guide to Understanding the Brain*, Alpha, 2002.
2. Frédéric Sedel and Olivier Lyon-Caen, *Le Cerveau Pour les Nuls*, First, 2010.
3. Pierre-Marie Lledo, quoted in Victoria Gairin, "Les vrais pouvoirs du cerveau," *Le Point*, vol. 2160 (February 6, 2014), pp. 60–67.
4. Patrick Georges, Anne-Sophie Bayle-Tourtoulou and Michel Badoc, *Neuromarketing in Action: How to Talk and Sell to the Brain*, Kogan Page, 2013.
5. Stanislas Dehaene, *Consciousness and the Brain: Deciphering How the Brain Codes Our Thoughts*, Penguin Random House, 2015.
6. Alan Turing, "Computing Machinery and Intelligence," *Mind*, vol. 49 (1950), pp. 433–460.
7. Bard and Bard, *The Complete Idiot's Guide.*
8. Andréas Kleischmidt, quoted in Gairin, "Les vrais pouvoirs du cerveau."
9. Paul Molga, "Notre ventre, une intelligence supérieure," *Les Echos*, May 2014.
10. Michael Gershon, *The Second Brain*, Harper Paperbacks, 1999.
11. Ibid.
12. Giulia Anders, *Gut: The Inside Story of Our Body's Most Under-Rated Organ*, Greystone Books, 2015; Emeran Mayer, *The Mind–Gut Connection: How the Hidden Conversation Within Our Bodies Impacts Our Mood, Our Choices, and Our Overall Health*, Harper Wave, 2016.
13. Cécile Denjean, *Le ventre notre deuxième cerveau* [Film], Arte, September 4, 2015.
14. See www.inserm.fr/en/health-information/health-and-research-from-z/intestinal-microbiota-intestinal-flora.

CHAPTER 8

Is the brain free or programmed?

There is a lot of research and many publications and videos on how the brain works. They range from hypotheses to empirical studies confirmed or not by experiments. The aim of this chapter is to try to better understand how the neuro-consumer's brain works, particularly in automatic mode. We will draw on the various theories and experiments of scientists, and particularly neuroscientists, without forgetting the philosophers. Some approaches are the subject of debate or even disputes in the expert community. Ideas evolve along with the progress made by neuroscientists. As with medicine, we use current knowledge, even though we know that it is likely to be considerably improved upon in the future. It has already revealed types of behavior in neuro-consumers that were ignored by traditional marketing and communication studies.

The "triune" brain

In 1969, at the University of Bethesda, American neurobiologist Paul D. MacLean formulated his theory of the "triune" brain, which was the subject of a book.[1] The theory describes three distinct brains that appeared progressively during the course of evolution (Figure 8.1).

The first, the reptilian brain, is approximately 400 million years old. It dates back to the period when fish came out of the water and became amphibians or reptiles. It is the seat of reflex behavior such as aggression or flight. It controls homeostasis, the body's equilibrium, by regulating functions such as respiration, heart rate, blood pressure, body temperature, etc. It deals with satisfaction of the vital primary needs: food, sleep, reproduction, etc. It is conservative and has an instinct for imitation. It favors the sense of smell over the other senses. Its actions are primal, instinctive and rapid.

The second, the paleo-mammalian or limbic brain, appeared 65 million years ago with the first mammals. It is linked to memory. It is the seat of the emotions and triggers alarm and stress reactions. It separates the world into two: "I like" or "I don't like." Everything pleasant is registered as an

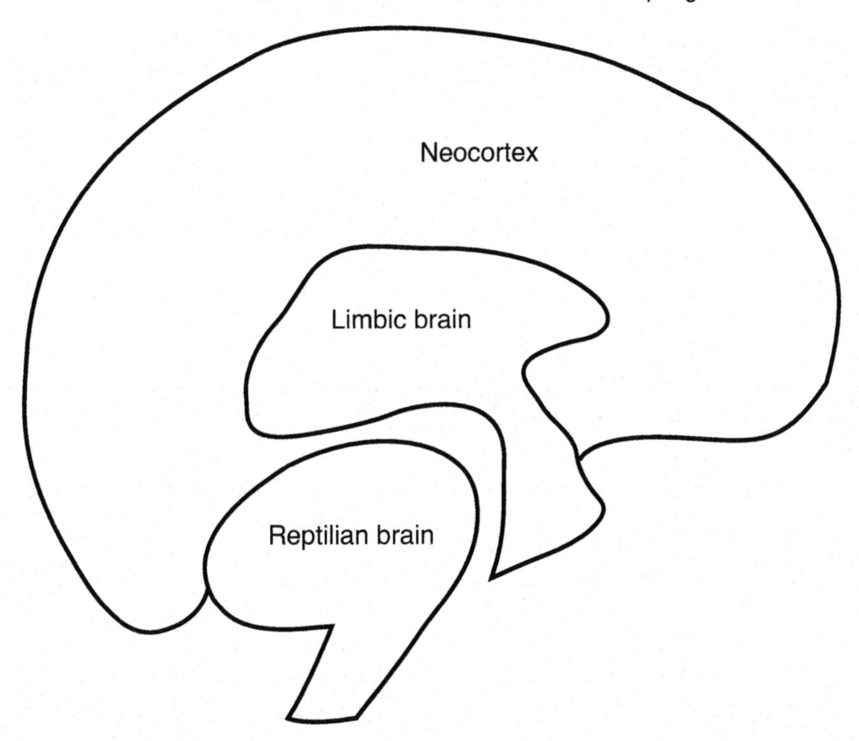

Figure 8.1 Representation of Maclean's "triune" brain

action to be repeated. Everything disagreeable is registered as needing to be avoided or fled from. It is the center of affectivity, encouraging us to look after our children and giving us a sense of family or clan. It compares everything with what has already been experienced. It favors the sense of hearing over the other senses.

The third, the neo-mammalian brain or neocortex, is the genuinely "human" part of the brain. It is only about 3.6 million years old. It appeared with the first bipeds, Australopithecus in Africa. It is capable of logical reasoning, abstract thought, language and anticipation. The neocortex or cortical brain analyzes, reasons, anticipates the future and takes decisions. It favors the sense of sight over the other senses. Relatively lacking in emotions, it functions like a computer. In a way, the neocortex renders the neuro-consumer intelligent.

The frontal lobes are the parts that make the neocortex human and differentiate man from animals. Some scientists consider them as a fourth brain. They enable man to be altruistic, to think of others, to be creative and to project himself into the future.

Neuromarketing takes a particular interest in the primitive brain when it leads the neuro-consumer into predictable, stereotypical behavior. As we will see in more detail in the following paragraphs, on the basis of the reflexive reactions of this part of the brain experts have developed approaches that help to improve the effectiveness of marketing, sales and communication.

Information moves around these three parts of the brain in a particular way. In an article, Belgian mental management trainer Rava-Reny[2] presents the ways in which information circulates according to the theories formulated by the American researcher.

Information arrives in the reptilian brain. If the survival of the human being or its essential needs (food, reproduction, defense) are not threatened, the information is sent up a level, to the limbic system. In certain cases (stress, alcohol, fear, strong irritation, strong irrationality, strong authority, atrophy of the frontal lobes, etc.) the reptilian brain can take decisions by itself.

The limbic system evaluates the information to determine whether it is agreeable or disagreeable. In the first case, it sends it on to the neocortex, which treats it with intelligence and often in a positive manner. In the second case, it does not pass it on. The individual then finds himself confronted with a negative perception. He is led to brood and to feel depressed. In order to "win," as Rava-Reny remarks, it is in an individual's interest to see the positive side of situations. "Not only does the limbic brain send the information to the neocortex but the latter gives it priority."[3] An old limbic proverb says: "If you start expecting to lose, you are sure to lose; start expecting to win if you want to win."

Paul MacLean's theories, popularized by Hungarian novelist Arthur Koestler[4] (1905–1983), are the subject of criticisms and objections from a number of scientists. The concept is contested in particular by French scientific journalist Michel de Pracontal.[5] The independence of the three parts of the brain is rejected by some neuroscientists who prefer to consider the different areas of the brain as interacting blocks. These criticisms do not prevent other parts of his theory from being justified, such as those dealing with paleontology and evolution.

The functional analysis of the "triune" brain put forward by Paul MacLean is used by neuromarketing experts. Patrick Renvoisé and Christophe Morin[6] have taken a specific interest in the functioning of the reptilian brain.

More generally, Paul MacLean's work has helped to understand the behavior of the neuro-consumer's brain in certain situations and in particular when it is subject to certain stimuli coming from its environment. His research has enabled progress to be made in different fields such as market segmentation of neuro-consumers, design, product presentation, training, motivation and management of sales teams, the attitude to adopt towards customers, communication, etc.

The brain and its automatic reflexes

For centuries, the decisions taken by the human mind were considered to originate from conscious thought. Gottfried Wilhelm Leibniz[7] (1646–1716) was one of the first philosophers and scientists to suggest that the mind is a combination of parts that are accessible and parts that are inaccessible to thought. A brilliant German mathematician, Leibniz wondered, like James Clerk Maxwell (1831–1879), Joseph Blake (1728–1739) and Johan Wolfgang von Goethe (1749–1832), if his mind did not contain deep caverns, inaccessible to his consciousness. He formulated the hypothesis that there are perceptions of which we are not aware. He called them "little perceptions." In his book, he suggests that we are unaware of some of our desires and tendencies that can nevertheless govern our acts. For the first time in history, subconscious urges were clearly brought to light.

In 1670, in his posthumously published *Pensées* (*Thoughts*), Blaise Pascal (1623–1662), who does not go as far as Leibniz, was already writing: "Man is incapable of seeing the nothingness from whence he comes nor the infinite where he is swallowed up."[8]

A century later, the German physiologist Johannes Peter Müller (1801–1858), following his experiments on the eye and light, deduced in his writings, published in 1833,[9] that man is not directly conscious of the world that surrounds him. His brain only processes the signals it receives from the nervous system. The relationship between the brain and perception is sophisticated. Many researchers have tried to find out more about this unknown element of us that guides most of our acts. Joseph Breuer, and above all Sigmund Freud, made an essential contribution to understanding the role of the subconscious. The brain's subconscious reflexes seem to have a particularly important place in creativity. Great artists in numerous fields have experienced their effects. Pablo Picasso was reported as often saying to his family and friends: "Painting is stronger than me. It can make me do what it wants."

Current technology enables researchers to go further in understanding the subconscious working of the brain. Among them, David Eagleman sheds new light: Free will was either an illusion or, at the least, much more widely constrained than we had ever thought. Towards the middle of the 20th century, intellectuals started to think about the fact that in reality human beings knew very little about themselves. We are not at the center of ourselves, but like the earth in the Milky Way and the Milky Way in the visible universe – far, far away on the edge. And very little of the information that circulates around our internal world actually gets to us.[10]

New information about how the brain works outside conscious control leads us to look with a new light at what conditions the neuro-consumer's behavior. It is probably this kind of thinking that made Gary Singer, co-founder of Buyology Inc., state in the lectures he gives all over the world: "Purchasing decisions are based 85% on the subconscious mind, traditional marketing analysis is based on the 15% that remain conscious."[11] In order to better understand the neuro-consumer's subconscious reactions, it is useful to examine his brain's automatic reflexes. To help us, we will look at research being carried out by today's neuroscientists. It is described in many books, including that by David Eagleman[12] and also in several films including *The Brain and Its Automatisms* (*Le cerveau et ses automatismes*).[13] It seems to us that there are three important questions about the neuro-consumer: What is his true perception of the real world? Is his brain free or programmed when it reacts to the demands of his environment? Why does his brain react as it does?

Our perception of the world: objective reality or simple construction by the brain?

Neuroscientists agree that our perception of the world is not the reality perceived by the senses, but a reconstruction created by the brain in accordance with its objectives. The neuro-consumer does not perceive what is in front of him. He perceives what his brain tells him to perceive.

In a dissertation dating from as early as 1847, the German physicist and physician Hermann Ludwig Ferdinand von Helmholtz (1821–1894) suggested that the brain makes suppositions based on the data that it receives.[14] These suppositions are based on information memorized from past experience. The brain makes the best possible hypotheses to transform external information into something bigger, which creates a perceived reality. Many facts and experiments attest to this reconstruction by the brain. They are widely illustrated in the above-mentioned lectures and films. They include optical illusions, false memories, the placebo effect and the ability to read words that are incomprehensible on paper but perfectly legible thanks to instantaneous reconstitution by the brain. The selectivity of the brain that can see only the essential elements of a scene according to its habits and its immediate needs is widely exploited by magicians.

In the 1960s, American neuroscientist Paul Bach-y-Rita from the University of Wisconsin became interested in enabling the blind to see again by sending information directly to the brain without using the sense of sight.

As David Eagleman[15] mentions, the blind, elite mountaineer Erik Weihenmayer has made many ascents and reached the summit of Everest. He is able to see thanks to a device called BrainPort placed on his tongue. The tongue uses its many nerve endings to transmit information to the brain. This experiment proves that we do not see with our eyes but with our brain.

The same is true for the other senses. The neuro-consumer must be aware that perceived reality is not necessarily veritable reality. One of the most important rules for fighter pilots refers to this phenomenon when it states: "Trust your instruments." This observation shows that, thanks to the way it functions, information can be transmitted to the brain that creates unconscious reflex reactions. This is important in the field of medicine. Hypnosis is used to reduce the level of certain types of pain during chemotherapy, but also to replace certain light anesthetics.

Knowledge of the neuro-consumer's unconscious behavior and objective reasoning has many lessons for marketing and communication. Researchers[16] have shown that simply being exposed to certain data has a very strong impact on positive memory. If an individual's face has already been seen on a photograph, that individual is judged to be more attractive when encountered later in real life. This is true even if the subject has no recollection of the photo. This phenomenon is called the "mere-exposure effect." It is not surprising that the simple repetition of images or messages has so much importance in election campaigns, marketing and communication. It aims to produce positive feelings in the neuro-consumer. This effect has led certain famous people to joke: "There's only one negative kind of publicity and that's no publicity" or "I don't care what the newspapers say about me as long as they spell my name right."[17]

Another observation relates to repetition.[18] It suggests that the more something is repeated, the more it appears to be correct. The role of the senses, which send information directly to the brain, can be used to positively influence the neuro-consumer's purchasing behavior. Sensory marketing seeks to better understand the power of their influence. It is possible to give the brain false information. Such information can be processed as if it were true.

In a European supermarket, which has asked to remain anonymous, the fish counter was failing to increase its turnover. When consumers were questioned in a survey, they said they felt that fish in a supermarket could not be very fresh. Neuromarketing experts suggested chemically diffusing a smell of the sea around the counter as well as adding the sound of gulls. Fish sales increased by 25% over the following months.

There is a significant risk of manipulation when sensory marketing is used. Actions implemented in this field must have a strong framework of ethical rules.

Is the brain free or programmed?

According to many neurology researchers, such as David Eagleman,[19] our innate behavior is the incarnation of ideas that are so useful for the species that they have ended up registered in the cryptic, minuscule language of DNA. This has occurred through the effects of natural selection over millions of years. Individuals who possessed instincts favorable to their survival and their ability to reproduce multiplied.

Authors such as Charles Darwin[20] (1809–1882) and Paul MacLean[21] have made us aware of the phenomenon of the development of the human brain over the centuries. Some of our instincts, those that lead to the most automatic and natural behaviors, have been programmed in our brains for thousands of years. They include sexual desire, food, fear of the dark, empathy, anger and the identification of different facial zones. The huge networks of neurons that underpin these actions are so well organized and regulated that we are unaware of their normal functioning.

Many neuroscientists[22] have shown that the criteria of feminine beauty and masculine handsomeness are programmed in our brains and linked to the need to optimize the reproduction of the species and the search for a partner in good health. Our nerve networks related to sexual desire are not attracted by the sight of a naked frog because we could not reproduce with it. On the other hand, the male brain responds to the dilatation of a woman's pupils because, according to the experts, this signal apparently delivers important information about her sexual availability.[23]

We are able to catch a moving ball because the laws of physics are deeply preprogrammed inside us. These internal models allow us to predict the position of the ball in space and where it will land. The parameters of our predictive system are calibrated by our experiences throughout our lives: partly instinctive and partly through learning. Swimming, skiing, cycling and all other sports seem difficult the first time we try. When these activities have become programmed in the brain, they are performed instinctively without our even being aware of it.

One of the first examples of the conceptual framework for the programming of the brain was produced by neuroscientist Donald MacKay in 1956.[24] He put forward the idea that the visual cortex is a machine whose fundamental job is to make representations of the world for the brain.

Studying the preprogramming of the brain opens up significant perspectives in the analysis of the neuro-consumer's attitudes, influenced by certain of his fundamental instincts such as short-term survival, reproduction and reaction to danger.

> *This can help explain the fact that the number of food items purchased in shops is greater before lunch, that advertising related to sex accounts for such a high proportion of advertising budgets in the majority of countries throughout the world and that advertisements offering solutions that deal with a fear often meet with success.*

The brain's automatic reflexes and the neuro-consumer

Two main reasons are commonly given for the brain having automatic reflexes. The first is speed of execution.[25] Studies using magnetic resonance imaging (MRI) show that an action is anticipated and carried out in the brain before being transmitted to the muscles to be carried out in reality. The brain's reaction time is faster than the reaction time called for by the conscious mind. A tennis player could never return a 230 km/h service if he had to think about it. In order to return the ball, he has to let his brain act unconsciously, using the program for this gesture acquired by practice after a long learning phase. What we call "concentration" in a top-level sportsperson is in fact "absence of concentration." The same is true of musicians. A virtuoso pianist would be likely to play badly if he started thinking about how his fingers move on the keyboard. In many situations involving danger or aggression, survival may be a question of reaction time. The brain's circuits are designed to produce behavior adapted to survival.

The second reason concerns energy efficiency. By automating its functions, the brain minimizes the amount of energy it needs to solve the problems facing it. When it accomplishes tasks automatically, it consumes much less energy than when it has to think. Using cerebral imaging, researchers[26] have shown that subjects in the process of learning how a video game like Tetris works consume a very large quantity of energy. Consumption becomes very low when they have become experts.

The process of "engraving" tasks to be accomplished into the brain's circuits is a fundamental aspect of cerebral function. This phenomenon may be dangerous for the neuro-consumer if he "engraves" into his brain the automatic reflexes of certain repetitive tasks, e.g., the habit of always playing the same number in the lottery, smoking, watching the same television series, regularly playing the same video games, never being without one's mobile phone, etc. These habits, categorized by the brain as automatic reflexes, can become addictions and lead to repeated purchases.

How the brain's "bipartite" system manifests itself in the neuro-consumer

Paul MacLean described a brain in three parts. The detail of his theories no longer seems to be favored by neuro-anatomists. The central principles remain. The parts of the brain are composed of subsets that compete with each other. The brain contains two distinct systems. One is rapid, automatic and below the surface of the conscience. The other is slow, rational and conscious. To simplify matters, the first is primary and emotional while the second is more rational. In a way, the neuro-consumer has two types of intelligence: instinctive intelligence and shrewd intelligence.

Instinctive intelligence Instinctive intelligence is primary but quick. It triggers immediate reflexes that are essential when faced with danger.

A stressful situation may be considered to be dangerous or problematical for the brain, which seeks a quick solution. Buying something is sometimes seen as a means of relieving stress. Neuromarketing experts such as professor Patrick Georges[27] have sought to calculate the ideal stress level of a population of neuro-consumers that will put them into the best frame of mind to make a purchase.

Instinctive intelligence acts on the basis of mental reflexes provided by genetics and learning that is experienced, generational or cultural. It is strongly conditioned by the emotions. Among the reasons why it acts are: too much information, fear, stress, being faced with strong authority, atrophy of the frontal lobe, alcohol, drugs, etc.

> *Unexpected price reductions are among the types of situation that bring the neuro-consumer's instinctive intelligence into action. The decisions are rapid but of poor quality, because they are based on primary reflexes such as the fear of not having something, fear of competition or the desire to get a good deal at any price. What is more, primitive reactions can be predicted, which makes the neuro-consumer vulnerable to well-informed professionals.*

Shrewd intelligence Shrewd intelligence comes from the brain's conscious, rational system, situated in the neocortex. The brain acts like a slow computer. It can only deal with one-fifth of the information that it receives. It is incapable of thinking and making good decisions at the same time. In this context, shrewd intelligence acts continuously when the neuro-consumer gives it time. In this case, it will take control of instinctive intelligence. To make a good decision, the neuro-consumer absolutely must give his brain the time to think. It is not recommended to make quick decisions in meetings or in the presence of a salesman if those decisions have not been previously thought over and prepared. If they have not, it is wise to request an interruption of about half an hour. Everyone knows the popular saying: "Sleep on a decision." This is particularly effective because while the subject is asleep his brain sifts through information. For important purchase decisions, it is recommended to relax for even longer, for example by taking a long weekend away. This allows the intelligence system to rest, enabling it to optimize the quality of its decision.

Another researcher, Daniel Kahneman, professor at the University of Princeton and awarded the 2002 Nobel Prize in Economics, has described the human brain's two ways of working in a book that has become famous.[28] In system 1, it works on automatic pilot and unconsciously. It does not require any attention and takes quick decisions. It requires very little energy to function. System 2, on the other hand, is slow but conscious and rational. It requires effort and attention. It seeks to make a positive or negative evaluation of the choices with which it is faced. It consumes a lot of energy. Depending on the nature of the decisions that it has to make, the brain moves from one system to the other. For important decisions – whether or not to have a child, buy a house or choose a car – it uses system 2. For the majority of decisions, using this system requires effort associated with consuming a lot of energy, something that the brain does not like. Daniel Kahneman describes the brain as a "lazy controller." Preferring to avoid too much effort, it uses system 1 to take the majority of everyday decisions. This system works on the basis of routines, mental shortcuts, memorized associations and approximations. It simplifies and has no doubts. Rapid but of poorer quality, its decisions are, moreover, predictable. Authors such as professor of neurosurgery Patrick Georges and professionals like Patrick Renvoisé and Christophe Morin are studying the brain's instinctive decision-making processes to better analyze the neuro-consumer's subconscious perception. Their objective is to make marketing, sales and communication more effective.

How the brain behaves when using its instinctive intelligence

When his brain uses instinctive intelligence, the neuro-consumer may have primitive, predictable reactions. This is not the case when his shrewd intelligence is called upon. A number of experts are working to try and better understand the reason for the neuro-consumer's behavior when he finds himself in the former situation. Their objective is to use that understanding to improve the effectiveness of marketing, sales and communication. By way of illustration, we will explain the approach used by Patrick Renvoisé and Christophe Morin in the United States and Patrick Georges in Europe. We are also interested in the brain's important function related to maintaining equilibrium, known as homeostasis, as presented by Bernadette Lecerf-Thomas.

The six stimuli of decision-making

In the United States, Patrick Renvoisé and Christophe Morin,[29] two neuromarketing experts who founded the SalesBrain company in San Francisco, refer to six stimuli that condition consumer behavior in relation to purchasing decisions. They base their thinking on neuroscience specialists such as Robert Ornstein, Leslie Hart, Bert Decker and Joseph Ledoux. According to Renvoisé and Morin, the stimuli come from reflex reactions of the primitive or reptilian brain when it is solicited in a commercial context. This analysis has led them to suggest an approach aiming to make salespeople more effective. The six stimuli are:

- *Egoism*: The primitive brain is egocentric. It is only interested in, and sympathetic towards, that which directly concerns its well-being and survival.
- *Contrast*: The primitive brain is sensitive to contrasts. Contradictions allow it to make quick, risk-free decisions.

- *Tangibility*: The primitive brain likes tangible information. It is constantly seeking the familiar and the amicable, that which can be quickly recognized and is concrete and unchanging. It appreciates ideas that are concrete, simple and easy to grasp.
- *Beginnings and ends*: The primitive brain remembers the beginning and the end of an event, but forgets more or less everything in between. This limited ability to maintain attention has a significant impact on the manner in which a project or a sales argument should be presented.
- *Visualization*: The primitive brain is visual. The optic nerve is physically connected to the primitive brain and transmits 25 times as much information to it as the auditory nerve. The visual channel provides a rapid and efficient connection to the true decision maker.
- *Emotion*: The primitive brain reacts strongly to emotions.

The traps of intelligence

Patrick Georges[30] has brought to light certain weaknesses in the neuro-consumer's intelligence.

Through his lectures, training courses and consulting activities, he helps companies to organize their marketing, sales and communication activities so as to become "brain compatible" and has highlighted the following "traps" of intelligence:

- Attention and perception are limited. We cannot do two things well at the same time, for example think and decide. Salespeople try to take advantage of this weakness by pushing customers towards an immediate purchase.
- Short-term memory is limited. The seller needs to know how to prepare a presentation so that the audience retains what is essential.
- Language makes communication possible, but distorts it as well. You should not hesitate to get your interlocutor to repeat what he has understood.
- The brain can only deal with one-fifth of the information that it receives. Judgment and decisions can be distorted.
- We have two intelligences in us: a shrewder, slower intelligence that, after analysis, can tell us the opposite of what our automatic intelligence tells us, generating potential internal conflicts, and an ancestral intelligence that is reflex, rapid and automatic. It obeys nine simple rules:
  - What is beautiful is good. We intuitively judge people who are beautiful, talkative, tall and slim to be more intelligent.
  - What is different is dangerous: we are suspicious of what we don't know.
  - According to evolutionary principles, Georges contends that to reproduce, a woman will subconsciously prefer a man with a flat belly. Her subconscious tells her that he is strong and will offer her

better protection. A man will tend to prefer a plump woman. Wide hips and a full bosom will seem to him to guarantee easy births and a well-nourished progeny.[31] According to Ramachandran, she should preferably be blonde and fair-skinned, because she cannot conceal her state of health as easily as a brunette.[32]

- The more visible an object, the more important it is considered to be.
- The more permanent an object, the more important it is considered to be.
- The bigger an object, the more important it is considered to be.
- The more frequently something is repeated, the more it is considered to be true.
- The more accessible an object, the less important it is considered to be.
- That which is placed in first place is considered to be important.

Our environment can favor use of one or the other. We activate our shrewd intelligence when we have time and when the environment is favorable to this type of intelligence. It is sometimes sufficient to increase the speed of information, and thus the stress level, for a person who was in "shrewd judgment" mode to move into "automatic intelligence" mode. When fully understood, these ideas are important for better analyzing our own behavior and that of others.

Homeostasis to ensure the brain's internal behavioral equilibrium

To end this chapter, let us mention Bernadette Lecerf-Thomas, who emphasizes the importance of homeostasis, from the Greek *hómoios*, "similar," and *stásis*, "stability, the action of standing upright," i.e., "stay constant."

When our internal parameters are normal, we say that our body is in equilibrium. This state of equilibrium corresponds to homeostasis, an idea introduced by French physiologist Claude Bernard[33] (1813–1878). Homeostasis is the ability to maintain equilibrium of function in spite of external constraints. According to Bernard, "Homeostasis is the dynamic equilibrium that keeps us alive." Every living system must simultaneously satisfy conditions of stability and movement to remain alive.

If marketing specialists wish to change the environment and be able to act on it, they must change this equilibrium. Those who succeed know how to change homeostatic equilibrium using innovations that anticipate neuro-consumer's needs. In order to avoid creating stress that is too invasive for those who are subjected to the change, they must prepare and lead it when they propose it. When faced with an imbalance, the brain sets a great deal of store by what seems to be harmonious in order to reduce stress. On the basis of studies of homeostasis, Bernadette Lecerf-Thomas[34] *presents a number of recommendations enabling company management to be adapted in order to satisfy the expectations of employees, partners and customers when they are confronted with change.*

Notes

1. Paul MacLean, *A Triune Concept of the Brain and Behavior*, Toronto University Press, 1974. See also Paul MacLean and Roland Guyot, *Les Trois Cerveaux de l'homme*, Robert Laffont, 1990.
2. Frédéric Rava-Reny, "Le cerveau triunique de MacLean," 2007,www.rava-reny.com/Auteur_Rava-Reny/Le_cerveau_triunique_de_Mac_Lean.pdf.
3. Ibid.
4. Arthur Koestler, *The Ghost in the Machine* [1968], Calmann-Lévy, 1994.
5. Michel de Pracontal, *L'Imposture scientifique en dix leçons*, Le Seuil, 2005.
6. Patrick Renvoisé and Christophe Morin, *Selling to the Old Brain*, SalesBrain, 2003.
7. Gottfried Wilhelm Leibniz, *New Essays on Human Understanding* [1704/1765], Cambridge University Press, 1981.
8. Blaise Pascal, *Pensées* [English ed.], Pinnacle Press, 2017.
9. Johannes Peter Müller, *Des manifestations visuelles fantastiques* [1826], L'Harmattan, coll. "Psyché de par le monde," 2010.
10. David Eagleman, *Incognito: The Secret Lives of the Brain*, Vintage, 2012.
11. Gary Singer quoted in Patrick Capelli, "Le marketing s'invite dans nos cerveaux," *Libération*, May 13, 2012.
12. Eagleman, *Incognito*.
13. *Le cerveau et ses automatismes: Le pouvoir de l'inconscient* [film], Arte, Part 1: December 1, 2011.
14. Ferdinand Von Helmholtz, *Science and Culture: Popular and Philosophical Essays*, University of Chicago Press, 1995.
15. Eagleman, *Incognito*.
16. G. Tom, C. Srzentic and C. Nelson "Mere Exposure and the Endowment Effect on Consumer Decision Making," *Journal of Psychology*, vol. 141 (March 2007), pp. 117–125.
17. Quotation attributed to the actress Mae West.
18. L. Hasher, D. Goldstein and T. Toppino "Frequency and the Conference of Referential Validity," *Journal of Verbal Learning*, vol. 16, no. 1 (1977), pp. 107–112.
19. Eagleman, *Incognito*.
20. Charles Darwin, *On the Origin of Species* [1859], Macmillan Collector's Library, 2017.
21. Paul MacLean, *A Triune Concept of the Brain and Behavior*, Toronto University Press, 1974.
22. In particular, Vilayanur S. Ramachandran, "Why Do Gentlemen Prefer Blondes?" *Medical Hypotheses*, vol. 48, no. 1 (January 1997), pp. 19–20.
23. D.W. Yu and G.H. Shepard, "Is Beauty in the Eyes of the Beholder?" *Nature*, vol. 396 (1998), pp. 321–322.

24. Donald M. MacKay, *The Epistemological Problem of Automata*, Princeton University Press, 1956.
25. Eagleman, *Incognito*; *Le Cerveau et ses automatismes.*
26. Eagleman, *Incognito.*
27. Patrick Georges, Anne-Sophie Bayle-Tourtoulou and Michel Badoc, *Neuromarketing in Action: How to Talk and Sell to the Brain*, Kogan Page, 2013.
28. Daniel Kahneman, *Thinking, Fast and Slow*, Penguin, 2012.
29. Renvoisé and Morin, *Selling to the Old Brain.*
30. Georges et al., *Neuromarketing in Action.* See also Patrick Georges, *Gagner en efficacité*, Éditions d'Organisation, 2004.
31. Ibid.
32. Ramachandran, "Why Do Gentlemen Prefer Blondes?"
33. Claude Bernard, *Introduction to the Study of Experimental Medicine* [1865], Dover Publications, 1957. See also Gregory Bateston, *Mind and Nature: A Necessary Unity*, Bantam Doubleday Dell, 1988.
34. Elizabeth Lecerf-Thomas, *Neurosciences et management: Le pouvoir de changer*, Eyrolles, 2009.

CHAPTER 9

How age and gender condition the brain

Elements internal to the brain, such as age and the sex hormones, significantly condition neuro-consumers' behavior. According to A.K. Pradeep, who has contributed to a great number of studies based on medical imaging, "age and gender affect how the brain is wired. Environment, upbringing, culture and experience also affect it. But they are learned. Age and gender are not."[1] In order to better understand how the structure of their brains affects neuro-consumers' behavior, new forms of segmentation of the population could be developed based on age and gender.

The age of the neuro-consumer's brain

Since the work of the famous Swiss professor, epistemologist and psychologist Jean Piaget (1896–1980), it has been commonly thought that the brain is fully formed and its functions almost completely developed by the age of 12. Research based on cerebral imaging shows that in fact the organ is not mature until the subject reaches the age of 20–25.

Paul MacLean[2] describes progressive formation of the brain in three parts: first the reptilian brain, then the limbic brain and finally the neocortex. The volume of young neuro-consumers' heads occupied by the three components of the brain changes with their age before the organ reaches complete maturity. This development partly explains the significant changes that may be observed in their behavior.

The reptilian brain and its dominant influence in young children

The reptilian brain is the center of instincts, of the satisfaction of the primary needs enabling us to survive and reproduce. It gives priority to the family or external group. It respects the "leader," often represented by the father or the mother, but also by the strongest, likely to be able to provide protection against external dangers. This part of the brain is dominant in young children, up to the age of 8–12 depending on the individual. The disparity varies according to the family and the social, cultural and environmental context in which the child lives.

During the first five years of his existence, the young child's memory undergoes very significant development. His capacity for learning is considerable. He "records" objects, images and events. Different parts of his brain develop depending on what it is trying to learn. This phenomenon explains young children's great capacity for assimilating foreign languages. The areas of the brain related to language develop most quickly between 6 and 13 years of age.

Play is a fundamental aspect of learning where the child acquires the most important knowledge. It is not surprising that the young neuro-consumer shows a lively interest in all forms of physical and electronic games, particularly those that encourage learning and imagination.

Products from brands such as Lego, Fisher-Price, Playmobil, Barbie, Hello Kitty, among others correspond pretty well to the child's expectations and have tapped into the child's need to belong to a group and the desire to acquire the same objects as other members. Such products are all the more appreciated because they encourage relationships within the group, be it family or friends. Collections inviting exchanges and games for all the family to play are particularly attractive to these new consumers. Monopoly, among many other family games, benefits from this enthusiasm and the brand Nintendo owes a great part of its success to its Wii console, designed for use in the family.

The children's market is particularly important for marketing professionals. Roy Bergold, vice president of communication and creativity at McDonald's, frequently says in his lectures: "If you have a single dollar to spend, spend it on children." McDonald's is well-known for having put the emphasis on the children's market very early in its development, particularly with its famous clown but also with other measures specifically aimed at children.

The limbic brain conditions adolescent behavior

The limbic or "mammalian" brain is the center for the emotions, stress and instinctive behavior such as desire and aggression, but also for memory. It records experiences by changing them into personal memories. It recalls memories charged with emotion. It conditions learning and reflexes in relation to past experience. Professor of psychiatry David Servan-Schreiber[3] (1961–2011), from the University of Pittsburgh, USA, explains that with the onset of puberty, at around 12 years of age, the ovaries and the testicles start to work at full capacity. The hormones that they release into the bloodstream bathe the neurons of the emotional brain and stimulate the need to assert oneself, to be taken seriously, to discover what lies beyond boundaries and to test one's membership of the group. The discrepancy between hormonal maturation and the region of the brain, the neocortex, that allows the individual to think before launching into something, explains certain types of adolescent behavior that adults consider to be "immature." According to doctor Jay Giedd of the United States National Institute of Mental Health, "The wiring of the white matter, the bundles of neurons that ensure reliable transmission of nerve impulses, does not become mature before the age of 20 on average."[4] The definitive formation of the prefrontal cortex, responsible for controlling urges and the ability to project yourself into the future, only appear after that age. This maturation of the brain strongly conditions the adolescent neuro-consumer's behavior. This may explain the misunderstandings that can arise between parents and their adolescent children, which are natural, given that their brains are not at the same stage of development. This is why professor Servan-Schreiber advises adults to take more interest in what is bothering the adolescent rather than concentrating on what is worrying the adults. Parents' worries are, however, not unfounded, given that the two biggest causes of mortality among adolescents are accidents and suicide.[5]

Adolescent neuro-consumers, unlike their younger brothers and sisters, are mostly attracted by new products or brands, new fashions that can distinguish them from, or even oppose, adult tastes. They often readily express their emotions and reveal and share them on social networks.

> *Some brands, such as Converse shoes and the Abercrombie & Fitch chain of shops understand this attraction and try to respond to it. Adolescents often show an interest in causes and subjects that generate emotions (e.g., social, humanitarian, concerning the future of the planet, sustainable development, fair trade, etc.) and give priority to emotional communication rather than rational information. They like risk, particularly boys, who undergo significant increases in testosterone levels during this period.*

Aging of the brain and its consequences among older people

Around the age of 20–25, neuroscientists consider that the brain's development is complete with the formation of the neocortex. The neocortex is the "logical" part of the brain. It is the center for language, anticipation of acts, decisions, it allows us to make choices, take more rational decisions and to manage the future. It makes the neuro-consumer more "intelligent," even if, as we have seen, other parts of his brain make him more likely to commit impulsive and irrational acts.

From the age of 20, brain cells start to die at the rate of 10,000 a day and the ability to learn and acquire new physical or mental skills begins to deteriorate. At 30, the neuro-consumer still has little awareness of this, but at 60, he might start to be aware of his limits. At 70, the brain has lost 5% of its mass: this loss can reach 20% by the age of 90. According to some neuromarketing experts, such as A.K. Pradeep, after the age of 60, the consumer experiences significant changes in his expectations and behavior when confronted with demands from his environment. The amygdalae of older people, and consequently their memories, appear to be more active when they receive positive messages. The brain, having far fewer neurons, is more sensitive to direct messages with simple, clear images without anything superfluous. It memorizes communication (in the form of sounds, songs, texts or images, etc.) that recalls past experiences. According to Pradeep, it has a tendency to consider familiar, repeated information received through different media (press, radio, television, etc.) to be true.[6]

"Digital natives" and "digital immigrants"

One of the big inter-generational differences among neuro-consumers is their use of the Internet. In his book, *Don't Bother Me Mom, I'm Learning*,[7] Marc Prensky puts forward the idea that this new form of communication changes people's emotions and perception of the world. He makes a clear behavioral distinction between neuro-consumers born after 1990, whom he calls "digital natives," and those born earlier, whom he calls "digital immigrants." According to Prensky, digital natives prefer information in staccato form, with neither verb nor object. Hence their taste for rap and hip-hop. They can follow different pieces of information from different media at the same time. They do not feel the need to structure their thoughts, and often read in a random manner. They feel more emotion from colors and graphic design than from the organized writing of a text. They want everything to happen quickly. They are the greatest enthusiasts of online purchasing, but also of fast food, speed-dating, drive-through retail outlets, etc. For them, intelligence is in the mind, in the speed of processing information, not in a sense of hierarchy or maturity due to age. These ideas are reflected in now world-famous start-ups like Facebook, Apple and Google. Above all, Prensky suggests that digital natives appreciate information that is amusing and love sending humorous "viral films" to their community. They feel a great need to be regularly connected and to communicate on social networks. He suggests that for some digital natives, the need for permanent connection can become an addiction, regularly using their mobile phones to send text messages, take selfies and send them immediately to their online community, etc.

In contrast, Prensky argues that digital immigrants tend to prefer linear processing of information. He suggests that they accord importance to the logic of a text. They prefer a message that arrives slowly and that is coherent in its structure. They prefer to protect their privacy, are careful about sending information via social networks and sometimes prefer to work alone. They consider that intelligence is hierarchical, but also linked to experience.

The arrival of the Internet and the increasing presence of digital natives are in the process of profoundly changing the design of training and communication. As a medium, the Internet seeks to be more efficient with respect to this new group. In the United States, new approaches to training such as "serious games" and "digital games-based learning," which question classical teaching methods, are obtaining good results.

The significant emotional differentiation that exists between these two categories of neuro-consumers requires different approaches to communication, adapted to their ways of thinking and feeling. At the time of publication of this book, a large proportion of neuro-consumers will be digital natives, but the purchasing power will still belong to the digital immigrants. In this context, as Georges Chétochine[8] (1938–2010) has pointed out, it has become essential to create Internet sites that are suitably adapted to these two populations.

The brain influenced by its generation's experiences

Belonging to a generation that has experienced significant events has been found to have a deep influence on the brain. It can emotionally condition people who have been actors in or witnesses to such events, leaving indelible traces in their brain's emotional memory that remain throughout their existence. In *Contributions to Emotion Research and Theory*, Magda Arnold[9] suggests that emotions are related to our experience and that it is particularly important to include this idea in the process of population segmentation. American sociologists William Strauss and Neil Howe[10] distinguish four main generations corresponding to this segmentation, with neuro-consumers who belong to each group showing similar intra-generational behavior, but different behavior from one generation to another: seniors (born between 1901 and 1946), baby boomers (1946–1958), generation X (1958–1975) and generation Y (1975–1994). Bernard Préel has also discussed this generational segmentation in relation to European society, in his book *Le choc des générations* (*The Clash of Generations*).[11]

The "seniors" (1901–1946) The "seniors" generation is marked by memories of two World Wars, including the sacrifice of their families or friends. According to Strauss and Howe, this generation believes in morals and religious or personal ethics. It grew up with small shops, the butcher and the grocer in the center of the village or quarter. It is much more used to the radio, the newspaper and the local bar than the television or, to an even lesser extent, the Internet. Doctors and schoolteachers benefited from an aura of wisdom and were respected. It believes in the values of marriage and other institutions. It remains faithful to its youthful political ideas and to its employer, but also to the local doctor, baker and butcher. It regrets that they retire without being replaced by someone from their own family.[12]

The "baby boomers" (1946–1958) The "baby boomer" generation is also marked by memories of the Second World War, experienced and related by their parents. In countries like France, the United Kingdom and Belgium, they experienced the end of empire and the associated loss of prestige, often occurring under dramatic circumstances. Other countries, like Germany and Japan, underwent the drama of occupation, the destruction of their country and often a sense of shame at the totalitarian regimes that preceded them. During this period, the United States was experiencing strong economic growth and pride at its greatness and its emergence as a world power. This was also the beginning of the Cold War. The baby boomers also lived at a time of reconstruction and economic development in Europe. Purchasing power was gradually increasing. The growing number of consumer products attracted them. This was the "marketing generation." It was the generation that thrilled to the May '68 civil unrest in France and the hippie movement in the United States. It was the time of sexual liberation and the introduction of the contraceptive pill. The baby boomer generation made Club Med a success but also widely conserved an attitude similar to that of the previous generation: fidelity and respect for institutions. Less austere than its elders, it developed a heightened sense of hedonism, a marked taste for consumption and liberalism and tended to be less strict in the upbringing of its children.

Generation X (1958–1975) Generation X was born in a period of transition and decline. It was the end of imperialism and the Cold War. In the United States, it was the time of the Vietnam War, strongly contested by youngsters on university campuses. This generation felt the full effects of the economic crisis. It saw its parents made redundant and a marked increase in divorce. But it was unprepared for any of this. The birth rate fell markedly, leading to it often being referred to as the "baby bust" generation. Generation X tended to reject values that were held dear by the two previous generations. It had a certain taste for adventure, anarchy, rejection of institutions, cynicism and the counterculture. British journalists Charles Hamblett and Jane Deverson wrote: "This was the generation that slept together before marriage, didn't believe in God, didn't like the Queen and didn't respect its parents."[13] It may be characterized by pragmatism and individualism but also by a degree of pessimism about the future. It is computer literate. It was raised on television and did not consider work to be an essential value, but rather a difficult challenge. It attaches importance to music and sport, particularly extreme sports. In his book, *Generation X: Tales for an Accelerated* Culture, Douglas Coupland[14] describes its intrinsic characteristics.

Generation Y (1975–1994) Generation Y has been described in several books[15] and, according to these authors, unlike the previous generation, which was surprised by numerous crises, generation Y had time to adapt. They are often self-confident, optimistic, independent, educated and perspicacious. Having lived in situations where both parents worked or were divorced, the children often found themselves alone. Relationships with the friends that they have chosen are often important for them. They tend to be more tolerant and less radical than their elders. Competent in using personal computers and the Internet, they know their personal worth and know how to negotiate it with their employers. They like working in teams, but consider that with modern means they are not obliged to stay in their offices to accomplish tasks that can be carried out at home. They do not feel at ease in hierarchical conflicts and sometimes prefer to resign rather than face them or when their job no longer interests them. They do not treat either work or hierarchy as a priority. They want to make the most of their leisure time and request time off to unwind. Mental and physical health is important for them. They seek a good quality of life that can reconcile work and other interests. They are mobile, keen to discover the world and to establish professional or personal relationships in other countries or cultures than their own. They are close to the generation of digital natives that has emerged after them and will perhaps form generation Z.

According to the above-mentioned authors, the particular generation to which a neuro-consumer belongs plays an important role in the unconscious behavior of his brain. Because the emotions felt by different groups may be very different, understanding them is useful both for developing the types of products and services that may satisfy them and for determining the distribution and communication methods best adapted to their desires. Appropriate marketing approaches are required for each segment.

The brain and the gender of the neuro-consumer

Neurologists are unanimous that the brain as such does not have a gender, but this does not prevent some authors from distinguishing net differences in the behavior of male and female neuro-consumers. Many publications in the United States and Europe attest to this difference. Among them is John Gray's well-known book, *Men Are from Mars, Women Are from Venus.*[16] Other books on the subject have been written by Claude Aron, Doreen Kimura, Simon Le Vay and Serge Ginger.[17]

It should be noted that this approach is often controversial and has been partially called into question by the "queer theory" proposed in the United

States by Judith Butler,[18] which has also been taken up in Europe by Marie-Hélène Boucier,[19] sociologist and assistant professor at the University of Lille III. "Queer" (or "gender") theories consider that the male–female dichotomy is too simplistic. They support the principle of a multiplicity of genders with several variants – heterosexuals, bisexuals, gays, lesbians and different categories of transsexuals. Their approach contests the existence of a typically male or female gender, preferring to opt for a "unisex" gender. Queer theory has recently become fashionable again and is reflected in the development of lesbian, gay, bisexual and transgender (LGBT) movements across many European countries.

While remaining aware of the above-mentioned caveats, in our research we refer to psychologists, clinicians and Gestalt therapy psychotherapist-educationalists such as Serge Ginger, John Gray, Doreen Kimura, Simon Le Vay and Claude Aron and A.K. Pradeep.[20] Their research and publications are based on converging studies from various fields, including psychology, genomics, neurology, medicine and religion.

More recently, in experiments carried out using magnetic resonance imaging (MRI), Ragini Verma and her colleagues from the University of Pennsylvania have shown the existence of differences in connections in the male and female brain. They found that the male brain has stronger connections within each hemisphere, whereas in the female brain, the connections are strongest between the hemispheres. Ragini Verma notes in her lectures: "It is fascinating that we can see structurally some of the functional differences between men and women."[21]

According to Serge Ginger:

> At this time, when we have just finished the first phases of deciphering the human genome, we have been able to show that man and apes possess a basic gene pool with 98.4% in common, i.e., leaving only 1.6% that is different ... compared to a genetic difference of about 5% between men and women. Thus, a male human being is physiologically closer to a male ape than to a woman while naturally, female apes are closer to women.[22]

But that is not enough to determine the behavior of a neuro-consumer.

Most researchers currently consider that a person's character is about one-third inherited, one-third congenital and one-third acquired through culture, upbringing and life experience. The epigenesis of the brain, i.e., its growth in relation to its cultural environment, was presented by Pierre Changeux as early as the 1980s, in his book *L'Homme neuronal* (*Neuronal Man*).[23]

The behavior of male and female neuro-consumers is largely considered to be conditioned by preexisting biological dispositions onto which are grafted influences from upbringing and culture. The suggested distinction between the "female brain" and the "male brain" is interesting in obtaining a better understanding of the differences in behavior between individuals of the two genders. This view must, however, be moderated by two factors. First, differences between individuals are often greater than those between genders. Second, according to Serge Ginger, it is estimated that about 20% of men have a "female" type brain and 10% of women have a "male" brain.[24]

The behavior of the female neuro-consumer's brain

A large majority of neuroscientists agree that the left hemisphere of the brain is more strongly developed in women, who are subject to the influence of hormones such as estrogen (the primary female sex hormone that cause development of the breasts, etc.) and oxytocin (the hormone that develops the need for tenderness, attachment, understanding, trust and love). This phenomenon is observed to an even greater extent when she becomes a mother. The female brain is also considered to be conditioned by genetic transmission between generations. According to evolutionary principles, this goes back to the Stone Age when people lived in groups and the main role of women was to look after the children with the other females in the clan. Serge Ginger contends that women are often more inclined than men to verbal sharing and communication. He suggests that they like talking and being listened to, that they need to share their ideas, feelings and emotions. Adult women spend an average of 20 minutes per telephone call, compared to 10 minutes for men.[25] According to Doreen Kimura, women often tend to be well orientated in time, often poorly so in space. She argues that contrary to certain widely held beliefs, women are actually less emotional than men, but that this is not always apparent because women tend to express their emotions more.[26] On the other hand, Ginger suggests that women are more sensitive, because more of their senses, particularly hearing, smell and touch, are more strongly developed than in men.[27]

Studies show that women hear things approximately 2.3 times louder than men. According to Serge Ginger:

> Women listen with both hemispheres, which means they hear a speech colored with emotions, perceived subjectively through their desires, fears, ethical and social values (sometimes feminist). They hear what I say, but above all how I say it. They are sensitive to the inflections of my voice and the rhythm of my respiration. They are sensitive to voices and music. Statistically, women sing more in tune than men.[28]

Research suggests that women's sense of smell is also more refined than that of men and that it can be 100 times more developed at certain periods in their menstrual cycle.[29]

Nineteenth-century glove makers knew how to make the most of this highly developed female sense to increase their sales by including perfumes in the design of gloves for women.

According to Serge Ginger, women's Jacobson's organ, considered to be a "sixth sense," enables them to perceive pheromones, the only neuro-transmitter hormones that can be transmitted from one person to another, which would explain women's apparent "intuitive" abilities.[30]

Finally, the sense of touch. Women possess ten times as many touch receptors in the skin. Their secretions of oxytocin and prolactin (hormones that favor attachment and the need to be hugged) increase their need to touch and be touched.

The brand Antoine & Lili has developed the texture of its textiles to make them agreeable to the touch and includes perfumes in them. Thanks to this, they have met with real success among female neuro-consumers.

An understanding of the specific attributes of the female brain is essential to better comprehend psychology and sexology. It is also useful to an understanding of female neuro-consumers' expectations and perceptions.

The behavior of the male neuro-consumer's brain

The right hemisphere of the brain is more strongly developed in men, who are subject to the influence of testosterone (the hormone regulating aggression, sexual desire and military and romantic conquest). According to Serge Ginger, men are more emotional than women, but express their emotions less. He contends that men are focused on action and competition, are well orientated in space, know how to find short-cuts and have a "clearer vision" than women. Evolutionary speaking, these faculties are believed to stem from the Stone Age, when men's main role in society was to hunt for food and defend the clan. Ginger suggests that the secretion of testosterone accentuates men's taste for adventure, new experiences, risk-taking and the instinct for domination. It provides men with endurance and tenacity. The male brain has good spatial memory whereas the female brain has better temporal and verbal memory and better remembers colors and the location of objects.[31]

The need to address the two brains differently

The apparent distinction between the male and female brains calls for careful differentiation in approaches to marketing, communication and sales. It enables responses to be found to male and female consumers' deepest needs. It appears to be easier to meet the male neuro-consumer's needs than those of the female.

> *This is, for example, the case for cars, luxury products, good restaurants, etc. Patrick Georges, Anne Sophie Bayle-Tourtoulou and Michel Badoc suggest that men are particularly receptive to the sight of products or publicity referring to sex. They note:*
>
> > *Advertising referring to sex can't fail to attract men. One-fifth of the advertising in the world is related to sex. The turnover of the sex industry is greater than that of the automobile industry. One of the most-often re-transmitted viral films in the world shows Paris Hilton washing a car while wearing particularly suggestive clothes and pausing suggestively, in order to promote a brand of hamburger that is generally forgotten.*[32]
>
> *It is, however, worth noting that although advertisements referring to sex play an important role in attracting men's attention, this does not translate so easily into purchases. Some may even prove to be counterproductive, because of over-seductive visions that hide the advantages of the product or service*

According to Roger Dooley,[33] the male neuro-consumer is more receptive to products or services offering a short-term advantage. He suggests that men prefer publicity that is simple and direct and avoids verbiage, that they prefer images to text and attach more importance to price than women do.

Dooley contends that female consumers have more complex needs and process information both rationally and emotionally thanks to better connections between the two hemispheres of their brains. According to Dooley, the sight of nudity attracts women less. He suggests that their attraction towards a salesperson of either sex depends on a range of factors perceived by their dominant senses: the voice and the smell, but also the salesperson's facial expression, their ability to listen and the quality of their answers to the female consumer's questions. According to Dooley, women prefer written, documented communication.[34] A.K. Pradeep argues that women like the social media that enable them to express themselves, to communicate their ideas, where they can meet comrades who share their tastes. He suggests that women filter rational messages through their emotions, that they are interested in stories relating to a brand or a product and prefer positive communication.[35]

According to Pradeep, before choosing a product or service, women tend to listen to the advice of others who have already tried it, as well as to that of experts. They are interested in the testimonial of recognized testers. Their sense of communication leads them to obtain information from friends and colleagues, but also to compare products or services before buying. Pradeep suggests that women are less impulsive than men, even if buying can often be a therapeutic activity for them, and that tactile and olfactory qualities can influence women in certain purchases, particularly where textiles are concerned.[36]

Important changes when a female neuro-consumer becomes a mother

In the United States, authors such as Katherine Ellison[37] and Michael Neman and Thomas Insel[38] have demonstrated that female neuro-consumers undergo profound behavioral modifications when they become mothers. They suggest that these changes result from significant increases in the secretion of hormones such as oxytocin, prolactin and cortisol, which lead to a change in their brain's priorities.

According to these authors, when a female neuro-consumer becomes a mother, one of her brain's main preoccupations is her offspring, making her very receptive to offers for products and services, but also communication related to the health, safety, well-being and happiness of her children and family. They are very receptive to testimonials, to the experiences of other mothers and child specialists, who they consult on social networks or online discussion forums. They prefer messages that are positive and optimistic. They more easily memorize messages that are easy and agreeable and that they feel to be authentic and honest. They are subconsciously attracted to shapes that they can identify as resembling a baby's face. Because their sense of smell has become more strongly developed, they avoid smells they consider to be noxious. They like products and shops that have an agreeable smell and produce a feeling of cleanliness. They find places with a calm, serene musical atmosphere agreeable.[39]

Psychological and sociological studies of the differences between the two sexes are numerous, even if they are sometimes disputed. There are fewer studies in marketing and communication in relation to these differences. They should be developed to take more account of gender-based behavioral differences, but also of the events that create endogenous factors that could lead to changes in the perceptions of neuro-consumers' brains. Today, researchers are interested in maternity. Tomorrow, they will perhaps find specific modes of behavior in the brain related to paternity or becoming a grandfather or grandmother.

Notes

1. A.K. Pradeep, *The Buying Brain: Secrets for Selling to the Subconscious Mind*, Wiley, 2010.
2. Paul MacLean, *A Triune Concept of the Brain and behavior*, Toronto University Press, 1974.

3. David Sevan-Schreiber, *The Instinct to Heal: Curing Depression, Anxiety and Stress Without Drugs and Without Talk Therapy*, Rodale Books, 2005 and *Anticancer: A New Way of Life*, Michael Joseph, 2011.
4. Jay Giedd, Jonathan Blumenthal, Neal O. Jeffries, F.X. Castellanos, Hong, Liu, Alex Zijdenbos, Tomáš Paus, Alan C. Evans and Judith L. Rapoport, "Brain Development During Childhood: A Longitudinal MRI Study," *Nature Neuroscience*, vol. 2 (1999), pp. 861–863.
5. Melonie Heron, "Deaths: Leading Causes for 2017," *National Vital Statistics Reports*, vol. 68, no. 6 (2019), pp. 1–77.
6. Pradeep, *The Buying Brain.*
7. Marc Prensky, *Don't Bother Me Mom, I'm Learning*, Paragon House, 2006.
8. Georges Chétochine, *Le Marketing des émotions: Pourquoi Kotler est obsolète*, Eyrolles, 2008.
9. Magda Arnold, *Contributions to Emotion Research and Its Implications*, Psychology Press, 2006 and *Emotional Factors in Experimental Neurons*, Sagwan Press, 2015.
10. William Straus and Neil Howe, *Generation: The History of America's Future*, Morrow, 1991.
11. Bernard Préel, *Le Choc des générations*, La Découverte, 2000.
12. Straus and Howe, *Generation.*
13. Charles Hamblett and Jane Deverson, *Generation X*, Tandem Books, 1965.
14. Douglas Coupland, *Generation X: Tales for an Accelerated Culture*, St Martin's Griffin, 1992.
15. Buddy Hobart and Herb Sendek, *Gen Y Now: Millennials and the Evolution of Leadership*, John Wiley & Sons, 2014. See also Olivier Rollot, *La Génération Y*, PUF, 2012; Florence Pinaud and Marie Desplats, *Manager la génération Y*, Dunod, coll. "Best Practices," 2011.
16. John Gray, *Men Are from Mars, Women Are from Venus*, Harper Collins, 1998.
17. Claude Aron, *La Sexualité: Phéromone et Désir*, Odile Jacob, 2000; Doreen Kimura, *Sex & Cognition*, MIT Press, 2000; Simon Le Vay, *Le cerveau a-t-il un sexe?* Flammarion, coll. "Bibliothèque scientifique," 1994; Serge Ginger, *Gestalt Therapy: The Art of Contact*, Routledge, 2007.
18. Judith Butler, *Gender Trouble*, Routledge Kegan & Paul, 1990.
19. Marie-Hélène Bourcier, *Queer Zone: Politique des identités sexuelles et des savoirs* (2001), Amsterdam Poche, 2011.
20. Pradeep, *The Buying Brain.*
21. Ragini Verma, "Brain Connectivity Study Reveals Striking Differences Between Men and Women," *Penn Medicine News*, December 2, 2013, www.pennmedicine.org/news/news-releases/2013/december/brain-connectivity-study-revea.
22. Serge Ginger, lecture in Vienna to the third World Psychotherapy Congress, 2002, published in Serge Ginger, "The Evolution of Psychotherapy in Western Europe," *International Journal of Psychotherapy*, vol. 8, no. 2 (July 2003), pp. 129–139.
23. Jean-Pierre Changeux, *Neuronal Man*, Princeton University Press, 1997.
24. Ginger, "The Evolution of Psychotherapy in Western Europe."
25. Ginger, *Gestalt Therapy.*
26. Kimura, *Sex & Cognition.*
27. Ginger, *Gestalt Therapy.*
28. Serge Ginger, "Female Brains vs Male Brains," paper presented at the 3rd World Congress of Psychotherapy, Vienna, July 15, 2002, pp. 2–4.

29. Ibid.
30. Ibid.
31. Ibid.
32. Patrick Georges, Anne-Sophie Bayle Tourtoulou and Michel Badoc, *Neuromarketing in Action: How to Talk and Sell to the Brain*, Kogan Page, 2013.
33. Roger Dooley, *Brainfluence: 100 Ways to Persuade and Convince Consumers with Neuromarketing*, John Wiley & Sons, 2012.
34. Ibid.
35. Pradeep, *The Buying Brain.*
36. Ibid.
37. Katherine Ellison, *The Mommy Brain: How Motherhood Makes us Smarter*, Basic Books, 2005.
38. Michael Neman and Thomas Insel, *The Neurobiology of Parental Behavior*, Springer, 2003.
39. Ellison, *The Mommy Brain*; Neman and Insel, *The Neurobiology of Parental Behavior.*

CHAPTER 10

How memory conditions the brain

Understanding how memory works leads to better knowledge of neuro-consumers' behavior. This is particularly true when the consumer is solicited by messages and subject to brand influence. In their book about the brain,[1] Doctors Arthur and Mitchell Bard devote a whole chapter to explaining how memory as a whole and individual memories work.

In *Descartes' Error*, neuroscientist Antonio Damasio[2] puts forward the theory of "somatic markers." For their part, Giacomo Rizzolatti and his colleagues from the Parma Faculty of Medicine are interested in a new theory that is revolutionizing our knowledge of the brain's behavior: "mirror neurons." Patrick Georges, Anne-Sophie Bayle-Tourtoulou and Michel Badoc[3] have thought about the applications of these theories to the behavior of the neuro-consumer confronted with messages and brand policy.

Memory and how it works

Contrary to widely held belief, memories are not stored in a specific zone of the brain like a computer file. The prefrontal cortex plays a specific role in the conservation of memory, but it shares that role with other zones such as the hippocampus and the amygdala. Memories are equivalent to pieces of information that enter the brain and can be recalled at need. They concern events but also texts, words, images, tastes, smells and feelings. When the neuro-consumer sees, hears or receives information, the brain creates new circuits in the form of synapses distributed among these structures. The objective of these circuits is to recall the appropriate memories. Forgetting occurs because circuits disconnect progressively when they are not used.

Researchers believe that there are several sorts of memory (see Figure 10.1). *Explicit memory* refers to mental representations solicited by the conscience. *Implicit memory* refers to automatic, subconscious processes such as getting dressed, brushing one's teeth or driving a car while automatically obeying the Highway Code.

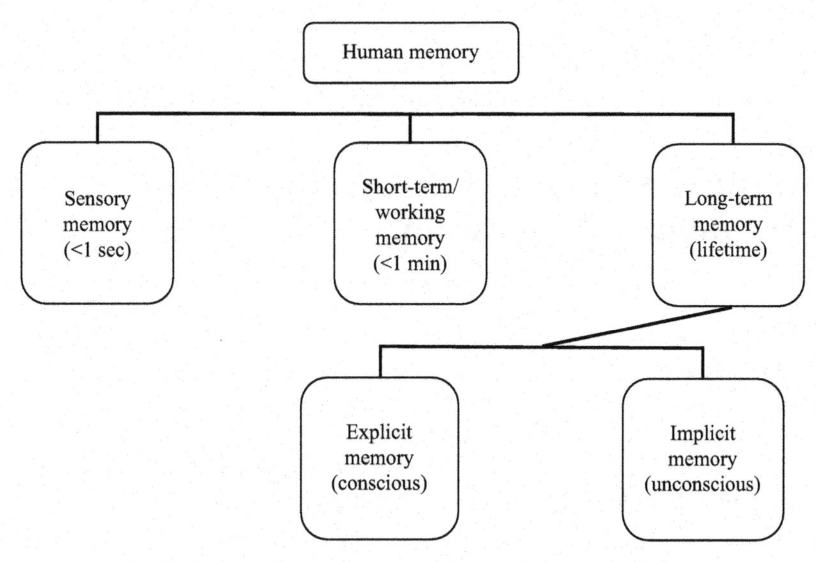

Figure 10.1 The different types of memory and how they work

Researchers often distinguish three levels of memory:

- *Short-term memory*: This hardly lasts more than 30 seconds. The words that we have read at the top of a page in a novel have already been forgotten by the time we read the bottom of the same page. Most of the sentences that we hear during a lecture, however interesting, are lost a few minutes after they have been spoken.
- *Working memory*: This type of memory maintains information for as long as it is useful.
- *Long-term memory*: This can last for a whole lifetime, even if it is often said that people cannot remember anything before the age of 3.

Arthur and Mitchell Bard use an approach enabling us to follow the pathway of information as the brain processes it. Information enters via sensorial memory, then becomes working memory. If it is learned, it can go into long-term memory. It appears, however, that short- and long-term memory can be independent of one another.

In order to be stored in long-term memory, information must be encoded. Only information that we need, or is of personal interest, producing a strong emotion, or relating to exceptional events, is transferred into long-term memory. The neuro-consumer essentially remembers messages and brands that meet these criteria. Others are forgotten soon after being seen. It is thus easy to understand the low rate of memorization of most television advertisements, even if repetition has a positive effect on the extent to which they are remembered. Numerous neuroscientific experiments have shown that the memory is reorganized by the brain to meet its own objectives. It can also prove to be wrong to a significant degree, even among sincere neuro-consumers. Sensorial perception plays an important role in memory recall.

Somatic markers and their influence on the behavior of the brain

The theory of somatic markers has been highlighted by neuropsychologist Antonio Damasio in several publications and particularly in his book *Descartes' Error*.[4] It has been the subject of several studies by Damasio himself and his colleagues, including Hanna Damasio and Antoine Bechara. His ideas are based on the fact that memories, often associated with emotions or sensorial stimuli, remain in the subconscious. They are somatic markers that can be reactivated when the person is confronted with the same stimuli. Neuropsychologist Nicolas Vermeulen summarizes the theory:

> The orbitofrontal cortex associates emotional sensations, which are implicit and automatic, with a stimulus. It records the relationship at the same time and is, moreover, able to reactivate the emotional sensations when it meets the conditioning stimulus again. Damasio (1995) uses the term "somatic markers."[5]

The neuro-consumer is subconsciously influenced by his own somatic markers.

Marketing professionals try to make the most of this by organizing the sensoriality of brands and sales outlets, with the help of specialists in cognitics. Through sensoriality, they try to associate the brand with the memory of a comparison that is favorable to it, stored in the neuro-consumer's memory.

> *When it hears a German-sounding brand name, the European neuro-consumer's brain often associates the name, without even thinking about it, with the idea of technical excellence and reliability. By choosing the name Weston, the well-known 125-year-old French brand of shoes evokes British style. By using the name K-Way, the windbreaker manufactured by Duhamel, also of French origin, associates itself with the idea of American efficiency, essential when confronting storms in the western United States.*
>
> *Just like the narrator of Marcel Proust's novel,[6] the smell in a sales outlet can give the customer a feeling of well-being by reminding him of times when he has been happy and making him predisposed to buy. Some sales outlets like Nature et Découvertes use sensorial ingredients that are agreeable to the neuro-consumer's senses.*

Some colors are also attributed precise meanings. The achromatic colors – white, black and gray – are often associated with an image of luxury, or the "top of the range." They can also have different associations depending on the culture or the country. White is associated with purity and happiness in Western countries, whereas it is linked with death in many Asian countries. Orange is considered festive in the Netherlands and appreciated in the Ukraine, where in 2004 it became the color of the Orange Revolution that led Yushchenko to the presidency. On the other hand, it may have a negative connotation in Ireland for members of the Catholic Church, where it can be associated with the Orange Order, a Protestant organization. Brands can seek to link themselves to somatic markers to project an emotive image. They can also try to create such images by telling real or invented stories, recalling happy or exceptional moments in neuro-consumers' memories. The story of the brand told as a saga invites consumers to associate themselves with the tale and to share the hero's feelings and values.

> *When he buys Chanel products, he is subconsciously part of the Coco Chanel saga (originally Mrs. Gabrielle Chasnel). When he buys a Ralph Lauren shirt, he is wearing part of the British Empire. With Lacoste, he is part of the world of tennis. In Shanghai, Dunhill has created a sales outlet in an old colonial house decorated to give the impression of being Alfred Dunhill's former home.*

Their story is told to create a feeling of nostalgic affection for the past among interested neuro-consumers.

Deliberately or otherwise, major brands often use ingredients that solicit somatic markers in the brain linked to elements that touch upon the collective subconscious.

> *For example, a secret, like that of Coca-Cola's recipe or Nike's air bubble, the life of an exceptional person, for example Kentucky Fried Chicken's colonel; or British and American Second World War aviators wearing Mac Douglas or Chevignon leather jackets.*

The brain conditioned by its mirror neurons

Professor Giacomo Rizzolatti's research team discovered mirror neurons by accident. They were first observed in the 1990s using cerebral imaging in the ventral premotor cortex of macaque monkeys. They were proved to exist in man a few years later. Several publications describe experiments, provide definitions and explain the usefulness of mirror neurons,[7] including the fundamental work by Giacomo Rizzolatti and Corrado Sinigaglia, *Mirrors in the Brain*.[8] Dr. Vilayanur Ramachandran, professor of neuroscience and director of the Center for Brain and Cognition at the University of California, San Diego, wrote:

> Discoveries about mirror neurons are the most important untold story of the decade. I predict that mirror neurons will do for psychology what DNA did for biology. They will provide a unifying framework and help to explain lots of mental states that up until now have remained mysterious and inaccessible to an empirical approach.[9]

Mirror neurons are a class of neuron situated in the brain's premotor cortex, which activate not only when an individual – human or animal – carries out an action but when watching someone else carry out the same action or even when imagining that action – hence the word "mirror." When reading a novel including characters or scenes that he finds moving, the reader can experience the same emotions or feelings as those felt by the hero. Giacomo Rizzolatti's important discovery was to demonstrate the existence of this activation linked to the simple observation of an action carried out by another individual.

For neuroscience experts, mirror neurons play an important role in social cognition, particularly in learning by imitation. It is partly the case in body language, nonverbal communication and learning sports. Watching how a top-level sportsman or woman performs generates mimicry likely to improve a beginner's game.

These neurons also play a part in emotional processes such as empathy. Professor Vilayanur Ramachandran then calls them "empathic neurons." Psychologists, including Frans de Waal, Jean Decety and Vittorio Gallese,[10] have carried out research into the role of mirror neurons in empathy, i.e., the ability to perceive and recognize another individual's emotions. Their work is based on the idea that there would appear to be a mirror neuron for emotions. When we observe an emotion in another person, we feel the same emotion albeit nuanced depending on the degree of affection we have for that person. Everyone has experienced the power of a simple smile to create a real feeling of sympathy in another. Mirror neurons thus enable communication with others. For A.K. Pradeep,[11] mirror neurons display stronger activity in women than in men. For Patrick Georges,[12] the tendency towards imitation thanks to mirror neurons can be triggered as soon as the premotor cortex, the zone of the brain providing the link between decision and action, is warmed up. If it overheats and its electrical activity spills over into the neighboring zone, action may result, for example when an animal starts twitching its tail before acting. Another example is a person sitting in front of his television watching a football match who starts shouting to encourage his team or instinctively kicking an imaginary ball. The reaction of mirror neurons is easy to observe. In a big railway station, everything is calm and the passengers patiently await information about the arrival of their train. Suddenly someone starts running. He is quickly followed by other people who start running after him without any real reason.

> *The effect of mirror neurons may have conditioned the exceptional mobilization of the French people after the terrorist killing of* Charlie Hebdo*'s journalists and cartoonists and the assassination of the customers at the kosher supermarket on Boulevard de Vincennes. There was also an exceptional mobilization among heads of state from all over the world who came to support the French president, as well as among the country's citizens. The killings, although highly symbolic, caused nothing like as many deaths as 9/11 or the Madrid railway station bombing. And far fewer than the several tens of millions who died during the Second World War. Nevertheless, according to the media, it led to the mobilization of millions of people, as many as demonstrated in France after the Allied victory and the capitulation of Nazi Germany. The newspaper itself, which normally sold about 60,000 copies per week, received thousands of additional subscriptions from one day to the next. Its print run went up to 7 million in just a few days. Some people stood in line at newspaper kiosks from 6 a.m. on the first day, even though they knew that the newspaper would be reprinted as often as necessary and that they would easily be able to get it in the days that followed. Apart from the important role of emotion in attributing a value to an event, the phenomenon of "Je suis Charlie" illustrates the importance of mirror neurons in guiding the behavior of people in gatherings and of consumers making purchases.*

The empathy effect explains why a person watching a film or reading a novel can suddenly burst into tears while watching a scene or reading a passage that has no link or relationship to himself.

Again, according to Patrick Georges, some phenomena like stress, or the visualization of strong authority can increase the imitation effect created by mirror neurons.[13] Knowledge of their effects is particularly important in understanding certain instinctive and sometimes irrational types of behavior among neuro-consumers.

> *An empty restaurant does not attract customers, who often prefer to stand in line to get into one that is full without knowing anything about the quality of its menu. Some chains have learned how to generate stress by organizing artificial lines or orchestrating rarity. In Zara clothes shops, owned by the Spanish group Inditex, products are not restocked. If a customer does not buy the item that appeals to him there and then, he cannot be sure of being able to find it, or even one like it, if he comes back another day. Abercrombie & Fitch organize artificial lines by filtering people at the entrance to its boutiques. Employing false customers to buy designated products is another well-known trick used by a few shops to artificially increase their sales. The presence in films of*

products used by favorite actors also contributes to increased sales, but only if they are well displayed and have a significant role in the screenplay.

Methods aimed at persuading customers to buy by imitation are implemented, in a pragmatic and empirical way, by the marketing, sales and communication departments of a number of companies. The techniques have been developed from practice and observation, often with no understanding of how mirror neurons contribute to the process. Future research enabling us to better understand the role of mirror neurons in the cognitive processes of buying, as well as in the processes of perception of information and messages, will significantly improve knowledge of certain hitherto unexplained instinctive types of behavior among neuro-consumers.

Notes

1. Arthur S. Bard and Mitchell G. Bard, *The Complete Idiot's Guide to Understanding the Brain*, Alpha, 2002.
2. Antonio Damasio, *Descartes' Error: Emotion, Reason and the Human Brain*, Vintage, 2006.
3. Patrick Georges, Anne-Sophie Bayle Tourtoulou and Michel Badoc, *Neuromarketing in Action: How to Talk and Sell to the Brain*, Kogan Page, 2013.
4. Damasio, *Descartes' Error*.
5. Nicolas Vermeulen, "Les émotions en tant que 'marqueurs somatiques' (Antonio Damasio)," *Le Psychologist.be*, July 5, 2014, www.lepsychologue.be/articles/damasio_marqueurs_somatiques.php.
6. Marcel Proust, *In Search of Lost Time* [1913–1926], Everyman, 2001.
7. Vittorio Caggiano, Leonardo Fogassi, Giacomo Rizzolatti, Peter Their and Antonio Casile, "Mirror Neurons Differentially Encode the Peripersonal and Extrapersonal Space of Monkeys," *Science*, vol. 324, no. 5925 (2009), pp. 403–406; Jean Decety "Naturaliser l'empathie," *L'Encéphale*, vol. 28 (2002), pp. 9–20; Vittorio Gallese and A.I. Goldman, "Mirror Neurons and the Simulation Theory of Mind-Reading," *Trends in Cognitive Sciences*, vol. 2, no. 12 (1998), pp. 493–501; S.D. Preston and F.B.M. de Waal, "Empathy: Its Ultimate and Proximate Bases," *Behavioral and Brain Sciences*, vol. 25 (2002), pp. 1–72.
8. Giacomo Rizzolatti and Corrado Sinigaglia, *Mirrors in the Brain: How Our Minds Share Actions, Emotions, and Experience*, Oxford University Press, 2008.
9. Vilayanur Ramachandran, *Le cerveau fait de l'esprit: Enquête sur les neurones miroirs*, Dunod, 2011.
10. Gallese and Goldman, "Mirror Neurons."
11. A.K. Pradeep, *The Buying Brain: Secrets for Selling to the Subconscious Mind*, Wiley, 2010.
12. Georges et al., *Neuromarketing in Action*.
13. Ibid.

CHAPTER 11

The influence of emotions and desires

The role played by emotions

Emotions are difficult to define. There is nothing palpable about them. They are a particular state that can be felt inside oneself or result from a situation capable of inducing them. The way in which they manifest themselves is easier to identify.

The origin of the word "emotion" comes from the Latin *emovere* or *emotum*, which may be translated as "pick up" or "shake," and *movere*, which means "move." For the ancients, emotion was movement in the sense of "movement of the soul." Emotions produce psychic or behavioral changes inducing a positive or negative reaction that in turn leads to behaviors of approach or flight. Bryan Kolb and Ian Q. Whishaw define them as "a cognitive interpretation of subjective feelings."[1] In *The Chemistry of Our Emotions*,[2] French journalist, columnist and writer Sébastien Bohler summarizes current knowledge of their role in people's lives.

Emotions are closely linked to our interactions with the world. They relate to the social brain that allows us to live in a community. Their expression is easier to describe.

Psychologist Paul Ekman[3] lists primary emotions – joy, disgust, fear, anger and sadness – to which he adds variations related to their intensity. Emotions concern love, euphoria and optimism. Other psychologists add manifestations such as jealousy and embarrassment.

Neuropsychologist Antonio Damasio[4] distinguishes three ways in which emotions can be triggered:

- *primary emotions*, which represent an immediate adaptive and objective response, e.g., hunger, thirst, fear, desire, etc.;
- *secondary emotions*, calling on memory and trying to avoid a future problem, e.g., not to stroke an animal, not to stare directly at certain people, etc.;
- *tertiary emotions*, assimilated with empathy, i.e., the ability to understand the feelings of another person.

The behavior of the brain linked to emotions: the case of Phineas gage

In 1848, Phineas Gage was working on the construction of a railway line in the United States. As he was preparing to blow up a rock, an explosion went off unexpectedly and an iron bar went through his skull. It damaged the left frontal lobe of his brain. Despite the apparent seriousness of his injuries, after being treated by Doctor Harlow, the patient managed to survive. After his treatment, Phineas Gage got back to an almost normal physical state. He had no paralysis and recovered a major part of his intellectual faculties. But the doctor noticed a significant change in his mood and behavior. Before his accident, he was considered as a serious, sociable, reliable companion with good judgment. After his accident, his emotional, social and personal behavior became unstable and profoundly asocial. His mood became changeable, his temperament was unstable, capricious and rude. He died 12 years after his accident.

In 1867, Doctor Harlow had Phineas Gage's brain exhumed in order to study it. He did not obtain any conclusive information. Conserved at the Warren Anatomical Museum, it was once again studied in 1994 by neuro-anatomists Antonio and Hanna Damasio. They used a computer to reconstitute the trajectory of the iron bar through the skull. They compared Phineas Gage's case with other patients who had undergone profound changes of behavior after comparable lesions in the same part of the brain. There was the case of Elliot, a well-known business lawyer who had had the same part of the brain removed after suffering from a tumor. After his operation, he underwent a profound personality change comparable to that suffered by Phineas Gage.

Antonio Damasio drew the conclusion that feelings play a decisive role in the incessant flow of personal decisions. In some areas, reason becomes blind without feelings.

Formal logic will not enable us to choose a job or a partner or to know whether we can trust someone. Similarly, it will not help a neuro-consumer to know whether he should trust an offer, a salesperson or a message.

The intuitive signs that guide the neuro-consumer in these decisive moments come from visceral impressions of limbic origin. They come from somatic markers, instinctive feelings. Antonio Damasio established a close link between emotions conditioned by the brain and the physical behavior of patients. The work of this researcher and his team are the subject of many publications including the best-seller *Descartes' Error*, but also in more recent works.[5]

Emotions occupy a fundamental place in the conditioning of the behavior of the neuro-consumer. From a positive point of view, they are a source of inspiration for creation. This is particularly the case in artistic fields such as poetry, cinema, painting and innovation and creativity in publicity. From a negative point of view, they can lead to stressful situations that can cause profound depression, obsessive fears and pathological anger, etc.

The use of fear in certain messages can prove to be efficient in relation to neuro-consumers who cannot easily master their emotions. It increases the sale of products or services that promise to remove the source of the fear.

The physical manifestations of emotions (sweating, increased heart rate, facial expressions) are identifiable signs that are difficult to hide. They can be seen and analyzed by neuroscientific techniques and by well-informed professionals. Their use enables the emotional state of neuro-consumers to be detected.

Emotional intelligence

In his famous story *The Little Prince*, aviator and writer Antoine de Saint-Exupéry (1900–1944), wrote: "We only see well with the heart. The essential is invisible to the eyes."[6]

Now, at the beginning of the 21st century, several authors have called into question the predominant role played by rational intelligence, as measured by the intelligence quotient (IQ) and its famous test, designed in 1912 by German psychologist William Stern (1871–1938). Psychologist Howard Gardner argues that people are more or less intelligent: "They're born like that and there's nothing much anyone can do about it."[7] He shows that there is not one single form of intelligence but a large range of intelligences and particularly what he calls "personal intelligences." Studying former American university students, he was surprised to discover that those with a high IQ but low relational intelligence succeeded less well than those with a lower IQ but who were better equipped with relational intelligence. In Gardner's opinion, "We should spend less time classifying children and more time helping them to recognize their natural gifts and cultivate them."[8] There are an infinite number of routes to success. Social perspicacity is an aptitude that a good education should cultivate instead of ignoring or suppressing it.

A new form of intelligence that is not rational plays a predominant role in people's choices and behavior: emotional intelligence. It is defined as a person's ability to use emotions and feelings in making decisions.

Harvard psychologist and *New York Times* journalist Daniel Goleman has written a fundamental book, *Emotional Intelligence*.[9] For him, the real questions are: How can we make our intelligence match our emotions? How can we render daily life its civility? Community life its humanity? The author shows that although emotions have acted as good guides throughout the evolution of humanity, civilization's new realities appear so quickly that the slow march of evolution cannot keep up.

The first laws and moral codes – Hammurabi's laws, the Ten Commandments, the edicts of the Emperor Ashoka – were written as attempts to channel, dominate and domesticate the power of human passions emanating from the emotions.

In *Civilization and Its Discontents*,[10] Sigmund Freud suggests that society is obliged to impose rules on the individual aimed at containing his all-too-ready emotional outbursts. Despite social rules, it should be noted that passions often take precedence over reason and that rational intelligence is no use when it is under the control of the emotions. The stronger a feeling, the more the emotional spirit dominates and the less effective rationality becomes.

The amygdala, also the seat of emotional memory, controls the emotions. Joseph Le Doux,[11] a psychologist and neuroscientist at the University of New York, was the first to demonstrate the fundamental role that it plays in the emotional activity of the brain. He explains how the amygdala manages to determine our actions even before the thinking brain, the neocortex, can make a decision. In its interaction with the neocortex, the amygdala is at the heart of emotional intelligence. In an emergency, its vast network of neurons enables it to take over the greater part of the brain, including the rational mind. Joseph Le Doux manages to show that the amygdala possesses a bundle of neurons linking it directly to the thalamus. The existence of this direct link, short-circuiting the neocortex, partly explains how the emotions manage to overcome reason.

The amygdala is also the center for emotional memories. Although the hippocampus enables us to recognize our cousin, the amygdala reminds us that she is unpleasant. The more the amygdala is stimulated, the deeper the information is stored in memory. Memories of frightening events are particularly strongly imprinted. All the senses may also be associated with an emotion in the memory.

These interactions are of great importance when dealing with a neuro-consumer. Birdsong could give him an impression of security. Nostalgic music could put him in a good mood or enable him to remember the relevant advertisement better. The odor of a waffle or of chewing gum could take him back to a happy moment in his childhood. The message is memorized better when it is based on emotions. Brands that are identifiable through several senses are not only more credible, but also more easily remembered.

When he can no longer control his emotions, the brain of the neuro-consumer is highly vulnerable, particularly in a context where he is strongly solicited by marketing and its messages.

The aim of acquiring emotional intelligence is to help the subject to control his emotions. In 1989, Doctor Peter Salovey,[12] chancellor of Yale University, and John D. Mayer, a psychologist at the University of New Hampshire, were already thinking about the ingredients of emotional intelligence.

They distinguish five areas that define emotional intelligence:

- *Knowledge of emotions*: Knowledge of oneself. Socrates' (470–399 BC) famous injunction, "Know thyself," which is engraved as an epitaph on the temple of Apollo at Delphi, refers to this keystone of emotional intelligence. Anyone who is blind to his own feelings remains under their control. People who are capable of achieving this conduct their lives better.

- *Mastery of one's emotions*: The ability to adapt one's feelings to each situation depends on one's self-awareness. A set of methods can help to pacify the mind, to liberate the subject from the control of anxiety, sadness and anger.
- *Self-motivation*: This leads the individual to put himself into a state of "psychological fluidity." To defer the satisfaction of his desires and suppress his urges, to know how to channel his emotions in order to concentrate, to control and motivate himself.
- *The perception of other people's emotions*: Empathy is a vital element of interpersonal intelligence. Empathic people are particularly receptive to subtle signals indicating the needs and desires of others. Such people are often good at communication and commerce.
- *Understanding of human relations*: The ability to maintain good relations with others is largely about being able to manage one's emotions. People who have this quality often succeed in any enterprise that is based on harmonious relationships.

The neuro-consumer's brain under the influence of his emotions and his desires

While deciding whether to purchase something, the neuro-consumer is fully exposed to the influence of marketing that targets emotions and desire. Although he thinks, in all good faith, that he is making conscious, reasoned choices, he is largely guided by subconscious urges. They operate quite independently of reason. Knowledge of them can be used to make powerful tools of persuasion. Alain Berthoz, a professor at the Collège de France and the author of several works on the role of emotions, writes: "The fact of buying or not buying a product, the choice of a brand ... are nothing more than the result of starting the emotional motors in our brains."[13]

The neuro-consumer confronted by his emotions

Addressing the emotional memory of the neuro-consumer is a good way of getting his sympathy. Political marketing advisors know this well. They use, and sometimes abuse, the fact by having their candidates visit people who are victims of tragedies, such as serious illness, fire, flooding, accidents, commemoration of wars, etc. Great historical characters, from Alexander the Great through Napoleon and Louis XIV to General de Gaulle, have known how to orchestrate their people's emotions. Some people write their story, or have it written, to make it more romantic than historical reality.

The effects produced by the emotions on neuro-consumers are used in marketing and communication.

The neuro-consumer confronted by pleasure and emotional attachment

In 1954, two American psychologists, James Olds (1922–1976) and Peter Milner, identified a "pleasure center" in the brain.[14] The pleasure center lies in the accumbens nucleus and when stimulated by different factors like the contemplation of a moving scene, the reminder of a happy moment included in the somatic markers, a sexual relationship or the consumption of drugs, the nucleus triggers the production of dopamine, one of the neurotransmitter hormones, which produces a sensation of pleasure.

Starting with the brand, marketing and communication seek to communicate this sensation to the neuro-consumer in order to make him more open to their offers. The pleasures can be very different.

> *The pleasure of taste in drinking an exceptional coffee at Starbucks. Visual pleasure, followed by taste when children watch their parents preparing a Nutella sandwich. The pleasure of accompanying a retailer in the fight against the high prices imposed by manufacturers or their competitors and, more recently, in the fight to save the environment (like French retailer Leclerc). The pleasure of driving a Porsche, a BMW, a Mercedes, a Tesla. The pleasure of being at the height of fashion with Abercrombie & Fitch.*

Pleasure may also be felt in sales outlets that appeal to the neuro-consumer's senses, where the welcome and the layout of the shop are agreeable.

Big brands often draw their strength from the unique pleasure they bring to their followers.

> *Nostalgia stored in the somatic markers brings moments of pleasure. Advertising sometimes tries to bring them to the surface. To do this, they use well-known characters, living or dead. Christian Dior's advertisement for Eau Sauvage presents film actor Alain Delon at the time when he was considered to be a great seducer. Songs that moved consumers at another period are used to create a feeling of nostalgia and to transfer it to the product or brand being promoted.*
>
> *The attachment becomes even stronger when the brand manages to create a community. Some, like Apple, Nespresso, Converse, Google and Microsoft, make big efforts to create and maintain an emotive community feeling among their customers, particularly through the social media.*

The neuro-consumer subject to fear and anxiety

The amygdala is the brain's center of fear and stress. When activated, it stimulates the liberation of neurotransmitters, including the stress hormone, cortisol, produced by the adrenal glands, but also the fight-or-flight hormone, adrenalin.

Fear produces a strong reaction in the neuro-consumer, leading him to make a purchase with the objective of reducing or eliminating the fear. The feeling of fear for the neuro-consumer can take many different forms:

- *Fear for oneself*: To lose one's good health, to die, to get old, to put on weight, to no longer please others, to have yellow teeth, dry skin, bleeding gums, incontinence, premature ejaculation, etc.
- *Fear for others*: To have a baby with a wet nappy, a partner or children inadequately protected in case of death, a poorly nourished pet, etc.

Advertising managers know that fear plays an important role among neuro-consumers and often successfully use anxiety-producing messages to increase the sale of products or services aimed at reducing or eliminating that fear.

Fear can become anxiety and even dread if it is too intense or if it creates a major dilemma in relation to a choice.

A widely developed idea in the Western world is that freedom makes people happy, because they can choose anything they want. Some psychologists, such as Professor Barry Schwartz from the University of Pennsylvania, strongly refute this idea, and consider that "the more choice there is, the more the consumer is disappointed, sad and unhappy."[15] His experiments show that the fact of being presented with a choice creates more problems than satisfaction.

Two other psychologists, Amos Tversky (1937–1996), Israeli professor from the Universities of Michigan and Jerusalem and lecturer at Stanford, and Daniel Kahneman, Israeli-American professor at the University of Princeton,[16] explain this phenomenon on the basis of an experiment. They show that the consumer is more sensitive to losses than to gains. After making his choice, the consumer thinks that he has perhaps missed a better deal, which makes him frustrated. This frustration is all the greater because people find it difficult to acknowledge their mistakes and admit that they have made a bad choice.

The realization of having made a mistake creates a phenomenon known as "cognitive dissonance" in the neuro-consumer's brain. This is explained by Leon Festinger (1919–1989),[17] a professor at Yale University, who suggests that, in a way, we suffer from our own foolishness, from the emotional impulse that has deprived us of the best choice. To reduce cognitive dissonance, we try to convince ourselves that we have nevertheless made the right choice. We produce arguments where we are dishonest with ourselves, and end up believing them. According to Festinger, and other researchers such as Barry Schwartz from Yale University,[18] it is not the incessant waves of advertising that make things worse for consumers. First and foremost, it is the fact of having to choose, sometimes to regret the choice that one has made, and finally to suffer from the effects of cognitive dissonance that produce the main frustrations in today's civilization of hyper-choice.

Several companies have been thinking about ways to deal with neuro-consumers' fear or anxiety by limiting their frustration. Among their ideas is to identify the optimum number of products in the same category that should be placed on a shelf to protect consumers against the negative effects of cognitive dissonance.

> *Other companies are trying to limit their customers' anxiety through innovations and better organization. Examples from Europe include Dyson, Décathlon and Darty. Dyson has tried to deal with neuro-consumer frustration with vacuum cleaner bags by inventing a bagless machine. Sports equipment chain Décathlon is trying to deal with the anxiety generated by too much choice by employing specially trained salespeople whose job is to provide genuine advice in relation to customers' needs rather than necessarily trying to sell the most profitable products. Darty tries to reassure its customers by offering them a "contract of trust" alongside a quick, efficient after-sales service.*

The neuro-consumer influenced by his desires

In a famous speech, Cicero (106–43 BC) said: "If you want to convince me, you should think my thoughts, feel my emotions and speak with my words."[19]

For Doctor Pierre Sedel and Professor Olivier Lyon-Caen,[20] the driving force behind desire comes from factors that add to the emotions, such as pleasure, love and sex.

For neuropsychologist David Eagleman,

> Urges and desires are programmed in our brains. The greater part of our personality is defined neither by our opinions nor by our choices. Can you imagine changing your ideas in relation to beauty or seduction. Could you do that? No doubt you could not. Your fundamental urges are imprinted in the circuits of your neuronal tissue and you do not have access to them. Some things attract you more than others, but you do not really know why.[21]

In this view, the objectives of our species, fixed by evolution, orientate our desires.

All these factors explain why our desires are sometimes uncontrollably strong. Novelist Ken Follett writes of one of his characters:

> For the moment, however Woody found it difficult to think about the threat of war in Europe. His feelings for Joanna overwhelmed him, intact ... He'd been out with other girls in Buffalo and at Harvard, but none of them had ever inspired the consuming passion that he felt for Joanna.[22]

Desires can go as far as changing our perception of the world. The philosopher Baruch Spinoza (1632–1677) was one of the first to recognize this force when he wrote: "I don't desire things because they are beautiful and good. They are beautiful and good because I desire them."[23]

Desire may be triggered by subconscious signals detected by the brain. Several of the authors mentioned above, for example Sedel and Lyon-Caen and Eagleman, describe a subconscious attraction creating a profound desire in men for women with dilated pupils.[24] They suggest that dilated pupils seem to exercise a real power of attraction and that in the inaccessible depths of the brain, "something knows" that, in women, dilated pupils are an indicator of sexual arousal and availability. According to Sedel and Lyon-Caen,

> Experiments have shown that women, like men, consider photographs of people of the opposite sex whose pupils are dilated to be more attractive. Belladonna (a substance that stimulates the sympathetic nervous system, leading to dilation of the pupils) was used by Renaissance women to seduce. Belladonna means "beautiful woman."[25]

Desire is sometimes so irresistibly strong that it can lead us to commit reprehensible actions such as murder, rape, theft, etc. Most philosophies and religions and some schools of thought like Buddhism draw up sets of principles, codes or laws to try and channel desires.

Therapeutic methods involving meditation, such as those developed by doctors like Jon Kabat-Zinn at the University of Massachusetts[26] based on the search for "mindfulness," aim to help us master our desires.

> *Means of subconscious substitution exist to satisfy certain desires that cannot be satisfied fully at the time. For example, the thumb or the dummy to replace the desire for the mother's breast, the e-cigarette when faced with the need to smoke or the sex toy to satisfy a sexual urge. The search for different types of unsatisfied desires among neuro-consumers is an important driver of marketing innovations for companies wanting to invent and offer products or services as substitutes.*

Towards marketing of the emotions and desires

Nowadays, marketing and communication professionals are increasingly interested in research into emotions and desires.

Marketing of the emotions

Georges Chétochine (1938–2010), professor of marketing at the University of Paris IX-Dauphine and international consultant, describes the changes linked to the emergence of "marketing of the emotions."[27] He suggests that today, despite companies' marketing words and actions, too many customers remain "victims" of what is offered to them and the way it is supplied. His aim is to develop marketing based on "zero frustration" for the customer. His approach begins by detecting all the emotions felt by a consumer before and after the buying process. Next, he proposes a policy based on a new state of mind, by encouraging all of a company's employees to recognize and admit the various frustrations felt by customers throughout their relationship with the company, and then to do their best to reduce them. The elimination of emotional frustrations is thus added to the satisfaction of needs. All aspects of marketing (product, price, distribution, sales, communication and e-communication) must become imbued with this mentality. This approach, known as customer-related marketing (CRM), is designed to anticipate and understand all of a customers' individual frustrations. Companies are beginning to develop permanent "relational marketing" with customers in accordance with the principles set out above.

Marketing of desire and inbound marketing

A small number of authors are interested in the marketing of desire. In Europe, Jean Mouton, consultant and lecturer at the University of Paris XII and the ESCP Europe Business School, suggests that desire cannot exist without a "lack" accompanied by an "expectation": "The customer desires a product because he lacks it."[28] Desire becomes the emergence of a need, but also of a recognition, both glorified by the imagination. Most of the time, traditional marketing is only interested in need and gives priority to the offer of products and services. For his part, the customer silently has the feeling of being acknowledged, respected and loved. The customer and his desire must be placed in the center of the company's strategy. Mouton suggests a method that seeks and awakens potential consumers' desires in order to win them over, then reactivates their positive feelings to gain their loyalty. He has developed the double interaction of desire

(DIAD) concept. The first interaction tries to satisfy the customer's need using the product. The second tries to respond to his request for recognition. Mouton defines this concept as follows: "The double interaction of desire means that a company's overall strategy should take into account the intersection between an initial interaction customer-buyer-actor and company-actor-buyer, and a second interaction between customer satisfaction and the response to his search for recognition."[29] Keeping the promises made to a customer is an absolute condition for ensuring his loyalty to a company.

> *Jean Mouton illustrates his ideas with examples that include Club Med. In the 1960s, its slogan (Eros, Helios, Thalassa) became "Sun, Sex, and Sea." Its concept provided a good response to consumer desires at a period when sexual freedom was becoming widespread with the introduction of the contraceptive pill and the opening up of ideas. In May 1968 in France, freedom movements – based on American campuses and the hippy phenomenon – began to grow. The idea for the club was invented by the Belgian Gérard Blitz (1901–1979) in 1950. It was developed and organized by Gilbert Trigano (1920–2001), who became chairman and chief executive in 1963. The concept was a perfect response to the desires of a population, mainly European, in love with freedom. Idyllic camping pitches bathed by the sea and heated by the sun, big tables encouraging conviviality, the insistence on the use of first names only, the presence of "nice organizers" (NOs) welcoming "nice members" (NMs), the elimination of money as a discriminatory symbol. The response to everyone's desires was there, promises were kept, leading to customer loyalty and then commitment. Club Med's success was spectacular.*

Some years later, the desires of traditional customers, who were then older, changed. Those of the new customer categories, now international, were no longer the same. The formula had been – at least partly – copied by numerous competitors. Unable to find a new model meeting new consumer desires, Club Med's attraction has somewhat reduced, reflecting the fact that a company that is unable to anticipate emerging desires early enough runs the risk of failure if it does not manage to offer new orientations aimed at satisfying them.

Another European lecturer-consultant in communication, Gabriel Szapiro, has developed thinking about the marketing of desire to which he adds the idea of permission. He has put forward the concept of "marketing of permission and desire" integrated into that of "inbound marketing." He defines the concept thus: "Make the customers come to you instead of going to look for them ... by giving 'desire' priority over 'need.'"[30] His approach is partly inspired by five authors of different nationalities (American, Korean,

Canadian and Indian): Seth Godin,[31] Chan Kim and Renée Mauborgne,[32] Brian Halligan and Dharmesh Shah.[33] The method has several phases:

1. Stop any "interruption" marketing. This type of marketing saturates the consumer with messages he does not want. It is counterproductive, expensive for the company and irritating for the customer because it does not correspond to any desire that he has formulated.
2. Only communicate with customers who give their permission. To obtain this permission, you must find a "bait" that seems desirable to them.
3. When faced with intense competition, do not limit yourself to just trying to meet customers' needs, but make them aspirational by identifying, and then satisfying, their desires. At this stage the company must become indispensable and different. To achieve this, you may use techniques of "value innovation" based on the "Blue Ocean" strategy developed by Chan Kim and Renée Mauborgne, professors at the INSEAD.
4. Anticipate and satisfy new desires in neuro-consumers' brains, but also win them over repeatedly to gain their loyalty. The use of CRM proves to be very useful in this case.
5. Create and maintain a "loving" relationship with customers who should become veritable partners. Permanent media pressure must be put on the customer, using all the available means of developing relationships, particularly e-communication and social networks.
6. Based on this permanent, emotional relationship with people who have become partners and given their permission, use "buzz marketing" or "viral marketing" to obtain new customers. This technique is based on the selection of "opinion leaders" attached to the product or brand. They are transformed into "ambassadors" who communicate their affection to their community on social networks. The trust that comes from efficient and permanent responses to the satisfaction of desires is essential to ensure loyalty and conquest based on referral.

Without realizing it, the neuro-consumer is already subject to various marketing and communication techniques that go beyond his needs and favor his emotions and desires. A new form of marketing and communication, exploiting knowledge of neuroscience linked to emotions and desires, is being born. Its applications are multiplying with the considerable development of "sensory marketing."

Notes

1. Bryan Kolb and Ian Q. Whishaw, *An Introduction to Brain and Behavior* (2nd ed.), Worth Publishing, 2006.
2. Sébastien Bohler, *La Chimie de nos emotions*, Aubanel, 2007.
3. Paul Ekman, *The Nature of Emotion*, Oxford University Press, 2008.

4. Antonio Damasio, *Descartes' Error: Emotion, Reason and the Human Brain*, Vintage, 2006.
5. Antonio Damasio, *L'autre moi-même: Les nouvelles cartes du cerveau, de la conscience et des émotions*, Odile Jacob, 2010 and *Le Sentiment même de soi: Corps, émotions, conscience*, Odile Jacob, 2002.
6. Antoine de Saint-Exupéry, *The Little Prince* [1943], Egmont Books, 2001.
7. Howard Gardner, *Frame of Mind. The Theory of Multiple Intelligence*, Basic Books, 1983.
8. Ibid.
9. Daniel Goleman, *Emotional Intelligence*, Bantam Books, 1995.
10. Sigmund Freud, *Civilization and Its Discontents*, W. W. Norton & Company, 2001; James Olds, *Pleasure Centers in the Brain*, Scientific American, 1956.
11. Joseph Le Doux, *The Emotional Brain: The Mysterious Underpinnings of Emotional Life*, Touchstone, 1998.
12. Peter Salovey and John D. Mayer, "Emotional Intelligence," *Imagination, Cognition and Personality*, vol. 9, no. 3 (1989), pp. 185–211.
13. Alain Berthoz, *Emotion and Reason: The Cognitive Neuroscience of Decision Making*, Oxford University Press, 2006.
14. James Old and Peter Milner "Reward Centers in the Brain and Lessons for Modern Neuroscience," *Journal of Comparative Physiological Psychology*, vol. 47 (December 1954), Seminal Paper, McGill University.
15. Barry Schwartz, *The Paradox of Choice*, Ecco, 2005.
16. Amos Tversky and Daniel Kahneman, "Judgment Under Uncertainty: Heuristics and Biases," *Science*, vol. 185, no. 4157 (1974), pp. 1124–1131.
17. Léon Festinger, *When Prophecy Fails*, Pinter & Martin, 2008.
18. Schwartz, *The Paradox of Choice*.
19. Cicero, *Complete Works*, Delphi Classics, 2014.
20. Frédéric Sedel and Olivier Lyon-Caen, *Le Cerveau Pour les Nuls*, First, 2010.
21. David Eagleman, *Incognito: the Secret Lives of the Brain*, Vintage, 2012.
22. Ken Follett, *Winter of the World*, Dutton Penguin Group, 2011.
23. Benedict de Spinoza, *Ethics* [1675], Penguin Classics, 1996.
24. Sedel and Lyon-Caen, *Le Cerveau Pour les Nuls*; Eagleman, *Incognito*.
25. Sedel and Lyon-Caen, *Le Cerveau Pour les Nuls*.
26. Jon Kabat-Zinn, *Wherever You Go, There You Are*, Hyperion, 1994.
27. Georges Chétochine, *Le Marketing des émotions: Pourquoi Kotler est obsolète*, Eyrolles, 2008.
28. Jean Mouton, *Le Marketing du désir: L'indispensable stratégie*, Éditions d'Organisation, 2000.
29. Ibid.
30. Gabriel Szapiro, *L'Inbound marketing selon la stratégie du Sherpa*, Jacques-Marie Laffont, 2015.
31. Seth Godin, *Permission Marketing: Turning Strangers into Friends, and Friends into Customers*, Simon & Schuster, 1999.
32. Chan Kim and Renée Mauborgne, *Blue Ocean Strategy*, Harvard Business School Press, 2005.
33. Brian Halligan and Dharmesh Shah, *Inbound Marketing: Get Found Using Google, Social Media and Blogs*, John Wiley & Sons, 2009 and *Inbound Marketing Revised and Updated: Attract, Engage and Delight Customer Online*, John Wiley & Sons, 2014.

Points to remember

- Twenty-first-century researchers are deploying considerable means to understand how the brain works. Thanks to neuroscience, they are discovering a highly complex organ, endowed with unexpected powers: a very sophisticated system of communication with the rest of the body based on neurotransmitter hormones and a significant influence on thought, the senses, memory and conscience. Better than a powerful computer, the brain may be compared to an orchestral conductor of human behavior.
- Recent neuroscientific discoveries show that the digestive tract contains such a large number of neurons that it represents a second brain, responsible for specific tasks.
- American researcher Paul D. MacLean has hypothesized that the human brain is made up of three parts. He calls them the "triune brain." The preponderance of the three parts, related to genetic evolution over the centuries, varies with the age of the individual. They influence and explain the differences in behavior observed in children, adolescents and adults. The reptilian brain, which predominates in children, gives priority to reflexes, to the leader and to the importance of the group. The limbic brain, which takes priority in adolescents, strengthens the emotions and the sense of affectivity. The neocortex, which develops in adults, enables logical reasoning, abstract thought and anticipation of actions.
- An individual's perception of the world is widely considered to be more a construction of their brain than an objective reality. It is "programmed" to a large extent, leading to instinctive behavior, particularly where making purchases is concerned. Detailed study of "instinctive intelligence" provides understanding of neuro-consumers' often irrational acts.
- The neuro-consumer's brain is largely conditioned by its structure and its moods. Age and gender can influence specific behaviors and attitudes that are sometimes difficult to understand for someone who does not belong to the same category. Elements such as memory, mirror neurons and somatic markers strongly condition the behavior of the human brain. They are the subject of relatively recent studies that are useful to understanding the neuro-consumer's brain.
- Emotions and desires have a large influence on the reactions of the human brain. The many research projects undertaken on this subject have shown the importance of "emotional intelligence." Studies related to the neuro-consumer show that they require the development of specific marketing adapted to them, i.e., "marketing for the emotions and desires."

PART III

The neuro-consumer's brain influenced by the senses

Thanks to their direct access to different areas of the brain, unfiltered by the barrier of reasoning, the senses have a primordial influence over people's behavior. Skilfully orchestrated, their use contributes to improving the effectiveness of marketing, communication and sales, but also to establishing long-lasting links with brands. Sensory marketing is becoming an integral part of traditional marketing.

Introduction

The five senses – sight, hearing, smell, touch and taste – are essential to human life, providing the brain with information about its environment. They are keys for us to sense and then perceive the world in which we live and are fundamental to our understanding of that world. Linked to the brain's limbic system, the senses are the gateway to human emotions. They are linked to memory and to emotional memories provoked by sensory information. Their evocative power is immense.

The world of marketing was very quick to understand the incredible power of the senses. Their capacity to develop emotional links, at the very center of the brand concept, is exploited by companies. Although sight and hearing attract the most attention, the other senses are generating increasing interest. The core aim of sensory marketing is to establish a durable link with the brand through unique sensory experiences. The approach involves capitalizing on the links between the senses and the emotions – brought to light by scientific research – to make the brand stronger, contributing to its differentiation from competitors that will be noticed by neuro-consumers, thus increasing their loyalty and leading them to buy more. Companies that play the sensory card count on the fact that emotional memories linked with their brands are more likely to form and be stored and retrieved when a large range of the senses are involved.

Part III gives an in-depth analysis of each sense through a description of how it works, its specific characteristics and its cognitive, affective and behavioral influences, revealed by academic research. The senses are inseparable from the working of neurons and they are the key elements through which brands create distinctive and durable connections with the neuro-consumer.

CHAPTER 12

The neuro-consumer and the sense of sight

Our understanding of the world is very closely connected with what our eyes see. Sight is the sense that enables us to observe and analyze the environment from a distance using light radiation. It is of great importance for the human species. One-third of our brain is devoted to it,[1] Some 70% of the sensory receptors in our body are in our eyes[2] and 80% of the information that reaches us is visual.[3] The optimal method for capturing the attention of the human brain is through the visual system. Advertising, packaging, web pages, applications, etc. solicit it incessantly. Sight is the subject of a great deal of research for marketing and communication in many sectors.

The mechanism of sight

Sight is the function through which images, once captured by the eye in the form of light rays, are transmitted to and then interpreted by the brain. Although they go through the eye, the organ of sight, they also require the intervention of specialized areas of the brain that make up the visual cortex. These areas analyze and synthesize the information collected in terms of shape, color, texture, relief and distance. They produce the final interpretation of what is "observed" by the brain.

Despite its apparent simplicity, seeing is a very complex process. Light rays reflecting off objects in the environment enter the eye through the cornea and various transparent media such as the aqueous humor, the vitreous humor and the lens. The process of refraction is regulated by the iris (colored part of the eye) and the pupil (the black circle at the center of the iris). Depending on the lighting conditions, the pupil enlarges or shrinks to admit more or less light.

The information arrives on the retina as an upside-down image. It causes chemical reactions in certain nerve cells called "photoreceptors." There are two types of photoreceptor in the retina: cones and rods. The cones, of which there are 6 million and three different sorts, are active during the day. They enable us to distinguish millions of different shades of color. When light levels are too low, the retina's 120 million rods become active. They enable us to see in the dark, without distinguishing colors. Nocturnal vision is mainly in black and white.

The chemical reactions that take place in the cones and rods trigger electrical messages that are sent along the optic nerves. These are partially crossed over so that the right brain receives information from the left field of vision while the left brain receives that from the right field. The electrical signals go to the occipital visual cortex, at the back of the brain, which processes visual information.

This part of the brain is made up of several areas specialized in the management of nerve messages. Each area interprets one characteristic of the image observed: shape, color, movement. The different areas communicate with each other. From their combined information, the brain produces an overall interpretation in three dimensions.

The exchange of analyses and interpretations from the different visual areas enable the brain to produce an overall, unified perception in the form of a multidimensional final image. Because the two cerebral hemispheres receive information about the same image from a slightly different angle, the brain is able to estimate the volume of the object and its distance from the viewer.

Despite this extremely complex visual machinery, the time lapse between the moment when the light goes through the pupil and the moment when the signals arrive in the brain is only a few milliseconds.

Vision and the brain's interpretations

What we call "vision" is a cerebral interpretation of electrochemical signals. The real world is never "seen" just as it is, but reconstructed by the brain. Its perception is fashioned by the brain, which is subject to various influences: previous experience, context, the brain's objective that directs its attention and the phenomena of illusion.

Vision, previously acquired experience and memory

As neuroscientist David Eagleman explains, "The brain has to do a phenomenal job to unambiguously interpret the billions of photons captured by the eyes."[4] To accomplish this task, it activates the images already processed and stored in the memory. Previously acquired experience is essential. Seeing, through the eyes, is innate. Vision, on the other hand, is acquired. This is the painful and surprising experience of patients recovering their eyesight thanks to surgery after decades of blindness. The new visual information that

> they perceive is nothing but a whirling and highly disconcerting deluge of shapes and colors. They cannot suddenly see the world when they leave the operating theatre: they must learn, or re-learn,

> to see it. Even when their eyes work perfectly from the mechanical point of view, their brain must train itself to interpret the data that it receives.[5]

In a similar way, to be able to read a text it is first necessary to learn and memorize individual words. They enable the brain to reproduce a text that is written incorrectly. This is demonstrated by an experiment carried out by researchers at the University of Cambridge, which we suggest you try yourself by trying to read the following text:

> Acrocnidg to a sudty at the Urenvitisy of Crimabgde, the oderr of the lertets in a wrod has no ipmrotncae, the olny ipmrotnate tinhg is taht the fsrit and the lsat are in the rhigt pclae. The rset mhgit be talolty out of oerdr but you can slitl raed it woutiht any porlebm. This is busacee the hamun barin deos not raed ecah letetr ivniddliualy, but the wrod as a wolhe.

The text above is completely illegible. Nevertheless, the brain reconstitutes the correct text in a fraction of a second. To do so it uses knowledge acquired about writing stored in its memory and subconsciously corrects the errors that it perceives in the words.

Vision and context

As Eagleman puts it, "strictly speaking, all visual scenes are ambiguous."[6] He illustrates this with an image of the Tower of Pisa. Whether it is the actual monument or a cheap souvenir held at arm's length, the image projected onto the retina is the same. The eyes alone are not enough to recognize the difference between the real tower and its reproduction. The brain uses the context to interpret the information received. It formulates hypotheses then evaluates them before deciding on just one. It makes a lot of effort to "disambiguate" the data that arrive from the eyes, using contextual elements.

> *Rubin's vase or face was developed around 1915 by the Danish psychologist Edgar Rubin. The image of it, easily found on the Internet, may be interpreted in different ways by the brain's visual system: either as a representation of two identical black faces looking at each other, or as a white vase.*
>
> *The brain analyzes the two possibilities by passing from one to the other. To paraphrase David Eagleman, the faces and the vase are both present on the image, which does not change.[7] The change occurs in the brain as it oscillates between the two interpretations.*

Vision and attention

Fundamentally, the brain works according to its needs and objectives. This is because it is incapable – according to neuroscientists – of processing more than one-fifth of the information sent to it by the senses. It concentrates its attention on what is useful for it at that moment and subconsciously ignores many other external stimuli that may be present.

The philosopher William James (1842–1910), considered to be the father of American psychology, gives a definition of attention that has become classic:

> Attention is taking possession of the mind, in clear and vivid form, of one out of what seem several simultaneously possible objects or trains of thought ... It implies withdrawal from some things in order to deal effectively with others.[8]

The soliciting of the attention is often linked to specific objectives. Because the cognitive capacity of the brain is limited, such soliciting implies excluding some information in favor of processing that which is necessary to the achievement of its objectives.

In 1999, two researchers in cognitive psychology from the University of Harvard, Christopher Chabris and Daniel Simons, developed a test involving manipulation of visual attention called "the invisible gorilla test."[9]

The instruction given to the participants is to watch a video attentively. Two teams of basketball players, one in white strip, the other in black, are throwing a ball to each other. The participants in the experiment are asked to count the number of passes between the members of the team in white. During the match, a person disguised as a gorilla crosses the field of view from right to left, beating his chest with his fists.

The participants were asked about the number of passes they had counted and whether they had seen anything out of the ordinary. Approximately 50% of them did not see the gorilla going across.

Although this may appear to be counterintuitive, we do not fail to see something because of lack of attention, but because we place so much attention on one thing that we do not see another. Magicians know all about this process and exploit it to perform their tricks without the spectators understanding how they do it.

Not only is our vision of the world a construction, different from the real world, but it restricts itself to the minimum necessary to the brain to meet its needs and achieve its objectives.

The links between attention and memory are interesting for those working in marketing. An object to which we give our attention is memorized better. The fight for consumers' attention is justified and has been the subject of many research studies.

Vision and illusory phenomena

Optical illusions are numerous and remind us to what extent visual errors are frequent. Another proof that vision is a construction, sometimes erroneous, made from the ocular messages that we receive.

There are various tests involving optical illusions that you can easily access on the Internet.

> *A now-famous test of geometrical illusion was developed by the Italian psychologist Mario Ponzo (1882–1960) in 1911. It underlines the impact of perspective in visual perception. Two identical lines are drawn horizontally, separated by a space between them and flanked by a pair of convergent lines like a railway track. The effects of perspective influence our perception of the real size of objects, the upper line being incorrectly evaluated by the brain as being longer. Although the visual sensations caused by these two lines are the same, the brain's interpretation is not. They are perceived differently.*

Peripheral vision is, by its very nature, easily subject to illusions. Unlike central or "foveal" vision, which has great acuity and precision, peripheral vision gives blurred but rapid impressions from the wider visual field. It sends up to 100 images per second, instead of the 3 or 4 images per second sent by foveal vision. It allows ultra-rapid perception of movements in the periphery of the visual field. The more detailed and slower analysis of precise objects is left to foveal vision.

Thus, vision is a complex process. The data received are recognized by comparison with the images already contained in the visual memory. Shapes, outlines, distances and movements become a whole that the brain interprets in accordance with the different elements mentioned previously.

Visualization of products and services: logos and designs

The importance of the sense of sight in humans is widely exploited by marketing. Besides advertising, companies are constantly increasing the attention paid to visual logos and product design. The objective is to make an impression. Just like names and slogans, logos and designs have become elements that differentiate brands. They wield great power.

To increase the ease with which they can be memorized and associated, logos are tending to become simpler, like those of Coca-Cola and HEC Paris. The most marked changes are concentration on one or two colors, stylization and refinement of drawings and graphic design, leading to an overall simplification of logos.

Certain brands are recognizable by a single symbol, without any explicit mention of their name. This has long been the case with Nike and its "swoosh."

The strength of the link created between a color and a company, between a symbol and a brand, represents a stimulus capable of activating the relevant somatic markers stored in the neuro-consumer's emotional memory.

The same is true for specific designs, which can, by themselves, evoke brands.

This is the case with Coca-Cola and Orangina bottles and the Toblerone bar. No other bar of chocolate, deprived of its brand name on the packaging, would benefit from the same level of recognition as Toblerone. The triangle has become a distinctive sign, inseparable from the brand, like the red soles and high heels of Louboutin stiletto shoes and the very characteristic design of Apple's iPod. Colors can also become strongly linked with particular categories or brands. Black and gold are both often associated with luxury. There is also a color called "Ferrari" red. Exclusive patterns are another important visual basis for the identity and recognition of strong brands like Burberry and Vuitton.

Eye-tracking research into the visual marketing of advertisements

In 2015, on the HEC campus in Paris, Professor Michel Wedel from the University of Maryland gave a lecture on "visual marketing," a subject in which he is an acknowledged expert. He gave a summary of the main results of research into advertising carried out using the technique of "eye tracking." A major part of this work is described in detail in his book *Visual Marketing: From Attention to Action,*[10] coauthored with Rik Pieters.

Eye tracking includes a group of techniques enabling the recording of eye movement. The most common systems analyze images of the human eye recorded by a video camera. They measure the fixations of the eyes and the saccades or time lapses between two fixations.

The first application of eye tracking in marketing goes back to 1924. Nowadays it is used by companies like Procter and Gamble, IBM, Pepsico, Target and Google. It enables them to predict what consumers will look at, their choices and the purchases that might well follow. It helps to optimize not only the way in which products are presented on the shelves in shops, but also packaging, Internet sites and banners as well as advertisements.

Many visual elements receive only one ocular fixation at best. Nevertheless, consumers appear to be able to grasp the essence of an advertising poster in a single fixation (100 milliseconds). They do this all the more easily when the message is characteristic of the industry or the product concerned, e.g., a woman's face for beauty products, a car for the automobile industry. The association efficiently facilitates understanding.

As well as being brief, the exposure is often blurred. The use of colors and a characteristic central counteract the rapidity and blurring often associated with the visualization of advertisements.

The effects of exposure time vary with the type of advertisement. Traditional adverts are generally well identified. However, appreciation of them decreases with exposure time. "Mystery" communications where the brand is only revealed at the end improve their identification and their appreciation.

Strong brand presence is crucial to improving the effectiveness of a message. The bigger its typeface, the greater the amount of attention it receives and the amount given to the rest of the advertisement. The attention paid to the brand makes it easier to memorize.

Repeated exposure leads to less attention being paid. The length of visualization is reduced by approximately 50% for each new exposure while neither the place nor the sequence of fixations by the eyes, visualized by eye tracking, are modified.

Originality encourages attention and memory. It increases the total fixation time on the brand and the image. This is even more the case when the advertisement becomes familiar to the consumer. For fairly long and repeated exposure, the accent should be on originality. For very short exposure times, it is more effective to use traditional or already-known advertisements.

Complexity can have diametrically opposite effects on attention. Featuring a great density of attributes (colors, fussy borders) reduces the attention paid to the brand. On the other hand, complexity in the design (a lot of asymmetrical, irregular or dissimilar objects) leads to more attention being paid to the image and the message (but not necessarily to the brand).

Leaflets used today are far from optimal. Although their size facilitates processing of the brand name and the text by the brain, the same is not true for the images. Research carried out by Michel Wedel and Rick Pieters suggests that images should be smaller, unlike the brands and the prices shown, and that this would result in greater attention being paid both to the featured distributor and the manufacturer, thus increasing future sales.[11]

Finally, in television advertising, playing on the emotions is recommended. For example, joy increases the appreciation of a message while surprise enables the company to "grab" inattentive spectators.

The importance of sight for goods and services sales outlets

The sense of sight is strongly solicited anywhere that sells goods and services. Market traders in all civilizations have made use of this since antiquity. The arrangement of colors and the relationships between volumes are strong factors in attracting a clientele, and shops today work on them meticulously. They concentrate first on the esthetic appeal of their shop windows in order to increase their penetration rate, i.e., the number of visitors to their shops as a proportion of the total number of potential customers in the catchment area. They then work to improve their conversion rate, which represents the proportion of visitors to a shop who become buyers. Visual factors can help with this.

For example, research confirms the influence of color on customer behavior and perception in points of sale.[12] Warm colors, said to be "activating," have a strong power of attraction. It is often recommended to use them on the outside of a shop to attract customers. Warm colors favor impulse purchases and increase the rate of customer rotation, which explains why fast-food chains, like McDonald's and Subway, like to use them. At the other end of the scale, cool colors, which are more relaxing, favor well-considered purchases and increase the time that customers spend in the sales outlet. They also contribute to increasing customers' evaluation of a product or sales outlet as expensive. Pastel colors are chosen to amplify the impression of space. Depending on what behaviors or perceptions they are seeking, outlets for goods or services choose one color scheme or another.

It is important, however, to be careful when choosing colors not to forget their cultural dimension in the international context. Symbolic associations with colors can vary widely from one culture to another. In many cultures, black symbolizes mourning, whereas in others it is represented by white. Some colors are highly appreciated in certain parts of the world and much less in others.

As researchers Charles Areni and David Kim[13] have shown, intensity of light can also influence customer behavior. A stronger light contributes to approach behaviors and increases the degree to which customers examine and handle products.

Sight is also solicited by the use of plasma screens, whose objective is to showcase the brand or the company's products.

> *This is the case in many ready-to-wear clothes shops that broadcast their fashion parades, and sports shops like Décathlon, whose video displays alternate sporting exploits with promotional spots. Video screens are not only installed inside shops. They have proved to be very effective in catching the eye of passers-by – who are very sensitive to movement – thus increasing the attractiveness of shop windows.*

Companies also work on the visual appearance of their shop staff and sales representatives. The way they dress and their "look" can strengthen the desired atmosphere or image for a place or a brand.

> *The "look" of staff can be sophisticated in luxury boutiques, professional in some chains of opticians where staff wear white coats,*

relaxed and youthful in ready-to-wear shops aimed at adolescents, or even partly naked at Abercrombie & Fitch. The latter's selection of individuals with the flattering, muscle-bound physique that their target clientele aspires to seeks to make the most of the "halo effect" described in psychology. This effect is produced when one of the characteristics of an individual, such as their physical appearance, dominates the way in which they are perceived.

Robert Cialdini contends that we subconsciously associate people who have an agreeable physical appearance with qualities such as talent, kindness and intelligence without even realizing the influence of that physical appearance on our judgement.[14] He suggests that we tend to accord them, and the products they represent, more trust and the "halo effect" thus spreads to the products.

Marketing Professors JoAndrea Hoegg from the Sauder School of Business in Vancouver and Joseph Alba from the University of Florida show that the color of a drink can have more influence on our perception of its taste than the actual taste itself. In an experiment, the researchers manipulated the color and the juice in several pairs of orange drinks for which they asked the participants to evaluate the similarity in taste. The results revealed that the pairs of drinks presenting a similar color but with different juices were perceived to be closer in taste than those with an identical juice but different colorations. The color had more impact on the perception of taste than even the quality of the juice, underlining the superiority of the sense of sight over that of taste.[15]

Neurological research confirms this relative strength of the optic nerve, revealing that it transmits 25 times more information than the auditory vestibular nerve[16] and is physically connected to the primitive brain, therefore highlighting the importance for marketing specialists that their communication efforts are addressed first and foremost to the eyes.

Notes

1. David Eagleman, *Incognito: The Secret Lives of the Brain*, Vintage, 2012.
2. A.K. Pradeep, *The Buying Brain: Secrets for Selling to the Subconscious Mind*, Wiley, 2010.
3. Bernard Roullet, "Comment gérer les couleurs et les lumières?" in Sophie Rieunier (Ed.), *Marketing sensoriel du point de vente* (4th ed., pp. 129–166), Dunod, 2013.
4. Eagleman, *Incognito.*
5. Ibid.
6. Ibid.
7. Ibid.
8. William James, *The Principles of Psychology* (vol. 1), Henry Holt & Company, 1890.
9. "The Invisible Gorilla," *Le cerveau and ses automatismes: Le pouvoir de l'inconscient* [film], Arte, Part 1: December 1, 2011. See also Christopher Chabris

and Daniel Simons, *The Invisible Gorilla and Other Ways Our Institutions Deceive Us*, Crown Publishing, 2010.

10. Michel Wedel and Rik Pieters, *Visual Marketing: From Attention to Action*, Taylor & Francis, 2008.
11. Ibid.
12. Bernad Roullet and Olivier Droulers, *Neuromarketing: Le marketing revisité par les neurosciences du consommateur*, Dunod, 2010.
13. Areni Charles and David Kim, "The Influence of In-Store Lighting on Consumers' Examination of Merchandise in a Wine Store," *International Journal of Research in Marketing*, vol. 11, no. 2 (1994), pp. 117–125.
14. Robert Cialdini, *Influence: The Psychology of Persuasion*, Harper Business, 2006.
15. JoAndrea Hoegg and Joseph W. Alba, "Taste Perception: More Than Meets the Tongue," *Journal of Consumer Research*, vol. 33, no. 4 (2007), pp. 490–498.
16. Patrick Renvoisé and Christophe Morin, *Selling to the Old Brain*, SalesBrain, 2003.

CHAPTER 13

The neuro-consumer and the sense of hearing

Every day, we are bombarded by thousands of natural, mechanical or commercial sounds. They provide us with information about the outside world and help us to live in society thanks to the exchanges that they elicit. Their absence is considered worrying because it is often synonymous with abnormality or death. What parent has never wondered about the sudden silence from a child's bedroom? Who has not observed a minute's silence in memory of the departed?

Sounds are an integral part of our social life and are unique in that they do not require all our attention to be noticed. Whereas visual, gustatory or tactile attributes require direct interaction with the product in order to be perceived, sound is an easy way of reaching consumers even when they are passive. Whether we like it or not, we are all exposed to sounds and we do not need to do anything in particular to hear them.

Marketing communication professionals often solicit the sense of hearing. They do it through radio or television advertising messages, jingles, songs or slogans attached to brands as well as "background" music in shops, hotels and other public places. This substantial use of music and other sounds finds its justification in the academic literature, which underlines their affective, cognitive and behavioral impacts on individuals, including consumers.

Hearing and the processing of sounds by the auditory system

Hearing is the sense that enables us to perceive sound. Sounds are collected by the ears and analyzed by the brain. The organ of hearing consists of three parts: the outer ear, the middle ear and the inner ear. The outer ear is composed of the pinna and the external auditory canal. The pinna captures sound vibrations and conducts them towards the eardrum via the auditory canal. The middle ear is an amplifier that propagates the sound waves from the membrane of the eardrum as far as the inner ear, passing through the chain of ossicles. The inner ear consists of the vestibule, an organ of balance, and the cochlea with its 15,000 sensory cells. It is these cells that translate the vibrations into a stream of nerve impulses. These impulses leave the ear via the auditory nerve and are transmitted to the cerebral auditory centers. It is here that the initial vibrations are finally recognized and perceived as sounds. This long process occurs thousands of times every day in a fraction of a second.

Vibrating waves have several physical characteristics that may be sensed by the ears. One corresponds to the amplitude of the wave and is measured in decibels. Another is the frequency, which measures the number of cycles per second of a wave, commonly expressed in hertz. A third refers to the harmonics of the wave, perceived as the timbre of a sound.

Just as with sight, our ability to perceive sounds is fairly limited. Human beings can only hear sounds between 20 and 20,000 hertz, unlike many other species that can hear a wider spectrum, such as dogs and dolphins.

Sounds are the first pieces of information to reach the brain – as yet little formed – of a baby in its mother's womb. As shown by Finnish researchers, prenatal exposure to music can have long-term effects on the developing brain and enhance neural responsiveness to the sounds experienced in the prenatal period several months later.[1]

The use of sound in advertising

The use of sound to advertise a brand is not new. As early as 1922, a New York-based radio station invented the promotional announcement on the radio. Since then, sound-based publicity messages have multiplied and may be heard in many different formats: radio, television, call centers, Internet sites and applications.

The concept of linking a specific sound to a particular brand name is called "sonic branding," the title of a book by Daniel Jackson.[2] This term covers all audio signatures, adopted today by very many brands. They are based on audible logos and jingles as well as slogans.

> *One of the first brands to use a jingle as a signature was the American company NBC, in order to synchronize its radio stations in different regions of the country. The jingle quickly became a recognizable, distinctive sign of the brand.*
>
> *It is the same for the opening sound used by the Intel brand, now one of the most widely recognized audio signatures in the world. These examples are far from unique. Among the most famous commercial sounds, we can also cite that of CNN, the roaring lion of MGM studios and the characteristic noise of Nintendo.*

Alongside these audio logos and jingles based on short songs or characteristic sounds, whole phrases or slogans are also indissociably linked with certain brands.

In the consumer's mind, the slogan "Because you're worth it" is unequivocally associated with the L'Oréal brand. McDonald's "I'm lovin' it" and Nespresso's "What else?" have the same evocative power. The brand identity of all these major companies includes a characteristic sound personality, expressed through an audio signature.

In China, McDonald's launched a notable sound branding campaign by calling on famous Chinese pop star Leehom Wang to sing the "I'm loving it" slogan in Chinese.

The final category of audio logos consists of sounds from the product itself.

An example is the very distinctive sound of the Harley Davidson that the motor-bike brand sought to patent. A whole team of specialists worked for many years on each of the sounds emitted by the motor in order to arrive at the very characteristic sound of the Harley Davidson.

Audio marketing and the creation of audio signatures are the work of specialized agencies like the French company Sixième Son (Sixth Sound). The company was set up in 1995 by Michael Boumendil, himself a musician, composer and producer, and was the creator of the sound identity of two French giants, France Télécom in 2001 and SNCF in 2005, then of the Samsung brand in 2007. For Boumendil and his colleagues, the main characteristics they are seeking when creating a brand's sound identity are duration and clarity, singularity, the relationship to the product, agreeableness, familiarity and accessibility.

Today, more than 320 brands from all over the world have placed their trust in the Sixième Son agency including Aéroports de Paris, Alstom, Areva, Axa, Baccarat, Castorama, Chanel, Coca-Cola, FDJ, Fnac, France Télécom, Groupe Accor, Groupe Bel, Hammerson, ING Direct, Lancôme, Michelin, Nexity, Petit Bateau, Peugeot, RATP, Roland-Garros, SFR, SNCF, Société Générale, Total, Unibail-Rodamco and Royal Air Maroc.

During the reworking of its brand identity, Royal Air Maroc entrusted the sound aspect to Sixième Son with three major objectives: to better differentiate the airline, express its values and strengthen the impact of its communication. Five key values of the brand are at the heart of the musical composition. They can be summarized by the terms "Moroccan, majestic, magic, maternal and modern" in addition to respect for the brand's oriental roots. Since then, the audio signature has been used for advertisements on radio and television, on the company's Internet site, as a jingle in airport terminals, on CDs for customers and as a telephone ringtone.[3]

Academic research into sound and advertising reports numerous impacts on consumers. Music, with its ability to generate an atmosphere of energy or peace, has a strong emotional power, capable of producing joy or sadness. It very often calls to mind memories or past experiences, which represents a certain advantage when trying to build a strong brand in the minds of consumers.

Psychological research carried out at the instigation of marketing professor Richard Yalch[4] from the University of Washington, shows that music increases the extent to which advertising messages are memorized and that jingles are remembered more easily than spoken slogans. Associating music with an advertising message is thus a good way to help the consumer remember. When it obtains a high approval rating, music can improve attitudes towards the advertisement, the attitude towards the brand and the intention to purchase, as shown by research carried out by Professor Gerald J. Gorn[5] from the Polytechnic University of Hong Kong, and by Professors C. Whan Park, S.M. Young[6] and Deborah J. MacInnis[7] from the University of Southern California.

The development and impact of background music

In places where people buy goods or services, background music is among the elements contributing to atmosphere that is the easiest to control and the cheapest to develop. Its use has increased constantly since it appeared in the 1920s. Apart from the fact that shop employees and customers prefer it to silence, background music appeals to distributors because it enables them to influence consumers' emotions and behavior. In the absence of music, research shows that customers spend less time in the shop, spend less, are in a worse mood, talk less to sales staff and are more stressed.[8]

> *Background music helps to contextualize the sales outlet and to arouse emotions appropriate to the place: excitement and emulation at Abercrombie & Fitch with its techno music or well-being and relaxation at Nature and Découvertes with its sounds of water and birds.*

Besides these emotional effects, the style of music can have an impact on the perception of the price positioning of a shop, as shown by research carried out by Professors Richard Yalch and Eric R. Spangenberg.[9] When compared to middle-of-the-road music, classical music has been found to increase the perception of a shop being more at the luxury end of the market. Thanks to the pleasure it arouses, music has been found to improve consumers' evaluation of a shop and even their judgment of the sales staff.[10] When customers appreciate the music they hear in a shop, they are more likely to remain loyal to it.[11]

The role of background music in people's decisions to buy has been the subject of several interesting studies. In 1999, Adrian North, David Hargreaves and Jennifer McKendrick[12] from the University of Leicester in the United Kingdom demonstrated its influence on the choice of products bought.

> *In a wine shop, over several days, they programmed characteristically French music and characteristically German music on alternate days. On days when the music was French, French wine sold better than German, and vice-versa.*

A similar study was carried out in 1993 by Charles S. Areni and David Kim.[13]

> *In this case, the shop alternated classical music with music from the Top 40. More expensive wine was bought on days with classical music.*

Music can also help with "crowd management," by influencing the time spent inside shops. Two studies carried out in 1985 and 1986 showed that music with a rapid tempo pushed customers to leave more quickly.[14] On the other hand, slow music played at a low volume increases the time and the money spent. The same studies revealed that customers eat more quickly and consume less with fast, loud music, which is very often encountered in fast-food chains. Another piece of research from 1966 found that, in a sales outlet, the customer adapts his walking speed to the tempo of the music.[15] Background music thus offers marketers various ways of influencing the consumer during his visit to a shop.

The highly concentrated market for background music is shared by a few companies. Among them is the pioneer, Muzak, created in 1922 and since bought out by the global Mood Media Corporation. Equipping nearly 600,000 sales outlets throughout the world, this company offers, like its competitors, the creation of soundtracks unique to each of its client brands by professional DJs. It mixes music for many distribution and ready-to-wear companies including variations depending on the time of day and the day of the week. The playlists are generally calmer on Mondays and livelier on Saturdays. Broadcasting music in shops has been found to increase sales by 18%.[16]

> *Broadcasting of background music is not limited to shops. Internet sites appeal more and more to our sense of hearing: a pop-rock soundtrack consistent with Levi's laid-back rock image, the purring of the Cayenne motor for Porsche, the Zen atmosphere at Suzy Wan.*[17]

Congruence with the target and the objective sought

Audio marketing is based, like sensory marketing in general, on the idea of congruence. There is no good or bad music. There is music or a sound that is in congruence with the chosen positioning, the target customer and the objective being sought.

> *It would be just as incongruous to hear hard rock music at Nature and Découvertes or Yves Rocher as to hear the sound of water or birdsong at Abercrombie & Fitch.*

It is amusing to note that the choice of music and the volume at which it is played are sometimes guided as much by a target market that a shop wants to avoid as that which it is aiming to attract.

> *Techno played like in a nightclub at Abercrombie is seeking just as much to attract adolescents, the shop's core market, as to dissuade their parents from accompanying them. Free of their elders, the adolescents, in groups like in a discotheque, will supposedly lose their inhibitions more quickly and open their wallets more easily.*

Congruence is also sought with respect to other aspects of the purchasing environment: decor, lighting, texture, staff clothing. Interested in the coherence between smell and music, researchers Eric R. Spangenberg, Bianca Grohmann and David E. Sprott[18] found that customers' evaluation of products and of the shop as well as their intention to visit were increased when a smell evoking Christmas was diffused with music of the same genre.

Sound as an indicator of product quality

Just like Harley Davidson motorbikes, car manufacturers invest a lot in acoustic research in the laboratory.

> *At BMW, 60 people work to adjust the sound of each element of the cars, going as far as fine-tuning the "right sound" for the closure of the doors. At Audi, a team of 45 is dedicated to the phonic signature of the brand's models.*[19]

Consumers use sounds – often subconsciously – during the qualitative evaluation of a product and their absence (or their smaller amplitude) can sometimes be detrimental to that evaluation.

> *IBM learned this to its cost when it launched a silent typewriter. Perplexed, the consumers showed no interest in it and IBM chose*

> *to add artificial sound to replace the natural sound that it had gone to a lot of trouble to eliminate!*
>
> *For vacuum cleaners, consumers also require a minimum of noise to make a positive evaluation of the device cleaning power.*

These concrete examples from different industries lead to questions about the potential negative effect of the absence of noise in electric cars on customers' perception of their performance and relative attractiveness compared to petrol-driven cars.

Taste and hearing being linked, the food industry is working on the sounds heard while products are being consumed.

> *The famous "snap, crackle and pop" emitted by Rice Krispies when milk is poured over them was carefully developed in the laboratory by Kellogg's.*
>
> *In 2013, Coca-Cola gave Japanese DJ Jun Fujiwara the task of changing the "psscht" emitted when its bottles are opened. McDonald's is working on the sound that a hamburger should make when it is bitten into.*

Besides these examples of commercial exploitation, let us conclude this chapter with a discovery that is encouraging and brings hope with regard to the beneficial effects of music on severe cerebral cognitive and motor deficiencies. It has been made by researchers working under the direction of Emmanuel Bigand,[20] professor of cognitive psychology and director of the Laboratory for the Study of Learning and Development (LEAD) at the University of Burgundy. Cognitive neuroscience shows that music acts as a neurostimulator and a neuroprotector. Beyond the regulation of mood in certain patients, music stimulates cerebral plasticity and contributes, by the reorganization of the affected neuron circuits, to an improvement in the recuperation of motricity or the ability to speak.

Notes

1. Eino Partanen, Teija Kujala, Mari Tervaniemi and Minna Huotilainen, "Prenatal Music Exposure Induces Long-Term Neural Effects," *PLoS One*, vol. 8, no. 10 (2013), pp. 1–6.
2. Daniel M. Jackson, *Sonic Branding*, Palgrave Macmillan, 2003.
3. Vladimir Djurovic, "Branding auditif: la création d'une identité sonore," 2009,www.marketing-professionnel.fr.
4. Richard Yalch, "Memory in a Jingle Jungle: Music as a Mnemonic Device in Communication Advertising Slogans," *Journal of Applied Psychology*, vol. 76, no. 20 (1991), pp. 268–275.
5. Gerald J. Gorn, "The Effects of Music in Advertising on Choice Behavior: A Classical Conditioning Approach," *Journal of Marketing*, vol. 46 (1982), pp. 94–101.

6. C.Whan Park and S.M.Young, "Consumer Response to Television Commercials: The Impact of Involvement and Background Music on Brand Attitude Formation," *Journal of Marketing Research*, vol. 23 (1986), pp. 11–24.
7. Deborah J. MacInnis and C.Whan Park, "The Differential Role of Characteristics of Music on High and Low Involvement Consumers' Processing of Ads," *Journal of Consumer Research*, vol. 18 (1991), pp. 161–173.
8. Sophie Rieunier (Ed.), *Marketing sensoriel du point de vente* (4th ed.), Dunod, 2013.
9. Richard Yalch and Eric R. Spangenberg, "Using Store Music for Retail Zoning: A Field Experiment," in L. McAlister and M.L. Rothschild (Eds.), *Advances in Consumer Research* (Vol. 20, pp. 632–636), Association for Consumer Research, 1993.
10. L. Dubé and S. Morin, "Background Music Pleasure and Store Evaluation Intensity Effects and Psychological Mechanisms," *Journal of Business Research*, vol. 54 (2001), pp. 107–113.
11. Valerie L. Vaccaro, Veysel Yucetepe, Sucheta Ahlawat and Myung-Soo Lee, "The Relationship of Liked Music with Music Emotion Dimensions, Shopping Experience and Return Patronage Intentions in Retail and Service Settings," *Journal of Academy of Business and Economics*, vol. 11, no. 4 (2011), pp. 94–106.
12. Adrian North, David Hargreaves and Jennifer McKendrick, "The Influence of In-Store Music on Wine Selections," *Journal of Applied Psychology*, vol. 84, no. 2 (1999), pp. 271–276.
13. Charles S. Areni and David Kim, "The Influence of Background Music on Shopping Behavior: Classical Versus Top-Forty Music in a Wine Store," *Advances in Consumer Research*, vol. 20 (1993), pp. 336–340.
14. Thomas C. Roballey, Colleen McGreevy, Richard R. Rongo, Michelle L. Schwantes, Peter J. Steger, Marie A. Wininger and Elizabeth B. Gardner, "The Effect of Music on Eating Behavior," *Bulletin of the Psychonomic Society*, vol. 23 (1985), pp. 221–222; Ronald E. Milliman, "The Influence of Background Music on the Behavior of Restaurant Patrons," *Journal of Consumer Research*, vol. 13 (1986), pp. 286–289.
15. Patricia Cain Smith and Ross Curnow, "'Arousal Hypothesis' and the Effects of Music on Purchasing Behavior," *Journal of Applied Psychology*, vol. 50 (1966), pp. 255–256.
16. Emmanuelle Ménage, *Consommateurs pris au piège* [film], France 5, January 12, 2014.
17. Gabriel Dabi-Schwebel, "13 exemples de marketing sensoriel réussis and ratés," 2013, www.1min30.com.
18. Eric R. Spangenberg, Bianca Grohmann and David E. Sprott, "It's Beginning to Smell (and Sound) a Lot Like Christmas: The Interactive Effects of Ambient Scent and Music in a Retail Setting," *Journal of Business Research*, vol. 58 (2005), pp. 1583–1589.
19. "Feel It, the Blog That Develops Your Senses," https://feelitblogdotcom.wordpress.com.
20. Emmanuel Bigand (Ed.), *Le cerveau mélomane*, Belin, 2013.

CHAPTER 14

The neuro-consumer and the sense of smell

The sense of smell is vital for many species. It is the most elementary means of perceiving the world that surrounds us: it is useful for finding food, avoiding predators, identifying the members of one's group, marking one's territory and finding sexual partners. It contributes to the survival of species by helping individuals decide whether to accept or reject the close environment.[1] Its place in culture and religions remains significant even though it has lost importance to the benefit of vision in modern society.[2] This explains why smell is relatively little exploited in commercial and marketing terms compared to sight and sound. However, this is now changing because scientific research has demonstrated that the sense of smell is like an open door to our memory and emotions. Companies seek to make the most of this in developing emotional links with consumers. Scientific study of this "chemical sense" developed from 2004, when Richard Axel and Linda Buck, professors at the Universities of Columbia and Washington, received the Nobel Prize for Physiology and Medicine for their work on olfactory receptors.[3]

The mechanism of smell

Smell is the sense that enables us to analyze volatile chemical molecules – or odors – present in the air. There are two ways of detecting a smell: directly, through the nose (ortho-olfaction) and indirectly through the mouth (retro-olfaction). In both cases, the molecules reach the olfactory mucus membrane, situated at the top of the nasal cavity. It contains olfactory cells whose specialized receptors enable the detection of smell. Each molecule activates a unique group of receptors. The mucus membrane covers about 10%, or 2 cm^2 of the total surface of the nasal cavity. It includes 5 million cells and 400 different olfactory receptors. Thanks to these receptors, the mucus membrane can distinguish a trillion different smells. This phenomenal number was put forward by a study published in *Science* in March 2014.[4] Until that publication, the scientific literature considered that the sense of smell could only detect 10,000 different smells, making it one of our least developed senses. By way of comparison, the human eye perceives between 2.5 and 7.5 million different colors and the ear around 340,000 sounds. This recent discovery reveals the power of the human sense of smell.

How are odors perceived?

Odors are perceived via minuscule cilia on the outside of the olfactory cells that convert the chemical stimuli of smell into nerve impulses. The information is transmitted to the brain by the olfactory nerve. The flow of nerve impulses arrives directly at the olfactory bulb, in the prefrontal area of the brain where information about smell, as well as that related to taste, is processed. The bulb is the first stage in the processing of olfactory data by the brain. After the bulb, the information is sent to the olfactory cortex and then to the brain's limbic system. There it comes into contact with the areas dedicated to emotions and memory.[5]

Smell, along with hearing, is the sense organ that is spatially the closest to the brain. Unlike the other senses, it has direct access to the limbic system, responsible for emotions and memory. As neurosurgeon Patrick Georges puts it, "Smell is the sense that reason least dares contradict."[6]

Smell uses a processing system that is essentially hedonic: "When confronted with a smell, we ask ourselves first whether we like it before adopting a more analytical approach to determine what it is and where it comes from."[7]

The two perception thresholds for smells and individual differences

In practice, there are two thresholds for perceiving smells. The first, and lowest, or "detection threshold," corresponds to the level at which a smell can be detected in the air. The level of detection varies between species, individual organisms and the type of smell. Animals can detect smell at much lower thresholds than humans. Some smells are detectible at lower concentrations than others. Molecules such as pheromones are in fact perceived subconsciously by the olfactory system.

The second threshold involves recognition of the smell. It is called the "recognition threshold," and is higher than the first. It is difficult to attribute a given smell with descriptions such as floral, sensual, sweet, acid. It is even more difficult to identify the specific source of the smell: the precise nature of the flower or foodstuff. This difficulty with identification comes from a faulty memory and not from a deficiency in the sense of smell.

There are strong individual differences in the sense of smell that may be due to sex, age or culture. According to Virginie Maille, professor of management and expert in olfactory marketing, the main results brought to light by academic research thus far are:

- Women's sense of smell is considered to be more highly developed than that of men and research suggests that smells evoke more memories in women than in men.
- There are significant generation effects and the acuity of the sense of smell deteriorates over time.
- The ethnic origin, the geographical location of origin and the generation to which the person belongs would appear to affect their sensitivity to smell. The sea, for example, has been found to evoke different scents among people from Brittany and those from Mediterranean regions.[8]

Powerful though the sense of smell is, it appears that the brain processes it more slowly than the other senses. According to a study carried out by Rachel Herz, Professor of psychiatry and human behavior at the American University of Brown at Providence, and the Swedish Trygg Engen,[9] although it requires only 45 milliseconds for an individual to detect a visual object, it would take them about 450 milliseconds, i.e., ten times longer, to detect a smell. After a few minutes, the perception tends to disappear through a process of adaptation, the effect of which is to increase the initial detection threshold.

Smell, an acquired, persistent sense, with a direct link to the memory

Unlike the other senses, our knowledge of smells is mainly acquired. Humans do not have a preference for particular smells from birth.[10] However, the perception of some smells seems to be innate, particularly those connected with the survival of the species. The sense of smell develops over time, mainly by association. Many smells are directly linked to experiences, memories and the emotions felt on contact with them.

This link with memory is described by Proust at the beginning of the 20th century in *In Search of Lost Time*.[11] Moreover, neuroscientists have not only demonstrated that the sense of smell is closely linked with memory, but that smells are conducive to the recording of mental associations.[12]

Thus, there is a physical and neuronal proximity between the systems associated with the sense of smell and memory. According to researchers Rachel Herz and Trygg Engen, "There is a unique direct connection, with rapid synaptic transfers, between the olfactory cortex and the amygdala-hippocampal complex of the limbic system, seat of emotions and memory."[13] (It should be recalled that the amygdala is known to be the center for emotions and that the hippocampus is involved in processing and forming memories.) Whereas the thalamus sorts and sends on sensory information to the structures of brain, the sense of smell is an exception. Olfactory stimuli bypass the thalamus and go directly to the brain's limbic system to be processed by the structures associated with emotions and memory.

These strong links between smells and memory have been brought to light by many research studies, including that by Professor Aradhna Krishna,[14] a renowned researcher in sensory marketing at the University of Michigan. She reviews the main observations from studies on the subject:

- The ability to recognize smells encountered in the past persists over very long periods of time with minimal reduction in the precision of the recognition. This phenomenon may be observed whatever the interval between the first and later exposures to the smell, whether the new exposure takes place after a few seconds, a few months or even a few years. This is not the case with memories coming from other senses, which fade more quickly over time.
- The long-term olfactory memory proves to be stronger than other sensory memories. This is because memories linked to the sense of smell are initially more charged with emotion.
- The memorizing of information disseminated in the presence of smells persists for longer. This is a consequence of the direct link between smells and the memory.
- The presence of a pleasant background smell encourages memorizing.
- The sense of smell is also linked to the emotions. It is considered to be the most emotional of our senses.[15]

Rachel Herz[16] shows that autobiographical memories reactivated by smell are more emotional, more detailed and go back further than those reactivated by visual or auditory stimuli.

The close relationship between the emotions and smells shows that the latter are capable of influencing mood, generating a state of alertness or of relaxation and evoking emotional memories. Some smells act as physiological stimulants whereas others possess calming and relaxing qualities. When observed under magnetic resonance imaging (MRI), the smell of mint is capable of activating part of the brain responsible for alertness. It helps to improve performance and to increase the capacity to work longer and harder, be it in the context of sporting or intellectual activities. The scents of orange or lavender have been found to reduce anxiety.

One consequence of learning by association is that a smell is liked more if it is encountered for the first time in positive circumstances.

Research into marketing using the sense of smell

The research described above has aroused significant interest in the fields of business marketing and communication. In 2005, 99% of communication by brands was still concentrated on only two of the senses: sight and hearing.[17] The sense of smell, with its direct connection to human emotions and memory, has proved to be very promising. Over the last 20 years, we have seen the development of the discipline of olfactory marketing and the appearance of olfactory signatures for some brands. Research studies in marketing, of which a few are discussed below, have confirmed and expanded the results of the neuroscientific studies described earlier in the chapter.

As early as 1932, research by D.A. Laird[18] showed the positive impact of the presence of an agreeable smell on consumers' evaluation of the quality of a product. Articles of underwear perfumed with a floral scent received a more positive evaluation from the housewives interviewed in the study, without any reference being made to the scent.

More recent research has shown that the presence of an agreeable ambient smell has a positive impact on:

- the evaluation of the shop and its products;
- the intention to visit the shop, to make a purchase and the time spent in the shop, whether this corresponds to the customer's perception of the time he has spent[19] or the actual time spent,[20] the first being underestimated and the second increased;[21]
- the evaluation and memorizing of brand names (known and, to an even greater extent, unknown) by causing consumers to pay greater attention;[22]
- the amount of money spent in the sales outlet,[23] on condition that the smell diffused is consistent with the type of product purchased (for example a feminine/masculine smell in a zone of the shop selling men's/women's clothes).[24]

Individual differences have been observed. The capacity of an ambient fragrance to increase expenditure in a shopping center proved to be significantly weaker among older consumers.[25]

Concerning the link between memorizing, memories and smell, Aradhna Krishna, May Lwin and Maureen Morrin[26] carried out research showing that exposure to a smell at the same time as information is given about a product's characteristics improves the memorizing of that information, even when the smell in question was not reintroduced later, when the subjects were asked to recall the characteristics. The research was based on experiments carried out with pencils. Half of them were perfumed with the smell of an essential oil whereas the remainder were not perfumed. Each participant in the study was given a list of 11 characteristics of these products (environmentally friendly, easy to sharpen, stronger) at the same time as a pencil, perfumed or not. After several periods (25 minutes, 24 hours and 2 weeks), the participants were asked to list as many of the pencils' characteristics as possible. The subjects exposed to the perfumed pencils showed a memorizing capacity significantly superior to those having received unperfumed pencils. Beyond the "Proustian phenomenon," the research showed that the memorizing was better, not only when the smell of the essential oil was reintroduced, but also when it was not. As Aradhna Krishna points out, the experiments show that repeated exposure to a smell activates a link in the brain with the product attached to that smell, but also increases the capacity to recall other details about the product.[27] Olfactory signatures may prove to be more powerful than their visual or auditory counterparts. This observation is interesting when the objective of a brand's signature is to help consumers remember a product's characteristics.

Research into the perception of smell carried out by Rachel Herz and Julia von Clef[28] shows that perception can be manipulated by the name given to the smell. The same smell qualified as "vomit" in one case and as "parmesan" in the other is perceived in a much more positive fashion in the second case, including when the smell tested really is that of parmesan. The designation given to the smell, more than the smell itself, leads to a change in perception. As Aradhna Krishna explains, "Marketers can mask unpleasant smells by giving them attractive names."[29]

The relationship between the intensity of a smell and how much it is appreciated follows an inverted U curve for pleasant smells and a straight line for disagreeable smells. The evaluation of the smell of lilac (pleasant) tends to increase with its intensity up to a certain point where the relationship inverses and the greater the intensity the more the smell is perceived as disagreeable. A weak smell of fish (unpleasant) may be acceptable, but the more intense it becomes, the more its appreciation becomes negative.[30] The optimum intensity of a smell must therefore be determined before using it to avoid saturation of the sense of smell and a feeling of rejection. There are big differences between individuals. It is necessary to adapt the use of the smell to the target customer segment (thresholds may potentially be different between segments) or to prefer, in the absence of homogeneous olfactory behavior by the target, a threshold near the minima observed (the distribution curve must be studied carefully).

The results of this research confirm that the sense of smell is of great interest to marketers. Two additional reasons may be underlined. First, olfaction is an automatic process that we cannot completely prevent. Before we block our noses, a bad smell is effectively perceived by the brain. Second, according to a study by Lindstrom and the Millward Brown Agency in 2005,[31] when we look at the communication by companies, almost totally oriented towards sight and sound, it is appropriate to consider the fact that 75% of our emotions are generated by what we smell.

Because it began only recently, there is still relatively little research into marketing related to the sense of smell. It suffers from three main difficulties, underlined by Professor Virginie Maille: "The difficulty of manipulating olfactory stimuli, the degree of difference between individuals and the difficulty of putting into words what we feel."[32] Given the increasing interest in this sense, it is reasonable to think that research will continue to develop at a sustained pace.

Olfactory marketing and olfactory techniques

In parallel with this research, companies and brands have been developing so-called "olfactory" marketing for the last 20 years. Aware of the direct link between the sense of smell and the limbic system, it seeks to use the capacities of the sense of smell to attract and increase attention, encourage memorizing, create emotional links and provide a better ambiance. It is helped by the multiplication of new olfactory techniques that make it possible to solicit this sense in many different places, either actively or passively. Synthetic odors, micromolecular techniques, sprays, perfumed publicity vouchers, diffusers of background fragrances, etc. form a large panoply of tools aimed at companies and sales outlets.

A number of companies are contributing to the development of olfactory marketing. Air Berger, number one in Europe and the European division of the world leader Scentair, produces around a hundred different smells in its factory in Rouen, Normandy. They range from Christmas tree to Fraise Tagada and crème brûlée, including one called "Feet in the Water" (a mixture of salty, marine smells designed for surf and sailboard shops). By acting on moods, the diffusion of artificial fragrances results in a 38% increase in impulse purchases and an increase in perceived quality of 5% to 9% according to director Pascal Charlier.[33]

> *Air Berger's clients are numerous, for example the electrical appliance chain Boulanger. Several diffusers are distributed around each shop: a smell of coffee mixed with cappuccino in the coffee machine department, fragrance of freshly cut grass in the television department before the football World Cup and its beautiful turf pitches, clean smell in the washing machine department. The average increase in sales is estimated at 20%.*
>
> *A client who owns a car dealership asked the company to design a fragrance capable of "creating a feeling of joy and human warmth." He wanted to emphasize the euphoric aspect of buying a car and minimize any possible feeling of anxiety. The brief for the fragrance emphasized the desire to imprint the brand in customers' memories in order to develop their loyalty.*
>
> *The diffusion of specific fragrances is also used by the chain of toy shops Toys R Us in different parts of each shop. Just like butchers and bakers have done for a long time, since the 1990s, Disney has used smells to increase sales at its popcorn and other food stalls.*

During a television broadcast,[34] Stéphane Arfi, director of Emosense, testified to the effectiveness of using smells.

> *His company suggested that sales of chocolates could be increased by diffusing the smell of the product outside shops. It also presented*

travel agencies with the perfume Tiare Fleur de Tahiti, likely to boost sales of holidays. During the broadcast, we learned that smell can increase impulse purchases by something in the order of 30%.

Optician Caroline Rigondet uses an odor diffuser in her shop. She has noticed that as a result, customers stay longer but also that the odor conveys a feeling of luxury in relation to the products.

During the same television program, neuroscientist Idrisse Aberkane[35] *described the diffusion of the smell of coffee, much appreciated by the Koreans, in buses in Korea every time that an advertising jingle for the brand Dunkin' Donuts was broadcast. Dunkin' Donuts is a brand of an American restaurant chain specializing in the sale of doughnuts and coffee that has many outlets in South Korea.*

An agreeable smell can in no way compensate for a bad product or service. Management studies[36] confirm, however, its ability to modify attitudes and behaviors.

The use of a perfume diffuser in supermarkets by the hair products manufacturer Herbal Essences not only increases the extent to which their name is known, but increases their sales volume by nearly 30%, which would tend to confirm what was said in the television broadcast mentioned above.

To the use of pleasant smells, a source of differentiation often used by health and beauty products, may be added the design of veritable olfactory signatures, which are widely used by car manufacturers, ready-to-wear clothing chains and the hotel industry.

From 2003, Cadillac perfumed its car seats with a fragrance called "Leather Nuance." Today, Rolls Royce diffuses a smell of "Old Rolls" in its new cars. Victoria's Secret uses a smell of potpourri in each of its boutiques to increase brand recognition at each visit. Novotel hotels use an olfactory signature called "Cozy Lounge." According to a study carried out by BVA for Air Berger, there is an improvement in the company's image and in the level of satisfaction and customer recommendation when entrance halls are perfumed.

As well as Scentair and Air Berger, there are many other companies on the olfactory market such as Air Aroma (Australia), Firmenich (Switzerland) and Givaudan (France). They promise to put the customer in a good mood thanks to an agreeable background fragrance that makes him stay in the shop for longer, touch more products and be more likely to make impulse purchases. They do their best to create olfactory identities for brands on the basis of marketing "briefs."

There are two pitfalls to consider carefully. The first is the thorny – and hitherto unresolved – problem of patenting an olfactory brand and giving it legal protection. A solution is necessary to encourage growth in the marketing of smells. The second relates to the risk of having too many smells in one shop or making them too strong. This can result in consumers being indisposed, irritated or even worried, the complete opposite of what the olfactory approach is supposed to do. Carrying out careful, measured pilot tests can protect retailers against error and help them optimize the effects they are seeking.

Notes

1. Richard Axel, "The Molecular Logic of Smell," *Scientific American* (October 1995), pp. 154–159.
2. Mark M. Smith, *Sensing the Past: Seeing, Hearing, Smelling, Tasting and Touching*, University of California Press, 2007.
3. Linda Buck and Richard Axel, "A Novel Multigene Family May Encode Odorant Receptors: A Molecular Basis for Odor Recognition," *Cell*, vol. 65, no. 1 (1991), pp. 175–187.
4. Caroline Bushdid, Marcelo O. Magnasco, Leslie B. Vosshall and Andreas Keller, "Humans Can Discriminate More Than 1 Trillion Olfactory Stimuli," *Science*, vol. 343, no. 6177 (March 2014), pp. 1370–137.
5. www.ikonet.com/fr/ledictionnairevisuel/static/qc/smellat.
6. Patrick Georges, Anne-Sophie Bayle-Tourtoulou, Michel Badoc, *Neuromarketing in Action: How to Talk and Sell to the Brain*, Kogan Page, 2013.
7. Rachel S. Herz, "The Emotional, Cognitive and Biological Basics of Olfaction: Implications and Considerations for Scent Marketing," in Aradhna Krishna (Ed.), *Sensory Marketing: Research on the Sensuality of Products* (pp. 87–108), Routledge, 2010.
8. Virginie Maille, "L'influence des stimuli olfactifs sur le comportement du consommateur: un état des recherches," *Recherche et Applications en Marketing*, vol. 16, no. 2 (2001), pp. 51–76.
9. Rachel S. Herz and Trygg Engen, "Smell Memory: Review and Analysis," *Psychonomic Bulletin and Review* (September 1996), pp. 300–313.
10. Papadopoulou Maria, *Cultural Differences in Scent Preferences and Perceptions: An Overview of Scent Marketing and an Exploratory Research on Cultural Differences Between French and Chinese Consumers Regarding Scent Preferences and Perceptions in Fashion Retail Environments*, MSc dissertation, HEC Paris, 2014.
11. Marcel Proust, *In Search of Lost Time* [1913–1926], Everyman, 2001
12. Larry Cahill, Ralf Babinsky, Hans J. Markowitsch and James L. McGaugh, "The Amygdala and Emotional Memory," *Nature*, vol. 377, vol. 6547 (1995), pp. 295–296; Rachel S. Herz, James C. Eliassen, Sophia L. Beland and Timothy Souza, "Neuroimaging Evidence for the Emotional Potency of Smell-Evoked Memory," *Neuropsychologia*, vol. 42 (2003), pp. 371–378.

13. Herz and Engen, "Smell Memory."
14. Aradhna Krishna, "An Integrative Review of Sensory Marketing: Engaging the Senses to Affect Perception, Judgment and Behavior," *Journal of Consumer Psychology*, vol. 22 (2012), pp. 332–351.
15. Beryl Lieff and Joseph Alper, "Aroma Driven: On the Trail of Our Most Emotional Sense," *Health*, vol. 20 (1988), pp. 62–67.
16. Rachel. S. Herz, "A Comparison of Autobiographical Memories Triggered by Olfactory, Visual and Auditory Stimuli," *Chemical Senses*, vol. 29 (2004), pp. 217–224.
17. Martin Lindstrom, "Broad Sensory Marketing," *Journal of Product and Brand Management*, vol. 14, no. 2 (2005), pp. 84–87.
18. D.A. Laird, "How the Consumer Estimates Quality by Subconscious Sensory Impressions: With Special Reference to the Role of Smell," *Journal of Applied Psychology*, vol. 16, no. 3 (1932), pp. 241–246.
19. Eric R. Spangenberg, Ayn E. Crowley and Pamela W. Henderson, "Improving the Store Environment: Do Olfactory Cues Affect Evaluations and Behaviors?" *Journal of Marketing*, vol. 60, no. 2 (1996), pp. 67–80.
20. Nicolas Guéguen and Christine Petr, "Smell and Consumer Behavior in a Restaurant," *International Journal of Hospitality Management*, vol. 25 (2006), pp. 335–339.
21. Bruno Daucé, "*La diffusion de senteurs d'ambiance dans un lieu commercial: intérêts et tests des effets sur le comportement*," doctoral thesis, University of Rennes, 2000.
22. Maureen Morrin and S. Ratti Ratneshwar, "The Impact of Ambient Scent on Evaluation, Attention and Memory for Familiar and Unfamiliar Brands," *Journal of Business Research*, vol. 49, no. 2 (2000), pp. 157–165.
23. Maureen Morrin and Jean Charles Chebat, "Person–Place Congruency: The Interactive Effects of Shopper Style and Atmospherics on Consumer Expenditure," *Journal of Service Research*, vol. 8, no. 2 (2005), pp. 181–191.
24. Eric R. Spangenberg, David E. Sprott, Bianca Grohmann and Daniel L. Tracy, "Gender-Congruent Ambient Scent Influences on Approach and Avoidance Behaviors in a Retail Store," *Journal of Business Research*, vol. 59 (2006), pp. 1281–1287.
25. Jean-Charles Chebat, Maureen Morrin and Daniel-Robert Chebat, "Does Age Attenuate the Impact of Pleasant Ambient Scent on Consumer Response?" *Environment and Behavior*, vol. 42, no. 2 (2009), pp. 258–267.
26. Aradhna Krishna, May Lwin and Maureen Morrin, "Product Scent and Memory," *Journal of Consumer Research*, vol. 37 (June 2010), pp. 57–67.
27. Aradhna Krishna, *Customer Sense: How the 5 Senses Influence Buying Behavior*, Palgrave Macmillan, 2013.
28. Rachel S. Herz and Julia von Clef, "The Influence of Verbal Labeling on the Perception of Smell: Evidence for Olfactory Illusions?" *Perception*, vol. 30, no. 3 (2001), pp. 381–391.
29. Ibid.
30. Herz, "The Emotional, Cognitive and Biological Basics of Olfaction."
31. Michael Lindstrom, *Brand Sense, Sensory Secrets Behind the Stuff We Buy*, Free Press 2010.

32. Virginie Maille "L'influence des stimuli olfactifs sur le comportement du consommateur: un état des recherches," *Recherche et applications en marketing*, vol. 16, no. 2 (2001), p. 51–76.
33. Emmanuelle Ménage, *Consommateurs pris au piège* [film], France 5, January 12, 2014.
34. *La Quotidienne* [TV show], France 5, March 2, 2016.
35. Idrisse Aberkane, *Libérez votre cerveau!* Robert Laffont, 2016.
36. Sophie Rieunier (Ed.), *Marketing sensoriel du point de vente*, Dunod, 2013.

CHAPTER 15

The neuro-consumer and the sense of touch

Touch is a vital sense providing contact with the environment. It helps neuro-consumers to orient themselves in space, identify objects and detect obstacles. Thanks to touch, they can perceive heat, cold, pressure and pain. It acts as a natural alarm system and is widely considered to be the most indispensable sense for the survival of the human being. Linguistic metaphors for touch are legion and bear witness to its importance. For example, we hear that people are "touching," are told to "stay in contact," to "keep in touch" and to "take a problem in hand." On the other hand, there is still little research into its impact on consumer behavior, but what there is shows strong differences in people's need to touch others. It underlines the large degree of interaction between touch and the other senses and clarifies its importance in interpersonal relationships and in sales outlets. Given the impossibility of touching in online sales, it is interesting to consider using the other senses and new haptic technologies (defined as obtaining sensations by touch).

How touch works and types of touching

Dating back to antiquity, Aristotle[1] (384–322 BC) emphasized the importance of touch. According to his hierarchy, touch was at the summit of the five senses. The other four contribute to increasing the acuity of tactile perception. Because it enables living beings to perceive friction, pressure and other deformations to the outside of the body, touch is vital in many situations. It is part of defensive and offensive behaviors linked to the survival of species.

The sense of touch, also called the somatosensory system, includes three main divisions. The first two, called "proprioception" and "kinesthesis," concern the way in which we perceive our body and limbs in space or when they move. The muscles, joints and tendons provide information called "proprioceptive" and "kinesthetic," relating to the postures, movements and efforts carried out. The third division – more relevant from the point of view of marketing – involves the "cutaneous" system. This includes the perception of touch and of pain from the haptic receptors, mainly situated in the upper layers of the epidermis.

The density of touch-sensitive receptors varies a great deal depending on the part of the body. Fairly well scattered in less-sensitive areas such as the back, they are very numerous and close together in delicate areas, such as the tongue, the lips and the fingertips. Known as "mechanoreceptors," they include one or more neurosensory cells. When the cells undergo deformation following contact, nerve signals are transmitted by neurons to a particular zone of the cerebral cortex, the primary somatosensory cortex, where they are "interpreted."

Different types of touch coexist. The first difference concerns the active or passive character of touching. Active touching, also known as "haptic touching or perception," occurs when the individual deliberately decides to explore his environment seeking tactile information. On the other hand, passive touching occurs when the individual feels a tactile sensation without having sought it. Passive touching is frequently associated with the immobilization of the part of the body involved.[2]

A second difference relates to the objective sought. Touch is called "instrumental" when the objective sought is functional: we touch an object to obtain information about its properties. This information can relate to texture, hardness, temperature or weight. It is called "autotelic" when the objective is hedonic. It is then motivated by curiosity and the search for tactile sensations and stimuli through sensory experience.

The importance of touch in purchasing decisions and the appropriation effect

The observation of the importance of the sense of touch is not a new phenomenon, however, studies related to its impact on consumer behavior are few and recent. According to researchers Hendrik Schifferstein and Pieter Desmet,[3] professors of engineering and industrial design at the Technological University of Delft in the Netherlands, touch would appear to be, after sight, the second most important sense for the immediate evaluation of a product. Using experiments where participants were deprived of one of their senses, the researchers found that, when manipulating objects, people deprived of sight and touch suffered more than those deprived of hearing or smell. The absence of visual and then tactile information strongly reduced the quality of the experience.

More than any other sense, touch proves to be essential to judging the hardness, texture, temperature and weight of an object. These properties are called "material" by researchers Robert L. Klatzky from the University of Carnegie Mellon and Susan L. Lederman from the University of Kingston, Ontario, in Canada. Academic research has brought to light the importance of touch in evaluating numerous products through their material characteristics. This is particularly the case in the study carried out on sweaters by Professor Holbrook[4] from Columbia and that by Professors Deborah Brown McCabe from Merrill College in Atherton and Stephen M. Nowlis[5] from Washington University. These researchers found that products for which touching can reveal differences (such as towels or rugs) have a higher probability of being bought in commercial environments where they can be taken hold of compared to those where this is not possible. This difference is not observed, however, for product categories that do not show variation in their material properties (for example, DVDs, batteries, light bulbs). The difference in preference between the two types of commercial environment can be reduced if a verbal explanation of the material properties of the product is given. It is possible to compensate for the absence of opportunity to touch, as in the case of mail order for example, by a written description of the products on offer.

The reasons given to explain the role of touch in the evaluation of products, although as yet little studied, are linked to the fact that it affects consumers' cognitive (by giving information) or emotional (by arousing pleasant or unpleasant sensations) processes when they have direct contact with the object.[6]

Companies are making increasing use of the sense of touch to make the most of their products or their sales outlets.

> *Martin Lindstrom[7] describes the example of British supermarket chain Asda, which knows how to make the most of tactile information communicated to its customers. By giving them the opportunity to touch the different brands of toilet paper on sale in its stores, it recorded a significant rise in sales of its own brand, leading to an increase of 50% in its presence on the shelves.*
>
> *Apple, Orange and SFR shops are also good examples. All the products are freely available to be handled, despite their high unit value. This type of product had for a long time remained in locked cabinets, out of reach of consumers.*

The objective for these companies is to capitalize on the influence that touch can have on what is known as the "endowment effect": The simple fact of touching a product, of handling it, increases the feeling of ownership. If we refer to the principles of behavioral economics, Richard Thaler,[8] awarded the 2017 Nobel Prize for Economics, shows that the endowment effect relates to the hypothesis according to which people attach more value to things simply because they own them. The price that they ask when selling a product that they own proves to be higher than the price that they would be prepared to pay to buy it. This observation has been confirmed by research carried out by the American-Israeli 2002 Nobel Prize-winning psychologist and economist Daniel Kahneman and his colleagues Jack Knetsch from the University of Berkeley and Richard Thaler from Cornell University.[9] They showed that the owners of a cup ask significantly more money (about $7) to sell it than when they are potentially ready to buy it (about $3). The link between touching, feeling of ownership and readiness to pay is validated in research by Professors Joann Peck and Suzanne Shu.[10] They observed that the fact of touching an object (as opposed to being unable to touch it) results in a stronger feeling of ownership and a greater readiness to pay. According to a study carried out by IFM/MCA in 2003,[11] 87% of products handled in a shop are bought.

Individual differences in the need to touch

Although touch can have an impact on the decision to buy, it is not equally important for all consumers. Several pieces of research have measured individual differences in the need to touch and acquire tactile information about a product before deciding to buy it.

This research is based on two scales specific to the field of consumption dating from 2003 that measure preferences for haptic information. The most complete is the "Need for Touch" (NfT) scale, developed by Professor Joann Peck from Wisconsin University and Professor Terry Childers[12] from Iowa State University.

> *The NfT scale includes 12 criteria providing a measure of the individual need for touch. Examples of those criteria are: "When browsing in stores, I can't help touching all kinds of products"; "Touching products can be fun"; "I feel more confident making a purchase after touching a product." Each interviewee had to say to what extent he agreed or disagreed with each of the items on the scale using a scoring system going from –3 (totally disagree) to +3 (completely agree). Half of the criteria seek to measure the instrumental need for touch whereas the other half concentrate on autotelic need.*

The instrumental need comes from the desire to obtain information about the material properties of a product with a potential purchase in mind. The autotelic need is an end in itself, not necessarily linked to an intention to purchase but motivated by the search for tactile sensations.

Individuals with a high score on the NfT scale are more inclined towards autotelic or hedonic touching of objects and derive greater pleasure from it. They are also less confident in their judgment of a product when they are not able to touch it than when they can. On the other hand, this difference is not observed for individuals with a low score on the NfT scale.[13]

Suzanne Shu and Joann Peck[14] also show that the endowment effect is greater in individuals with a higher autotelic touch score when they have the opportunity to touch the product. This difference is not, however, observed in individuals with a strong need for instrumental touching.

Researchers Alka Varma Citrin, Donald Stern, Eric R. Spangenberg and Michael J. Clark[15] have observed a difference in the need for touch between men and women. They suggest that women appear to have a greater need to touch products than their male counterparts and that this touching increases their confidence in the product and their intention to purchase.

An individual's expertise in relation to a product generates differences in the need to touch it before purchasing. Research carried out by Atefeh Yazdanparast and Nancy Spears[16] shows that the expert has greater need than the nonexpert for tactile information in order to increase his confidence in the object and reinforce his intention to purchase.

Differences between products in the need for touch

Tactile information is essential to evaluate the properties of an object, e.g., its robustness, hardness, temperature and weight. Touching provides unique information about its properties that cannot be obtained by visual inspection. Tactile information is important in the evaluation of offers whose material properties are significant and highly variable.

In a study carried out in 2004 where participants' hand movements were filmed, Joann Peck and Terry Childers demonstrated strong correspondence between people's answers on the NfT scale and their actions.[17] Product categories that vary in their material characteristics like sweaters or tennis rackets were touched for a longer time than those with less variance such as calculators and mobile telephones and those where the physical aspects are not evaluated, such as breakfast cereals and toothpaste.

This difference between products led Citrin and colleagues to conclude that "haptic products" are less likely to be sold in an online environment.[18] This explains why consumers, when they evaluate "haptic products," are more frustrated and less confident in a situation where touching them is impossible.

The effect of touch would appear to be particularly positive for high-quality products, whereas tactile information generally has a negative effect on the evaluation of low-quality products.

When there is no opportunity to touch a product, a written description appears to be an adequate substitute for its functional characteristics – such as its weight – but inadequate for more hedonic characteristics – such as softness. For consumers with a weak autotelic need for touch, some researchers recommend supplying visual information to compensate for the absence of haptic exploration.[19]

Interaction of touch with the other senses

Aradhna Krishna, a specialist in sensory marketing, has worked on the interaction of touch with the other senses. Using experiments, she has demonstrated the influence of haptic properties on the perception of objects, based on other senses.

The taste of a drink may be affected by the haptic properties of its container. Wine is generally judged to be of better quality if it is drunk out of a glass rather than out of a plastic goblet. The rigidity of a cup containing water affects consumers' evaluation of the liquid. Not all consumers are influenced in the same way by haptic indices of no importance compared to the taste. Those with a strong autotelic need for touch are less influenced by these indices than those with a weak autotelic need.[20] The former make the most of the process of haptic exploration and know how to appreciate tactile indices when they are relevant to their judgment. On the other hand, they are better than consumers with a weak autotelic need at ignoring them when they are irrelevant.

Aradhna Krishna[21] has also studied the interaction between touch and smell. She confirms that certain haptic properties may be linked to types of smell. She has shown that a rough paper receives a more positive evaluation when it is accompanied by a masculine smell. On the other hand, smooth paper benefits from a better evaluation when it is linked to a feminine fragrance. The congruence between touch and smell depending on the sex of the individual can improve neuro-consumers' perceptions of the quality of a product. According to Aradhna Krishna, incorporating harshness or softness into an object or its packaging has a positive or negative impact on consumers' perceptions depending on the congruence of the texture and the sex of the individual interested in the item.[22]

Touch and interpersonal relations

Interpersonal touching is studied in many disciplines: cognitive and social psychology, neuroscience and cultural anthropology. As Alberto Gallace and Charles Spence remind us, studies have shown that women and young people have been found to be the most sensitive to touch.[23]

In the context of consumption, interpersonal touching between employees and customers has a positive influence on the latter's responses. According to Benoît Heilbrunn,[24] a professor at ESCP-Europe Business School and the author of many works, touch creates a feeling of intimacy and proximity with the person who takes the initiative. Thus, researchers April Crusco from the University of Mississippi and Christopher Wetzel[25] from Rhodes College have observed that a waitress who physically touches a customer will receive a higher tip, even if her service is not judged to be better. Research by Professor Jacob Hornik[26] shows that consumers are more inclined to test a new product in a supermarket when the sales assistant or demonstrator touches them. The physical contact causes the brain to release oxytocin, a hormone that produces a feeling of well-being and calm.[27]

On the other hand, when consumers are accidentally touched by other consumers while examining products, their evaluations of the brand and the product have been found to be more negative. They generally are less ready to pay and spend less time in the shop.[28] The stranger who is responsible for the accidental touching does not have the legitimacy of a sales assistant. His touch is perceived as a violation of personal space.

In some cultures, such as India and other Asian countries, touching may be considered to be ill-mannered, even impure, and should thus be avoided.

The implications of research into the sense of touch are interesting in terms of management of the sales outlet and training of sales personnel.

Tactile marketing

In parallel with this research, the sense of touch is increasingly being used in the marketing of products and companies.

The tactile aspect of many objects has been reviewed to provide consumers with more pleasure and sensations. "Soft-touch" textiles are becoming more widespread, be it for clothes, household items or the interior of cars. The softness of containers is also being improved, for example squeezable pouches for puree and many yoghurts. An unusual shape or very distinctive texture can become the characteristic elements of a brand.

> *This is the case with the Orangina bottle with its easily recognized shape and roughness. In a study of mobile phones carried out by Martin Lindstrom, 35% of consumers said that the feel of their phone is more important than its appearance.*
>
> *The ease with which products can be touched is encouraged. The piles of sweaters, prominent at Abercrombie & Fitch, are there to incite people to touch them and thus increase their desire to buy. Shops selling technical products that used to be displayed in closed showcases are pursuing the same objectives. Mobile phones, tablets, computers, cameras, vacuum cleaners and food processors are freely accessible so that consumers can judge their material qualities and appreciate how they feel in the hand.*

One of the material qualities whose optimization has been enabled by neuromarketing studies of touch is weight. When it sees a product, the brain automatically calculates the weight that it *should* have in order to be considered as being of good quality. If it fails to meet this criterion, it is very likely that the product will be put back on the shelf and not bought.

Following studies on the subject, a major mobile phone manufacturer decided to artificially increase the weight of its devices to better correspond to what its customers' brains expected.

The haptic quality of a product is sometimes highlighted in its name. The name of the iPod Touch emphasizes the new tactile interaction that it generates with its user although its features and its basic raison d'être are related to the sense of hearing. As Aradhna Krishna explains, this is an element of differentiation that makes the interaction between man and machine more personal and inviting.[29]

In many shops, the use of luxury materials, providing agreeable tactile sensations, is seen to increase the customer's feeling of well-being during his visit. The time spent wandering about the shop increases, producing favorable conditions for additional purchases.

The Sephora beauty products chain uses soft, red carpets, contributing to the feeling of voluptuousness that the company has taken the strategic decision to offer its customers. The women's underwear brand Princesse Tam.Tam has changed the textures in its dressing rooms to provide its customers with a sensorially agreeable haven. Velvet curtains, padded walls, velvet footstools and thick carpets seek to increase the sensation of cocooning that encourages people to let themselves go, including where their wallets are concerned.

Soliciting the sense of touch is widely considered to be all the more important when the target clientele is female. Research shows that women have ten times as many tactile receptors directly linked to their brains as men. During tactile contact, they also secrete more oxytocin and prolactin, hormones linked to affection and attachment, which lead to a feeling of well-being.

Very often associated with the use of other senses, tactile marketing continues to develop in the real world while awaiting new techniques to utilize in e-commerce.

What to do when customers can't touch products?

The fact of not being able to touch products reduces the consumer's confidence, particularly when he manifests a high level of need for touch. Confidence is lower when the product has material qualities that are important for its evaluation. What can be done to overcome this difficulty, particularly in e-commerce?

Images or photographs of products

Adding images or photographs of a product increases the confidence of consumers with a weak need for touch, for whom seeing, not touching, is the main source of information. However, when Joann Peck and Terry Childers studied this solution in 2003, they found that it does not work for consumers with a strong need for touch. The presence or absence of product photographs do not change their evaluation.[30]

Although for some consumers the presence of product images is insufficient to compensate for being unable to touch, it can nevertheless encourage them to imagine interacting with the product and to envisage themselves using it in the future. This observation refers to recent work inspired by psychological theories of "grounded cognition."[31] Based on our past physical experience and thanks to the mirror neurons brought to light by Professor Rizzolatti's team,[32] the simple fact of seeing a product in use can subconsciously lead us to imagine handling it, thus increasing the intention to purchase. Studying the influence of printed advertising material, Aradhna Krishna and her colleague Ryan Elder[33] show that this motor stimulation, suggested by the photograph, leads to a more favorable evaluation of the product and a greater intention to purchase.

> *In an advertisement for a yoghurt, when a spoon is placed in the pot and orientated towards the consumer's dominant hand, intentions to purchase are greater than when there is no spoon or when it is pointing towards the consumer's less dominant hand.*

Presenting haptic objects in natural situations where they are being used can make up for the absence of real handling and increase the feeling of appropriation of the product.

Descriptions and written comments

Roger Dooley[34] suggests that the comments posted on websites strengthen the appropriation effect and have a strong influence on the decision to purchase. Reading comments left by other users can lead the consumer to envisage himself using the product, to imagine himself interacting with it, thus increasing his confidence and his intention to purchase.

Haptic technology

The Échangeur by LaSer, a renowned center for technological innovation and marketing applied to the customer relationship, describes haptic technology as a small feedback indicated by a vibration of the device being used whose aim is to make the user *feel* that an action has been successfully accomplished. Used in the great majority of smartphones and electronic tablets, it enables vibrations to be generated in the joysticks of many games consoles (Wii, PlayStation, Xbox), to make the user feel the hardness of a blow during a combat game, for example. The next developments in haptic technology could involve the simulation of texture and relief and provide information leading to better interaction with tactile screens. In time, this technology could enable us to "feel" a piece of textile, a fruit or a work of art on a tactile screen, thus bringing to e-commerce what it lacks today: tactile information and its perception.

> *For the experts at L'Échangeur, use of this new technology is going to become generalized in video games. As with the Kinect or the Wii, a glove will enable the user to manage screen content and feel what the avatars are touching in their virtual worlds. Companies like Senseg in Finland are working on this technology. Haptics are going to revolutionize our interaction with the screen, just as a few years ago the first tactile screen on the iPhone revolutionized the telephone.*

Art, digital images and the sense of touch

Advances involving the digital world and the sense of touch can already be found today in the artistic field. During the fourth Digital Woman's Day, held in March 2016 in Paris, the painter, set decorator and graphic designer Claire Bardainne showed the fascinating realizations that she and her associate Adrien Mondot, computer scientist and juggler, have developed using digital images and the moving bodies of dancers. Extracts from their shows, including one called "Pixel," can be seen on YouTube or Vimeo.[35] The tactile sensation is very much present and astonishingly real, showing that digital images and touching can be part of the same world.

Notes

1. Aristotle, *On the Soul*, Peripatetic Pr, 1981.
2. David Katz, *The World of Touch*, Lawrence Erlbaum Associates, 1989.
3. Hendrik N.J. Schifferstein and Pieter M.A. Desmet, "The Effects of Sensory Impairments on Product Experience and Personal Well-Being," *Ergonomics*, vol. 50, no. 12 (2007), pp. 2026–2048.
4. Morris B. Holbrook, "On the Importance of Using Real Products in Research on Merchandising Strategy," *Journal of Retailing*, vol. 59, no. 1 (1983), pp. 4–20.
5. Deborah Brown McCabe and Stephen M. Nowlis, "The Effect of Examining Actual Products or Product Descriptions on Consumer Preference," *Journal of Consumer Psychology*, vol. 13, no. 4 (2003), pp. 431–439.
6. Alberto Gallace and Charles Spence, *In Touch With the Future: The Sense of Touch from Cognitive Neuroscience to Virtual Reality*, Oxford University Press, 2014.

7. Michael Lindstrom, *Brand Sense, Sensory Secrets Behind the Stuff We Buy*, Free Press, 2010.
8. Richard H. Thaler, "Toward a Positive Theory of Consumer Choice," *Journal of Economic Behavior and Organization*, vol. 5, no. 1 (1980), pp. 39–60.
9. Daniel Kahneman, Jack L. Knetsch and Richard H. Thaler, "Anomalies: The Endowment Effect, Loss Aversion, and Status Quo Bias," *Journal of Economic Perspectives*, vol. 5, no. 1 (1991), pp. 193–206.
10. Joann Peck and Suzanne B. Shu, "The Effect of Mere Touch on Perceived Ownership," *Journal of Consumer Research*, vol. 36, no. 3 (2009), pp. 434–447.
11. Sophie Rieunier (Ed.), *Marketing sensoriel du point de vente* (4th ed.), Dunod, 2013.
12. Joann Peck and Terry L. Childers, "Individual Differences in Haptic Information Processing: the 'Need for Touch' Scale," *Journal of Consumer Research*, vol. 13, no. 4 (2003), pp. 430–442.
13. Joann Peck and Terry L. Childers, "To Have and to Hold: The Influence of Haptic Information on Product Judgments," *Journal of Marketing*, vol. 67, no. 2 (2003), pp. 35–48.
14. Joann Peck and Suzanne B. Shu, "To Hold Me Is to Love Me: The Role of Touch in the Endowment Effect," *Advances in Consumer Research*, vol. 34 (2007), pp. 513–514.
15. Alka Varma Citrin, Daniel E. Stern, Eric R. Spangenberg and M.J. Clark, "Consumer Need for Tactile Input an Internet Retailing Challenge," *Journal of Business Research*, vol. 56, no. 11 (2003), pp. 915–922.
16. Atefeh Yazdanparast and Nancy Spears, "Can Consumers Forgo the Need to Touch Products? An Investigation of Nonhaptic Situational Factors in an Online Context," *Psychology and Marketing*, vol. 30, no. 1 (2013), pp. 46–61.
17. Joann Peck and Terry Childers, *Self-Report and Behavioral Measures in Product Evaluation and Haptic Information: Is What I Say How I Feel?* Association for Consumer Research, Working Paper Track, 2004.
18. Citrin et al., "Consumer Need for Tactile Input."
19. Joann Peck and Terry Childers, "To Have and To Hold: The Influence of Haptic Information on Product Judgment," *Journal of Marketing*, vol. 67, no. 2 (2003), pp. 35–48.
20. Aradhna Krishna and Maureen Morrin, "Does Touch Affect Taste? The Perceptual Transfer of Product Container Haptic Cues," *Journal of Consumer Research*, vol. 34, no. 6 (April 2008), pp. 807–818.
21. Aradhna Krishna, *Customer Sense: How The 5 Senses Influence Buying Behavior*, Palgrave Macmillan, 2013.
22. Ibid.
23. Alberto Gallace and Charles Spence, "The Science of Interpersonal Touch: An Overview," *Neuroscience and Biobehavioral Reviews*, vol. 34, no. 2 (2010), pp. 246–259.
24. Benoît Heilbrunn, *La Consommation et ses sociologues* (3rd ed.), Armand Colin, 2015.
25. April H. Crusco and Christopher G. Wetzel, "The Midas Touch: The Effects of Interpersonal Touch on Restaurant Tipping," *Personality and Social Psychology Bulletin*, vol. 10, no. 4 (1984), pp. 512–517.
26. Jacob Hornik, "Tactile Stimulation and Consumer Response," *Journal of Consumer Research*, vol. 19 (December 1992), pp. 449–458.

27. Bertil Hulten, Niklas Broweus and Marcus Van Dijk, *Sensory Marketing*, Palgrave Macmillan, 2009.
28. Brett A.S. Martin, "A Stranger's Touch: Effects of Accidental Interpersonal Touch on Consumer Evaluations and Shopping Time," *Journal of Consumer Research*, vol. 39, no. 3 (2012), pp. 174–184.
29. Krishna, *Customer Sense*.
30. Peck and Childers, "To Have and to Hold."
31. Margaux Limoges, *Dans quelles mesures le e-commerce souffre-t-il du déficit d'expérience tactile, et comment ce manqué peut-il être compensé?* MIM dissertation, HEC Paris, March 2013.
32. Giacomo Rizzolatti and Corrado Sinigaglia, *Mirrors in the Brain: How Our Minds Share Actions, Emotions, and Experience*, Oxford University Press, 2008.
33. Ryan A. Elder and Aradhna Krishna, "The 'Visual Depiction Effect' in Advertising: Facilitating Embodied Mental Simulation Through Product Orientation," *Journal of Consumer Research*, vol. 38 (April 2012), pp. 998–1003.
34. Roger Dooley, *Brainfluence: 100 Ways to Persuade and Convince Consumers with Neuromarketing*, John Wiley & Sons, 2012.
35. See https://vimeo.com/114767889.

CHAPTER 16

The neuro-consumer and the sense of taste

Taste receptors on the tongue

Until the appearance of the microscope and scientific analysis of the anatomy of the tongue, the sense of taste was poorly understood. The tongue is covered in many sense organs, called taste buds, each equipped with taste receptors. They were described in the 1960s by the American and Japanese anatomists and electron microscopy specialists Raymond Murray and Masako Takeda.

Having many cells that regenerate every 10 to 14 days, the taste receptors allow us to distinguish between four fundamental tastes: sweet, sour, salty and bitter. This system was first described by German physiologist Adolf Fick in 1864. All taste perception is described in terms of a combination of the four primary tastes, linked to four types of sensory receptor located in four zones of the tongue.

American Carl Pfaffman, the pioneer of the electrophysiology of the gustatory nerves, was the first to question this model. In 1980, French biologist Annick Faurion performed electrophysiological experiments that showed that each taste molecule has a particular flavor, recognized specifically by the brain. It is more accurate to say that there is a gustatory continuum around the four tastes, or elementary categories.[1]

At the same time, Japanese scientists brought to light a fifth fundamental taste, called "*umami*," meaning "tasty or flavorsome." This strong taste, close to that of a meat stock, is activated when the tongue comes into contact with two amino acids, aspartate and glutamate. They are present in protein-rich food such as meat, soya and mushrooms.

Contrary to what was taught for a long time, the gustatory receptors are located over the whole surface of the tongue (except for the center) and not in separate zones reserved for each flavor. When the receptors come into contact with food molecules through the saliva, they identify to which category of flavor the molecules belong and transmit this information to the nerve fiber. The signals are sent by three cranial nerves to the central nervous system. Certain regions of the brain decode the chemical information and translate it into a sensation of taste. All this happens in a few fractions of a second.

The tongue plays an essential role in classifying food based on the different flavors. It helps the brain to evaluate whether a food is dangerous or nourishing. A sweet taste signals the presence of carbohydrates with a significant supply of energy. A salty taste helps identify minerals that are indispensable to human metabolism. The *umami* is detected thanks to its proteinaceous elements that are nutritionally rich. Acid and sour substances, many of them toxic, are detectable by a greater number of receptors.

Working alone, the tongue is nevertheless incapable of telling the difference between a fruit juice and a soda or between an apple and a pear. In order for the complete sensation of a taste to form in the brain, the other senses must also be brought into play. More than a simple question of the tongue, taste must be viewed as an amalgam of the five senses.

Taste and olfaction

The nose makes it possible to distinguish between wine and coffee or between a shortbread biscuit and a cookie. A heavy cold can show how important this organ is in detecting flavor: When we are deprived of our sense of smell, foods can seem to have more or less the same, insipid taste.

Smell is essential in perceiving flavors. Its contribution to the perception of taste comes less from actively sniffing the smell from food than from internally inhaling volatile substances that arrive in the mouth via the larynx. Retronasal olfaction, at the back of the nose, is responsible for nearly 80% of taste. By comparison, the taste molecules liberated in the saliva and captured by the taste receptors on the tongue are responsible for about 10%.

The combination of olfactory and gustatory sensations in the brain provides the flavor of a food. Many experiments have registered a reduction in – or complete absence of – flavor when the sense of smell is blocked during the act of ingestion.

The fact that the flavor of a dish is a combination of its smell and taste explains why, in most cases, we like the taste of products whose smell we appreciate and vice-versa. There are certain exceptions to this, for example some of the famous French cheeses, the smell of which is often disagreeable but the taste widely appreciated.

Many companies, for example Starbucks, use smell as a vector of gustatory sensation for marketing and commercial purposes. The amplification of the pleasant aroma of coffee has the objective of increasing customers' gustatory satisfaction and hence the amount they consume. In their brains, the aroma becomes a distinctive part of the brand image. At the wine merchant, great wines are appreciated in the first place by the nose. Mid-20th-century oenologists first discovered the role of retronasal olfaction in taste.

Food manufacturers are constantly seeking new flavors, subtle combinations between the taste and the smell emitted by a product. Artificial aromas added to products with little taste enable such flavors to be generated. For example, surimi conveys the taste of crab or one fish or another depending on the aromas and colorants added to it. Another trend in the food industry is to use "taste amplifiers." They have the characteristic of enhancing the original flavor.

Taste and sight

Tasting food is not only a question of tongue and nose. Sight plays an important role. A product's visual aspect is essential to identifying and differentiating foods and drinks that cannot be distinguished by the tongue's tasting abilities alone.

The esthetic quality of products contributes to a better appreciation of their taste.

Truck farmers who emphasize beautiful colors and shapes on their stalls are well aware of this. The decoration of the fruit and vegetable department in supermarkets is another illustration. The lighting, carefully directed onto the products, showcases them. Water misting systems enhance their shininess. Attractive "palettes" of colors are created by playing with the nuances of the different varieties of fruit and vegetables, arranged into blocks.

Merchandizing techniques, or making the most of assortment in the customer's eyes, contribute greatly to the commercial performance of food departments.

The color of food items has an important impact on gustatory sensations. Researchers Cynthia Du Bose, Armand Cardello and Owen Maller[2] from the US Army's food science laboratory have highlighted this phenomenon. In experiments where participants were prevented from seeing the color of fruit-juice-based drinks, they precisely identified only 20% of the tastes given to them. They scored 100% when they could see the color of the drink being tested. In another experiment, cherry juice was manipulated to make it orange. Some 40% of the participants said that it tasted of orange juice. The perception of taste appears to be closely linked to the color stored in the memory. This might explain the failure of Crystal Pepsi, whose colorlessness is a complete break with the brown tone traditionally associated with cola. Another interesting result of the study by Bose et al. was that the intensity of the color appeared to increase the intensity of the flavor – perhaps an explanation of the widespread use of colorant chemicals in the food industry.

Taste and touch

An important part of the perception of taste relates to the consistency of foodstuff. This is detected when it is "touched" in the mouth.

> *Kellogg's considers that the crunchiness of its cereals is a determining factor in their success. The company's advertisements accentuate the sound made when the cereal is eaten.*
>
> *Sweetcorn producer Green Giant has two products distinguished by the texture of the niblets: ultra tender and ultra crunchy. Their tactile promises, very closely linked to their tastes, are clearly set out for consumers.*

As well as texture, touch conveys important information about temperature. Many dishes are preferred, according to different cultures or individuals, at a specific temperature.

> *Spanish gazpacho, unlike the majority of soups, is supposed to be eaten cold. Some people cool down their coffee by adding cold milk. Others can't imagine drinking it any other way but very hot or, on the contrary, iced. Presenting food at a temperature other than that intended often has a negative impact on the sensation of taste.*

As already mentioned, the haptic qualities of a container can affect the gustatory perception of the contents. Researcher Aradhna Krishna and her colleague Maureen Morrin[3] have found that the taste of a particular brand of water is perceived as being better when its container is more agreeable to the touch. Professors Carl McDaniel and Revenor Baker[4] from the University of Texas show that when a bag of potato crisps is difficult to open, people perceive the crisps as being better. Beyond its visual impact on the shelves, packaging plays an important role in the gustatory perception of products.

Taste and hearing

The influence of hearing on taste, although less obvious than that of smell or sight, exists for certain products. Brands use it in advertising, like Nestlé's Crunch cereal bars, Kellogg's Rice Krispies and Schweppes soda. Many foods have an auditory signature that neuro-consumers associate naturally with them: the crunchiness of an apple, a carrot, a dry biscuit. The sound produced when chewing has an impact on the perception of the product's freshness and tastiness.

Research carried out by Professors Massimiliano Zampini and Charles Spence[5] from the University of Oxford provides empirical proof of this. Using potato crisps as the study object, they equipped the participants with headphones. For some, the noise emitted as they ate the crisps was reduced, whereas for the others it was amplified. The participants were asked to score their perception of freshness and crunchiness. The scores obtained from participants where the sound was amplified were significantly higher. Potato crisp manufacturers have exploited this result by making their crisps bigger so that they have to be crunched. There is an impact on their perceived taste, and also on the desire of those around the eater to consume them.

Taste and the "happy hormone"

Serotonin, the "happy hormone," leads the brain to perceive the taste of foods with more intensity, whether they are bitter or sweet.[6]

> *Restaurants' efforts to increase their customers' well-being – the parking valet, the welcome from the staff, the general atmosphere – are fully justified. They help to put customers in a good mood and awaken their sense of taste.*

Gustatory marketing

Scientific research shows that taste relies heavily on the other four senses to be completely perceived. Beyond information about the five primary tastes, its perception results from the brain processing additional details provided by the other senses: olfaction (the smell of the food), touch (temperature, texture, the pain caused by certain spices), sight (the appearance of the food, its esthetic qualities, color), hearing (crunchiness, crispness, crustiness).

Many companies have adopted the theory that taste is an amalgam of the five senses in the marketing of food products, with more and more emphasis given to the other senses in order to render the perception of taste more agreeable.

> *Top restaurants take almost as much care over the presentation of their dishes as over their taste. The colors, textures, aromas and even the sound coming from a plate of food are considered to be very important. The esthetics of the presentation of dishes, the shape of the cutlery and the material from which they are made, even the bread rolls are studied carefully because they contribute to the overall perception of the meal. The stars awarded to great restaurants by the famous Michelin Guide accord almost as much importance to the quality of the ambiance as to that of the food.*
>
> *In the same way, shops selling food use the other senses to increase sales. The visual aspect of products, their texture, aroma and packaging are carefully designed. This is all the more important*

because consumers are not often allowed to taste a product before buying it. Tastings and samples are widely used by some companies to try to overcome this problem and increase sales. One third of those who taste a product buy it immediately afterwards and half of those who buy a product after trying a sample say they would buy it again.

The five senses may also be stimulated by advertising messages, which aim to improve consumers' perception of taste. In a series of experiments on chewing gum, crisps and popcorn, researchers Ryan Elder and Aradhna Krishna[7] show that an advertisement appealing to several senses significantly improves the scores for the gustatory quality of the products concerned.

Manufacturers of nonfood items also use the idea of flavors as a source of pleasure to strengthen the promises made about their products and increase their differentiation.

Lush, a British manufacturer of cosmetics made from essential oils taken from organic fruit and vegetables, associates some of its articles with the names of food products, for example Oatifix, Chocolate Whipstick, Brazened Honey, and products shaped like cakes or pastries.

Sephora sells a range of edible cosmetic products called Dessert Beauty with names referring to tastes such as Powdered Sugar Deliciously Kissable Body Shimmer. Sports chain Décathlon offers flavored snorkels and Colgate has developed bubble-gum-flavored toothpaste.

Gustatory marketing must also take account of the importance of the brand on the perception of taste. As early as 1964, an experiment[8] showed the impact of the brand name on taste preferences. In a blind test carried out with regular beer drinkers, they were unable to distinguish the different beers. When brand names were added, the drinkers gave their favorite beer a higher score for taste.

This experiment is reflected by a more recent study, already mentioned, carried out on colas by Read Montague[9] and his colleagues. Although preferences were equally distributed between the two brands in their blind tests, Coca-Cola scored higher on taste than its competitor when the brand names were added. The researchers were able to interpret this result using magnetic resonance imaging (MRI). Contrary to Pepsi, the brand knowledge of Coca-Cola activated two additional memory-related brain regions and biased preferences. Memories stored about the Coca-Cola brand during its many publicity campaigns dominated the pure sensation of taste and finally modified gustatory perception in its favor, thereby clearly illustrating the power of brands.

The five senses: congruence, saturation and sensory marketing

The use of the senses for marketing purposes is developing rapidly. It relies on the strength and singularity of the emotional links created by soliciting them and, to be effective, it emphasizes the congruence of the senses and avoids saturating them.

The congruence of the senses

It is essential to work with the senses in a congruent manner, both in terms of target and objectives. Achieving congruence between the different senses solicited leads to consumers giving better evaluations of products and shops and increasing their intentions to visit. On the contrary, inconsistencies in the stimuli used can generate cognitive or perceptional dissonances, sources of mental discomfort, potentially leading to negative feelings about products or shops.

The saturation of the senses

Another pitfall to avoid is saturating the senses. An inverted-U relationship has been found between the intensity of a pleasant smell and people liking it, while there is a linear function in the case of an unpleasant smell. Finding the optimum level for the intensity of a smell is essential to avoid potential rejection. This risk can apply to each sense, whether it is taken individually (light intensity, ambient temperature, sound level, intensity of a taste) or collectively (too much solicitation producing the opposite effect to that intended). The difficulty in trying to guard against this is that the level of saturation varies from one neuro-consumer to another. The more homogeneous the target customer segment the easier the task.

Sensory marketing

Although companies tend to give priority to the visual attributes of goods and sales outlets, many marketing experts, like Martin Lindstrom and Aradhna Krishna, recommend using the whole palette of the senses. In an era when the consumer is confronted with 5,000 brand images every day, continuing to favor the visual is no longer sufficient to ensure success. To make themselves more attractive, companies must consider all the senses of the consumer. Such multisensory exploitation can help to establish strong and lasting emotional connections with consumers.

Technological progress makes this sensory integration more feasible and affordable, particularly for the senses of hearing, smell and touch.

The senses influence neuro-consumers' emotions and the decisions they take. Each one alone and all of them collectively can play an important role in the formation of perceptions, attitudes and the ways that a product, service or experience is consumed. Sensory marketing, whose objective is the exploitation of the senses for commercial ends, represents a strong competitive advantage for companies in the current environment.

Notes

1. See www.larecherche.fr/idees/back-to-basic/gout-01-01-2002-88850.
2. Cynthia N. DuBose, Armand V. Cardello and Owen Maller, "Effects of Colorants and Flavorants on Identification, Perceived Flavor Intensity and Hedonic Quality of Fruit-Flavored Beverages and Cake," *Journal of Food Science*, vol. 45, no. 5 (1980), pp. 1393–1399.
3. Aradhna Krishna and Maureen Morrin, "Does Touch Affect Taste? The Perceptual Transfer of Product Container Haptic Cues," *Journal of Consumer Research*, vol. 34 (April 2008), pp. 807–818.
4. Carl McDaniel and R.C. Baker, "Convenience Food Packaging and the Perception of Product Quality: What Does Hard-to-Open Mean to Consumers?" *Journal of Marketing*, vol. 41, no. 4 (1977), pp. 57–58.
5. Massimiliano Zampini and Charles Spence, "The Role of Auditory Cues in Modulating the Perceived Crispness and Staleness of Potato Chips," *Journal of Sensory Studies*, vol. 19, no. 5 (2004), pp. 347–363.
6. Marion Hombach, "Les secrets du bon goût," in *Géo Savoir* édition spéciale, "Le fantastique pouvoir de nos 5 sens," vol. 7 (2014), pp. 96–99.
7. Ryan S. Elder and Aradhna Krishna, "The Effects of Advertising Copy on Sensory Thoughts and Perceived Taste," *Journal of Consumer Research*, vol. 36, no. 5 (2010), pp. 748–758.
8. Ralph I. Allison and Kenneth P. Uhl, "Influence of Beer Brand Identification on Taste Perception," *Journal of Marketing Research*, vol. 1, no. 3 (1964), pp. 36–39.
9. Samuel M. McClure, Jian Li, Damon Tomlin, Kim S. Cypert, Latané M. Montague and P. Read Montague, "Neural Correlates of Behavioral Preference for Culturally Familiar Drinks," *Neuron*, vol. 44 (October 2014), pp. 379–387.

Points to remember

- The senses are the gateway to the neuro-consumer. Any information received from the outside world passes through one or more of them. They are the origin of sensation, i.e., the reception and transmission of this information to the brain. They enable the information to be perceived, i.e., sensory data to be recognized and interpreted by the brain.
- Sight, which receives 80% of daily information, is the origin of our vision of the world. It is the interpretation made by the visual zones of the brain of the electrochemical signals transported by the optic nerves. It is subject to experiences acquired previously and stored in the memory. It varies with context. It is strongly limited by the visual attention span, itself defined by the needs and objectives of the brain. It can be subject to illusions. Sight is the sense that receives the most commercial stimulation. This may be explained by its frequent domination over the other senses, the brain often giving priority to its visual perception of a stimulus. The sheer number of solicitations that it receives tend to reduce their effectiveness. They are made to the detriment of the other senses, which receive less exploitation, perhaps with the exception of hearing.
- Hearing is the sense that enables us to collect and process the thousands of sounds coming from our environment. The fact that sound can be used easily and cheaply makes it the sense most used by marketing, after sight. Research shows that music increases the extent to which advertising messages are memorized. When there is a high level of accord, it improves our attitude towards publicity and the brand as well as increasing our intention to purchase. Customers and employees in sales outlets prefer background music to silence. It helps to contextualize the place of purchase and to arouse favorable emotional states. It has an influence on the perception of pricing in a shop and on the choice of products purchased. It contributes to crowd management by having an impact on the length of time spent in commercial premises.
- Along with hearing, smell is the sense organ that is physically closest to the brain. A chemical sense and considered to be the most emotional, it is increasingly used by brands to create strong cognitive and affective relationships with neuro-consumers. Its direct link to the brain's limbic system gives it strong connections to the memory and the emotions. According to scientific experiments, the presence of an agreeable smell improves the evaluation of the stimulus concerned, be it a brand or a shop. It increases intentions to visit and to purchase as well as the time spent in the shop. Brand names and advertising messages have been found to be better memorized in the presence of a nice smell. Use of the sense of smell's capacity to favor attention, memorizing, emotivity and

ambient comfort has given birth to olfactory marketing and the many olfactory signatures of brands.

- After sight, touch appears to be the second most important sense for a consumer during his immediate evaluation of a product. Studies show that, when manipulating objects, the quality of the experience is greatly reduced for people deprived of sight and touch compared to those who cannot hear or smell. Touch also contributes to a feeling of ownership of the handled product and increases its perceived value and the predisposition to buy it. Although touch can influence the decision to purchase, its importance varies according to the individual and the product. Its interaction with the other senses is significant as is its role in interpersonal relationships and in sales outlets. The commercial exploitation of touch is known as tactile marketing and has only recently been developed.
- Until the appearance of the microscope and scientific analysis of the anatomy of the tongue, the sense of taste was poorly understood. More than a simple affair of the tongue, taste should be understood as being an amalgam of the five senses. Beyond the gustatory information about the five primary tastes, its sensation is the result of processing by the brain of additional details provided by the other senses: the smell of the food by olfaction; the temperature, texture and sensation caused by certain spices through touch; its appearance, esthetic appeal and color through sight; crunchiness, crispness and crustiness through hearing. Stimulating the five senses, particularly through advertising messages, increases consumers' gustatory perception of a given product. Gustatory marketing works on this subtle alchemy of the senses. It cannot, however, always overcome the influence of a strong brand. Studies have shown that an extremely powerful brand, such as Coca-Cola, which is present in consumers' minds thanks to incessant communication, can often modify the perception of taste to its benefit.
- The stimulation of the five senses is at the heart of sensory marketing. Its job is to go beyond sight alone, traditionally overexploited by companies. The selection of the senses to focus on by companies must be consistent with – and in the service of – the brand's core concept. Multisensory exploitation increases the probability of creating strong, lasting emotional connections with the neuro-consumer. It must be careful to ensure the congruence of the chosen senses and to minimize the risk of saturation, both at the individual and the collective level.

[illegible] flavor has been [illegible] more [illegible] the [illegible] history of brands.

a After sight, touch appears to be the second most important sense for the consumer during his immediate evaluation of a product. Touch is more [illegible] than when manipulating objects, the touch of the [illegible] is greatly [illegible]. [illegible] people deprived of sight and touch, compared to those who [illegible] or smell. Touch, after evaluating a [illegible] of the [illegible] products and [illegible] [illegible] [illegible] [illegible] [illegible] [illegible] [illegible] [illegible] of touch is known as [illegible] has been [illegible] [illegible].

a Until the appearance of the microscope and [illegible] analysis of the anatomy of the tongue, the sense of taste was poorly understood. More than [illegible] [illegible] could be [illegible] as [illegible] of the five senses. [illegible] the anatomy [illegible] [illegible] [illegible], its perception is the result of processing by the brain of [illegible] of [illegible] by all [illegible] senses. The smell of the food, its texture, the shape, [illegible] [illegible] [illegible] [illegible] [illegible], its [illegible], [illegible] [illegible] [illegible] through [illegible], consistency, crispness, [illegible] through hearing. [illegible] the five senses, particularly through [illegible] [illegible] [illegible] [illegible] [illegible] [illegible] [illegible] [illegible] [illegible] [illegible] of the senses. It [illegible], how[illegible] [illegible] [illegible] [illegible] a strong brand. Studies have shown that [illegible] [illegible] [illegible] [illegible] [illegible] Coca-Cola, which [illegible] [illegible] [illegible] [illegible] [illegible] [illegible] [illegible].

[illegible] [illegible] [illegible] [illegible] [illegible] traditionally [illegible] by companies. [illegible] of the [illegible] to focus on [illegible] [illegible] [illegible] [illegible] [illegible] [illegible] [illegible] [illegible] [illegible] [illegible] [illegible] [illegible] the probability of creating strong, lasting emotional connections with the [illegible] [illegible] [illegible] careful to ensure the [illegible] of the chosen senses and to [illegible] the [illegible] [illegible] both at the [illegible] [illegible] [illegible] [illegible] level.

PART IV

The neuro-consumer's brain influenced by innovation

Developed with the aim of adapting to the subconscious behavior of neuro-consumers' brains, innovations in the presentation of products and services, in pricing and in sales techniques make them more effective. When sales outlets for goods and services are themed, staged and equipped with sensory experiences, customers spend longer in them and show more interest. Positive results are demonstrated by a significant increase in turnover and an improvement in customers' perception of the brand and the retail chain.

Introduction

Having looked at the subconscious behavior of the neuro-consumer's brain and the way in which it is conditioned by different internal factors and dominated by his emotions, desires and senses, we now consider the consequences of the neuro-consumer's perceptions when he is confronted with an environment devoted to consumption and purchasing. We will begin by looking at his reactions when confronted with policies of innovation, design and packaging of products and services used by brands to try and seduce him. We will particularly concentrate on the subconscious effects that can be caused by different shapes, colors, typefaces, etc.

Then, we will look at the obstacles the neuro-consumer experiences in trying to get an objective idea of the right price of a product or service and also at the difficulties in his brain's perception of value for money. Next, we will analyze the reactions of the neuro-consumer's brain when he visits a sales point for goods or services and examine his feelings when he is faced with deliberate and organized theming and staging in specific places with the objective of getting him to spend more money.

It is important for companies to understand what effect the development of experience-based and sensory marketing can have on the neuro-consumer's purchasing decisions. This is particularly true when the subliminal appeal to his senses is well adapted to the positioning chosen by the brand, coordinated with the retail chain's global strategy of communication, e-communication and its relationship policy and when there is harmonized congruence between the senses being solicited.

We will end Part IV by looking at how salespeople can influence consumers' subconscious attitudes and reactions and examine methods drawn from neuroscientific thinking that are likely to improve their ability to persuade.

CHAPTER 17

The influence of innovation, design and packaging

Companies spend large sums of money on marketing to ensure the success of their product development and launch policy among consumers. Such efforts concern different areas, including innovation, design and packaging. Many products launched on the international markets meet with failure just a few months after release, despite significant expenditure to promote them. Experts such as Emmanuelle Le Nagard-Assayag, professor at ESSEC Business School, and Delphine Manceau,[1] professor at ESCP Europe Business School, estimate the failure rate of new products two to three years after their launch to be between 70% and 90% depending on the product category and country. They believe that the failure of offerings in the "high-tech" and "mass-consumption" sectors is greater than in the industrial goods and services sector. They quote a joke made by the company 3M, which reveals the difficulties faced during a product launch: "You have to kiss a lot of toads before you find Prince Charming."

Faced with this fact and having exhausted all the traditional marketing methods, many companies are turning to research that provides a more acute understanding of neuro-consumers' irrational behavior in relation to current or future products and services.

They often contact agencies and professionals who propose studies and expert evaluation in neuromarketing, using neuroscientific techniques such as MRI and electroencephalography (EEG), or other simpler and less costly tools. In collaboration with their partners, client companies are particularly interested in what effects innovation, design and packaging produce in neuro-consumers' brains.

The neuro-consumer's brain faced with innovation

Despite the high failure rate among new products, the brain harbors a particular interest in innovation. For millions of years, innovation has allowed the human race not only to survive, but also to evolve at a faster rate than other living species. American Psychologist Robert Evan Ornstein[2] shows that the "primitive brain" considers our survival a priority, hence the particular attention afforded to innovations required to guarantee survival in the neuro-consumer's brain.

Unfortunately for companies, few inventions receive sufficient client interest to justify their development and turn them into real innovations. It should be noted that an "invention" is an offering considered as innovative by the company. It only becomes an "innovation" when it is considered new and interesting by clients.

Chan Kim and René Mauborgne[3] propose an approach that allows companies to find a "blue ocean" to create real "value innovations." The method is designed to help them offer original products and services by focusing on "being different" rather than trying to "be better." Their method is consistent with the brain's attraction to newness, but not just any kind of newness.

Neuroscience techniques now allow us to see whether inventions are in harmony with the impressions and expectations in neuro-consumers' brains.

Notions of beauty and ugliness, which are felt directly in the brain, generate a feeling of pleasure or disgust towards innovations. As described by two French researchers, Bernard Roullet and Olivier Droulers,[4] the regions of the brain involved in the evaluation of product innovations, and in particular their design or packaging, are the insula or insular cortex, the medial orbitofrontal cortex and the parietal cortex. The insula is not immediately visible, unlike the other cerebral lobes, because it is covered by the frontal, parietal and temporal lobes.

Roullet and Droulers argue that the insula plays an essential role in incorporating sensory information from the body in emotional processes. It is involved both when a person feels disgust and in recognizing an expression of disgust on another person's face. Using techniques drawn from neuroscience, activity in the insula has been observed when a neuro-consumer is presented with an innovation they do not like. It is also active when faced with stimuli that the neuro-consumer considers unappealing, as well as when confronted with unfair offerings, for example when it considers a price is too high with respect to the perceived quality of the item or service presented.

Hiro Kawabata, a Japanese professor of art and design at Nariyasu Gakuen University in Kyoto, and his Turkish colleague Semir Zeki,[5] professor of neuroaesthetics at University College London, show through scientific experiments that esthetic judgment correlates with activity in the aforementioned areas of the brain, namely the orbitofrontal cortex and the parietal cortex. The more "beautiful" the stimulus is considered to be, the greater the activity in the first cortex. If it is considered "ugly," the motor cortex becomes active and the orbitofrontal cortex ceases to be active.

The use of neuroscientific techniques helps detect which innovations please or displease the brain. With reference to results from the survey firm Novaction, Emmanuelle Le Nagard-Assayag and Delphine Manceau[6] suggest that 81% of projects generate negative results during pre-launch tests. It is therefore understandable that neuromarketing companies are looking to propose a range of neuroscientific methods and tools to complement traditional marketing research. Based on direct observation of the behavior of neuro-consumers' brains, these tools cast new light on how the very concept of innovation is perceived.

To do so, the techniques first try to explain why the neuro-consumer's brain does not adhere to certain concepts of innovation that have already encountered failure. Each failure is studied using neuroscience.

Nagard-Assayag and Manceau also aim to identify elements that may be accepted or refused by the brain with regard to a new concept to be launched. Aspects such as the concept idea, proposed positioning, formulation or functionality of the product or service, chosen name, brand used to support the offering, benefits perceived by the client, acceptability of the price, etc. are subject to neuroscientific investigations. Using MRI or EEG, characteristics that stimulate high levels of attention, pleasure, emotional engagement, memorization, a sense of novelty or intention to buy in the brain can be observed, as opposed to repulsion or indifference.

Neuroscientific verifications are also used during the transition from design to manufacture in order to measure interest, attractiveness, indifference, repulsion, etc.

> *All sensory data are subject to studies on the brain's subconscious reactions to the different components of the product. They concern a broad spectrum of fields such as the texture of creams or beauty oils, the noise of candies or chips in the mouth, the taste of snorkels for diving, the smell of coats or diving suits left in damp storage for several weeks.*

Shape and design are also subjects of neuroscientific interest. Neuromarketing experts such as A.K. Pradeep[7] consider that neuroscientific research, whether used alone or alongside traditional marketing surveys, provides major improvements to ensure the success of innovations leading to the launch of new products and services.

American professor Larry R. Squire and his colleagues write that:

> At the start of the 21st century, the Hubble space telescope provided us with information on previously unmapped regions of the universe and the promise that we could learn something about the origins of

> [the] cosmos. This same adventurous spirit is also turning towards the most complex structure that exists in the universe: the human brain.[8]

For Pradeep, neuromarketing is to marketing what the Hubble telescope is to astronomy. It provides "a quantum leap forward toward greater knowledge and a larger vision, achieved with scientific precision and certainty."[9]

The neuro-consumer's brain confronted with design

Success in design is of primary importance to companies. For some, it comprises an integral part of innovation.

The "neuroesthetics" of design, a new research focus for neuroscientific analysis

Bernard Roullet and Olivier Droulers[10] remark that esthetics, which are part of design, have long been considered highly cognitive, i.e., nonemotional, by art historians and researchers. Understanding and evaluating esthetics requires study, prior cultural knowledge and "objective" evaluation tools. Design efficiency is still evaluated using traditional marketing surveys with a conscious, rational basis and carried out through interviews, attitude scales and a range of conventional tools such as focus groups. Developments in neuroscience reveal the role of emotions and affect. Classical surveys are shown to be of limited interest in this field, and even more so when a consumer is asked to choose today the design of an item that will be brought to market in a few years' time. This is the case for cars, for example, where "a better understanding of the processes of esthetic judgment can also be achieved by taking account of the underlying cerebral processes."[11]

Semir Zeki[12] mentions the emergence of "bioesthetics" and "neuroesthetics." The first international conference on neuroesthetics was held in London in 2002.

For Roullet and Droulers,

> Today, the topic of design is a key element in differentiation and competitiveness for both manufacturers and distributors ... The neuroscientific approach is especially useful here because many design survey topics, such as esthetic evaluation (beauty and ugliness), the appreciation of innovative forms, the focus of attention on a given subject or the symbolic meaning of forms, cannot be easily addressed through direct questioning. Indeed, these processes are either hard to consciously access or difficult to describe verbally.[13]

Italian neuroscience researchers Cinzia Di Dio, Emiliano Macaluso and Giacomo Rizzolatti even claim that "some intrinsic factors in works of art (in particular the famous golden ratio) are able to trigger a specific neural pattern inherent in the sense of beauty in the observer."[14]

Pierre Boulez, Jean-Pierre Changeux and Philippe Manoury[15] believe that the brain's disposition allows us to recognize real harmony between a part and the whole of a work of art.

Research on the way neuro-consumers' brains are attracted to design now also uses neuroscientific techniques. These range from the simplest, such as combining eye tracking with electrodermal measurements or observation of the pupillary diameter to record the appearance of emotions or attention, to the most complex, such as MRI and EEG.

In the film called *Consommateur pris au piège*,[16] journalist Emmanuelle Ménage presents an experiment by Steve William, manager of the British firm NeuroSense, on the search for preferences in handbag design among consumers whose brain responses are studied using MRI.

Another company, NeuroFocus (Nielsen), studies neuro-consumers' reactions to a product design by observing waves emitted in the brain with its tool "Mynd," a sophisticated EEG connected to a computer via Bluetooth. The measurement of different designs, carried out by NeuroFocus and described by its founder, A.K. Pradeep,[17] complies with a specific protocol. It aims to understand a consumer's total consumer experience (TCE) by monitoring the responses in their brain at different stages in their relationship with the product design. TCE successively analyzes different moments in these relationships, from visual examination of the shape, handling of the product and anticipation of its use, removal of the product from its packaging, multisensory perception upon first contact with the product and responses from the neurological multisensory perception from the way the product is consumed. During these phases, experts try to identify precise elements that are of the greatest importance in the brain. These elements constitute the design's neurological iconic signature (NIS) that indicates which aspects produce the most powerful, striking or encoded evocations in their representation in the brain.

Additional tests relating to all the sensory components of the prototype, such as its form, ease of use, price and the value of the brand destined to support it, are also carried out based on analysis of the brain's responses.

Neuroscientific studies on design are carried out in most sectors and countries. Their use remains most common in Western countries and the mass-consumption industry, but they are also now starting to address industrial products. They concern a large number of items, from the most ordinary to the most complex such as perfume bottles, drinks, gas bottles, mobile phones, digital tablets, restaurant menus and cars.

> *Lead is added to some mobile phones to make their design more compatible with the idea of quality linked to the weight perceived by the brain. The company Bang & Olufsen add weight to their remote controls to reassure the consumer as they assess the quality of the product.*[18]

The brain forges a close connection between a food's taste and its color.

> *To inspire confidence among neuro-consumers, the food industry does not hesitate to use artificial colorants to make its products better suited to the brain's perception. For example, mint syrup, which is naturally colorless, is often colored green.*
>
> *Apple accords great importance to the design of its products and services. Their appearance is highly recognizable: the Mac has a specific design, the earphones are white and the iPods are made with bright colors. The App Store has an exceptional design in terms of practicality of use. Available for all Apple devices, it has a single point of entry destined for all iPhones.*

The automotive industry places great importance on successful designs for the interior and exterior of its cars.

> *In the early 2000s the company Daimler-Chrysler, in association with neuroscience researchers from Ulm University in Germany, became pioneers in the use of MRI to research attractive designs. Today, several manufacturers use neuromarketing and sensory marketing techniques to find a neuro-compatible exterior or interior look for their future cars. Toyota works in association with the Ricken Brain Science Institute in Japan.*
>
> *Béatrice Daillant-Vasselin, head of appearances for materials and interior and exterior ornamentation at PSA, describes the manufacturer's experience in looking for sensory designs.*[19] *Within the company, a number of professionals work constantly to compose the most favorable sensory ID cards for the brain for vehicles due to leave the assembly line. She says that,*
>
>> *We mainly work on the feel, look and sound. The texture of materials such as leather, the impression of solidity and safety of the dashboard, polyphonic sounds and indicators, the fastening of seatbelts, etc. are all integrated in the early stages of the vehicle development project.*[20]

Another field that is particularly favorable to neurosensory studies is product packaging.

The neuro-consumer influenced by packaging

Packaging has a specific appeal for neuro-consumers. It communicates the elements that the manufacturer wants to make them feel about the design or product inside the packaging. It is all the more important given that, according to experts,[21] between 70% and 75% of purchases of mass consumption goods are made in sales outlets and that 80% of products handled are then bought. However, if the feel of the product, such as its weight, does not match what is expected by the brain, it is quickly placed back on the shelf.

In an environment where multiple product offerings compete with each other on display shelves, packaging is designed as much as possible to meet the instinctive expectations in clients' brains.

In large French supermarkets such as Leclerc, Carrefour and Auchan, there can be over 100,000 competing brands. Product launch failures due to packaging that does not appeal to the responses in neuro-consumers' brains are common.

In the attempt to attract customers, packaging can be misleading. In some cultures, such as Japan, clients prefer transparent packaging that allows them to see the product inside. In this case, the inherent quality of the design and the presentation of the product are particularly important.

Through MRI studies, three researchers at Zeppelin University in Friedrichshafen, Germany, Marco Stoll, Sebastian Baecke and Peter Kenning,[22] have observed variations in brain activity when it is confronted with different types of packaging. Packaging that is considered attractive activates regions in the brain that process more visual stimuli and can also stimulate the reward system, encouraging the consumer to approach the product. Conversely, packaging that is considered unattractive activates the insula, mentioned above, and other regions predisposing clients to reject or leave the product.

Different techniques are used to try and evaluate the packaging's interest for neuro-consumers.

Éric Singler, chief executive officer at BVA, recreated convenience stores in Paris, the United States and Asia to carry out tests on packaging. The samples tested are presented on shelves surrounded by competing products, like in a real supermarket.

Neuro-consumers' attitudes are observed using oculometry techniques, sometimes in association with a tool for detecting emotion such as electromyography (EMG).

EMG, or EEG biofeedback, was first used in medical applications in the 1970s by the fathers of sophrology, American Edmund Jacobson (1888–1976) and the Colombian of Spanish-Basque origin, Alfonso Caycedo.[23] Using sensors placed on specific muscles, it records electrical currents produced by the brain during muscle activity. This generates an electrogram that can be used to produce an electrodiagnosis. An experiment using this process by Olivier Droulers, designed to detect the emotions in neuro-consumers' brains, was broadcast on television.[24]

To evaluate the appeal of a packaging, most neuromarketing consultancy firms prefer to use more sophisticated techniques such as MRI or EEG. NeuroFocus prefer to use its EEG "Mynd." The device is placed on the neuro-consumer's head and offers the advantage of allowing them to move around between shelf displays.

Properties of packaging that attract the brain's attention

The brain's attraction to packaging is analyzed using neuroscience techniques that assess the different properties of its design.

Images and iconography Images and iconography, displayed on the packaging, play an important role in triggering attractiveness in the brain, especially when they create an "emotional evocation" on display stands.

> *Depicting products' origins (olives or olive trees for olive oil, oranges or orange trees for orange juice, cows for milk, flowers for a beauty product) is likely to create an emotive feeling in the brain stemming from the natural appearance of the images. With other products, such as those for young children, the brain is attracted when it detects faces, because it is naturally programmed to identify them.*

Colors Some colors, such as red or black, attract the brain's attention because they activate somatic markers linked to danger or death. Others, especially when they have a cultural meaning, stimulate specific emotions in the subconscious in certain countries or populations, such as red and yellow in China, orange in Ukraine and Holland and green in some cultures linked to the Muslim world.

Blue is one of the most popular colors all over the world. It is particularly appreciated in Japan, especially Prussian blue, which was imported from Holland in the 1820s. It allows multiple shades to be used in prints and gives a dreamlike aspect to depictions of nature, which appear to be bathed in moonlight. The use of Prussian blue was made famous by Katsushika Hokusai (1760–1849) in his woodblock print series *Thirty-Six Views of Mount Fuji*, which includes the internationally iconic print, *The Great Wave*.

Text Text gives meaning to a product. The text must remain in harmony with the images. Terms such as "flavorsome," "new," "origin," "natural," "traditional," "fresh," "handmade," "farm produce" and others help generate emotion. They are carefully chosen to give the packaging value. The layout of the writing and images is designed to facilitate fast processing of the information in the brain, by placing images on the left and text on the right, for example. The form of the writing is also important. The brain seems to be more attracted when the text is slightly offset from the center. Curved lines also seem to be better visualized than straight lines.

The size and shape of packaging This field is known among professionals as "facing." Some bottle shapes are "optimized," meaning their size is increased to give the neuro-consumer's brain the impression that they have a larger content than competitors' bottles on the same shelf. Indications of the attractiveness of a particular packaging include whether a product's weight is considered pleasant by the brain and how easy it is to pick up or handle.

The brand featured on the packaging The brand featured on the packaging plays an important role in conferring qualities relating to a product's essence. All the elements depicted on the packaging must be consistent with the brand's core concept.

Most of the elements on the packaging are studied using different neuroscientific techniques, allowing observation of their compatibility with the brain's impressions.

Companies that use EEG, such as NeuroFocus, carry out tests to evaluate a design, to determine what effect the packaging has on the brain, whether it attracts attention, creates emotional engagement, is remembered, sparks an interest to buy, evokes an impression of newness, etc.

The brain's responses faced with different modes of packaging presentation

Neuroscience experts[25] have carried out experiments on multiple types of packaging and consider several types of behavior to be linked to responses in neuro-consumers' brains when faced with packaging design. For example:

- Reducing the amount of text and number of visuals makes it easier for the brain to understand the packaging. Too much text distracts attention. There should preferably be less than three images on the packaging.
- Images are more important than text in attracting the brain's attention.
- The brain memorizes images on packaging better when it has already seen them in ads on the television, online, on mobile phones, social media, flyers, in-store, external shop videos or on merchandise or presentation stands.
- Packaging is all the more effective when it allows sensory interaction with the product (sight, sound, taste, smell, touch). Some packaging is

transparent to allow the product inside to be seen. Others are designed to allow consumers to touch the product, smell its fragrance, etc.

- The layout and quantity of packaging designs on the shelves in sales outlets influence the level of attractiveness for the brain. Higher shelf positions, at eye level, have been found to boost attractiveness to the consumer.
- Too many items in the same product category can weaken sales.

  Experiments by Sheena S. Iyengar and Mark R. Lepper[26] at Columbia University show that a small range of products in a convenience food store performs better than a larger selection. Roger Dooley,[27] president of Dooley Direct LLC, indicates that removing two Walmart peanut butter brands bolstered sales in this category of products in his sales outlets. Procter & Gamble saw an increase in sales of skin creams and lotions by reducing the number of items on the shelves.

- Attractive packaging is essential to the success of numerous products on shelves with a large number of competing offers. This is why harmonizing artistic design with the emotive and sensory perceptions in neuro-consumers' brains is a constant goal of marketing.

Notes

1. Emmanuelle Le Nagard-Assayag and Delphine Manceau, *Marketing des nouveaux produits. De la création au lancement*, Dunod, 2005.
2. Robert Evan Ornstein, *Evolution of Consciousness: The Origins of the Way We Think*, Simon & Schuster, 1992; Robert Evan Orstein and Richard Thompson, *L'Incroyable Aventure du cerveau*, InterÉditions, 1991.
3. Chan Kim and Renée Mauborgne, *Blue Ocean Strategy*, Harvard Business School Press, 2005.
4. Bernad Roullet and Olivier Droulers, *Neuromarketing: Le marketing revisité par les neurosciences du consommateur*, Dunod, 2010.
5. Hideaki Kawabata and Semir Zeki, "Neural Correlates of Beauty," *Journal of Neuropsychology*, vol. 91, no. 4 (2004), pp. 1699–1705; Semir Zeki, *Inner Vision: An Exploration of Art and the Brain*, Oxford University Press, 1999 and *Splendors and Misery of the Brain: Love, Creativity and the Quest of Human Happiness*, Wiley Blackwell, 2008.
6. Nagard-Assayag and Manceau, *Marketing des nouveaux produits*.
7. A.K. Pradeep, *The Buying Brain: Secrets for Selling to the Subconscious Mind*, Wiley, 2010.
8. Larry R. Squire, Darwin Berg, Floyd E. Bloom, Sasch du Lac, Anirvan Ghosh and Nicholas C. Spitzer, *Fundamental Neuroscience* (4th ed.), Academic Press, 2012.
9. Pradeep, *The Buying Brain*.
10. Roullet and Droulers, *Neuromarketing*.

11. Bernard Roullet and Olivier Droulers, "Neuro-esthétique automobile: les neurosciences et le design," *Management et Sciences sociales*, vol. 6 (2009).
12. Zeki, *Inner Vision* and *Splendors and Misery of the Brain.*
13. Roullet and Droulers, "Neuro-esthétique automobile."
14. Cinzia Di Dio, Emiliano Macaluso and Giacomo Rizzolatti, "The Golden Beauty: Brain Response to Classical and Renaissance Sculpture," *PLoS ONE*, vol. 2 (2007), pp. 1201–1209.
15. Pierre Boulez, Jean-Pierre Changeux and Philippe Manoury, *Les Neurones enchantés*, Odile Jacob, 2014.
16. Emmanuelle Ménage, *Consommateurs pris au piège* [film], France 5, January 12, 2014.
17. Pradeep, *The Buying Brain.*
18. Martin Lindstrom, *Brand Sense: Sensory Secrets Behind the Stuff We Buy*, Free Press 2010.
19. Natacha Cygler, "Tous les sens," *Le Nouvel Économiste: Leadership & Management*, vol. 1584 (October 2011).
20. Ibid.
21. Pradeep, *The Buying Brain*; Martin Lindström, *Buy.Ology: How Everything We Believe About What We Buy Is Wrong*, Random House Business, 2009; Roger Dooley, *Brainfluence: 100 Ways to Persuade and Convince Consumers with Neuromarketing*, John Wiley & Sons, 2012; Patrick Georges, Anne-Sophie Bayle-Tourtoulou and Michel Badoc, *Neuromarketing in Action: How to Talk and Sell to the Brain*, Kogan Page, 2013.
22. Marco Stoll, Sebastian Baecke, Peter Kenning, "What They See Is What They Get: An fMRI Study on Neuronal Correlates of Attractive Packaging," *Journal of Consumer Behavior*, vol. 7, no. 4–5 (June 2008), pp. 342–359.
23. Paule Vern, *Force vitale: La sophrologie caycédienne*, La Méridienne Desclée de Brouwer, 2000; Edmund Jacobson, *Savoir relaxer pour combattre le stress*, Éditions de l'Homme, 1980.
24. Élise Lucet, *Cash Investigation*, France 2, May 25, 2012.
25. Vern, *Force vitale.*
26. Sheena S. Iyengar and Mark R. Lepper, "When Choice Is Demotivating: Can One Desire Too Much of a Good Thing?" *Journal of Personality and Social Psychology*, vol. 79, no. 6 (December 2000), pp. 995–1006.
27. Dooley, *Brainfluence.*

CHAPTER 18

The influence of the price of products and services

The price of a product or service is of great importance to the neuro-consumer. Prices that do not correspond to the perception that the customer's brain has of the item can kill a sale. When a product or service is considered to be too costly, preference will be given to a less expensive equivalent offered by a competitor, which accounts for the success of low-cost goods and services in many sectors. Consumers will refuse to pay more for a product whose intrinsic quality does not seem to justify the difference in price. On the other hand, if the product or service seems too cheap, it might also deter the neuro-consumer by making them think that the product is not of adequate quality. This is particularly the case with "top-of-the-range" or luxury products.

> *In Europe, Bic had problems when it offered a cheap perfume, despite the fact that experts and customers judged it to be of good quality during blind tests (when neither brand nor price were revealed). Sales did not meet expectations and the perfume was withdrawn from the market.*

Given how difficult it is for the brain to find a rational compromise between quality and price, research enabling it to determine a price that suits it has proved important in the decision-making process for purchases.

In many cases, the very idea of perceived quality is nothing more than a simple reflection of the stated price. This perception may also vary significantly depending on a number of associated factors. One of the most important is the brand. A book written by two directors of the international consultancy firm Simon-Kucher – Enrico Trevisan and Florent Jacquet – throws an interesting light on the influence of the brain's perceptions in developing a pricing policy.[1] Simon-Kucher & Partners Strategy and Marketing Consultants was founded in 1985 by the German professor, writer and businessman Hermann Simon. Today, it is one of the biggest international consultancy firms. It takes a particular interest in the psychology of pricing.

To try and understand consumer behavior in relation to price, marketing has long been interested in the customer's psychological perception of that price. By looking directly at the brain's perceptions, neuroscience provides complementary explanations that enable it to be better understood.

The psychological price perceived by the customer

Traditional marketing studies try to determine the psychological price that a consumer is willing to pay for a given product or service. The classic method is to simply present the product or a description of the service to a representative sample of consumers and to ask them three questions:

> "Above what price would you consider this product or service too expensive?"
>
> "Below what price would it seem to be of poor quality?"
>
> "At what price would you agree to buy it?"

With the results of this questionnaire, marketing studies experts can determine the psychologically optimum price, that which receives the greatest number of positive responses. To be valid, this type of study must be subject to certain precautions. If the product is sold in a shop, the study must be carried out at the sales outlet where it is sold. Marketing professionals have observed that the type of sales outlet can influence the consumer's perception of price. The same unbranded garment presented in the window of a lower-end shop is attributed a considerably lower psychological price than if it is presented in the window of a high-end luxury shop.

Other factors related to distribution, such as the position on a shelf, the proximity of other products, the immediate environment, etc. also have an influence on the perception of the psychological price.

Among the factors that condition the idea of the "right" price perceived by consumers, the brand plays a fundamental role. Chains that are well known and appreciated by a wide range of people, such as Apple, Nespresso, Starbucks, Abercrombie and of course the luxury brands, make the most of this. They are able to charge higher prices for products and services, the quality of which is not always necessarily better than those from competitors that are less well known or have a poorer image.

Apart from the brand and the sales outlet, the perception of the relationship between price and quality is influenced by other factors: the design of the merchandise, the type of communication associated with it and the comments about it on social networks, etc.

The customer's perception of a good relationship between the price of a product or service and its perceived quality remains highly subjective. This is a major concern for companies' marketing teams. Some of them now use neuroscientific research to complement and improve their knowledge in this field.

Price seen through the filter of neuroscience

In the United States, behavioral economists and neuro-economists are carrying out research related to the neuro-consumer's brain and the perception of price. Many research projects are being carried out at the universities of Stanford, Pennsylvania and Carnegie Mellon. Thanks to the use of neuroscience, publications from researchers such as Georges Loewenstein, Brian Knutson, Uri Simonsohn, Dan Ariely, Lisa Trei, Benedict Carey and others are pushing forward knowledge in this field.

Research carried out by Georges Loewenstein, Brian Knutson – professors at the universities of Carnegie Mellon and Stanford respectively – and their colleagues,[2] shows that the action of purchasing and paying for a good or service activates the zone of the brain linked to feeling pain. This feeling is strong when the neuro-consumer has to hand over his money and various means are used to try to reduce it, for example the use of credit cards, when the customer can either pay later or obtain credit for the purchase. The neuro-consumer will more readily accept a price when the salesperson manages to reduce the effect of all the elements that cause him pain.

The sensation of pain related to payment, felt by the brain, is compensated for by the immediate pleasure provided by possession of the product. Hence the importance of being able to explain the quality and advantages of a product or service to a customer who feels that the price is too high.

According to behavioral economics, the neuro-consumer considers a price to be appropriate when it favorably reconciles the conflict between the pain caused by payment and the pleasure of possessing or consuming the product or service in question. This resolves a conflict inside the brain between the regions involved in approach and avoidance processes. A positive assessment increases the activity of the nucleus accumbens, which plays a role in the reward circuit. When a price is considered excessive, there is an increase in the activity of the insular cortex, which is involved in feelings of disgust. According to Bernard Roullet and Olivier Droulers, "An excessive price is, in a way, a price that incites horror."[3] According to Brian Knutson and his colleagues,[4] simple observation of cerebral activity using magnetic resonance imaging (MRI) will tell us what a neuro-consumer has decided in relation to a price, without even having to ask him.

In many circumstances, the brain finds it difficult to determine the right price and establish a reference tariff. Dan Ariely, Israeli-American professor at Duke University, has studied many experiments in different fields that bear witness to this phenomenon.[5] Describing research carried out by Uri

Simonsohn and George Loewenstein, he shows that a real estate buyer keeps the former price of his house in mind for several years as a reference, even if prices in the real estate market change significantly.

With products or services for which the brain does not have a reference price, the first tariff encountered acts as a reference indicator. Having high-priced products on an end-aisle display makes other, similar products seem less expensive.

It can be useful to put the most expensive dishes such as lobster, foie gras and fillet of beef at the top of a restaurant menu. Whatever comes next, the brain has the impression that it is intrinsically better value for money.

Good realtors are instinctively aware that the perception of the first price encountered becomes fixed in customers' brains. In order to make the apartments they hope to sell seem cheaper, they start by showing their customers other apartments that are just as expensive, but that they know will be less to the customers' taste. They leave the visit of the properties they hope to sell to the end. The majority of potential purchasers base their idea of the "right" price on the first properties visited. They are in a better frame of mind to accept the prices of the last apartments visited, which seem naturally cheaper than those visited previously, considered to be of poorer quality.

Roger Dooley[6] *mentions an interesting example of this practice of fixing a high reference price in the neuro-consumer's brain used by Apple when the first iPhones came out. The price was considerably reduced afterwards, to give future customers the impression of getting a bargain. When it was launched in the United States, the iPhone was priced between $499 and $599, targeting early adopters. Only a few months later it was reduced to $200, which considerably stimulated sales to customers who were happy to make a good purchase. When the iPhone 3G came out, it was priced at $199, which enabled Apple to sell a million iPhones in only three days.*

When the neuro-consumer's brain lacks clear ideas about value for money, the price changes his perception of his experience of the product. A research study on wine carried out by Hilke Plassmann, John O'Doherty, Baba Shiv and Antonio Rangel bears witness to this attitude: Using functional magnetic resonance imaging (fMRI), they show that the part of the brain that experiences pleasure becomes more active when the drinker is enjoying the wine that they have been told is more expensive, even if in reality it is actually the cheaper wine.[7]

American journalist Benedict Carey[8] *describes this same perception with placebos. Also referring to scientific studies from the United*

> *States, he shows that 85% of patients who are given a placebo claim to feel better pain reduction when they pay $2.50 per pill whereas only 61% of subjects say that their pain is reduced when they pay merely 10 cents for the same product.*

Chains that sell luxury or top-of-the-range goods or services have long understood that a high price is necessary to give their products an image of exceptional quality but also to make them different.

The effects of subconscious mental strategies when faced with a price

Enrico Trevisan and Florent Jacquet[9] have studied the mental strategies employed when faced with pricing. They are interested in the different "effects" that operate and exert a subconscious influence on consumer choices. Some of them have already been described in this chapter. Their publication – inspired by the work of several psychologists and behavioral economists including Daniel Kahneman, Amos Tversky and Richard Thaler – provides responses enabling us to better understand and to develop appropriate strategies in relation to prices. Among the main effects felt by the brain when faced with pricing, they list the following:

- *Anchoring effect*: Experiments carried out by the Simon-Kucher consultancy show that the brain is subconsciously influenced by previous prices encountered in its environment.
- *Decoy effect*: Presented in 1980 by Richard Thaler and his colleagues in several articles, it shows that the way we determine our preference between two choices is modified if a third choice is offered.
- *Endowment effect*: The brain tends to accord greater value to a good that it possesses.
- *Hyperchoice effect*: If the brain is confronted with too many choices, it may defer the purchase of a good.
- *Sunk cost effect*: Money spent on irrecoverable costs influences choices related to further expenditure.
- *Payment decoupling*: The fact of uncoupling the moment and the location of payment for the purchase (bankers card, stage payment, credit, etc.) makes the brain perceive the item as less costly.
- *Self-control effect*: The brain lacks discipline in relation to purchases. It lets itself be easily influenced by subjective elements, even when it is fully aware of the fact.
- *Mental accounting*: According to Professors Tversky and Kahneman, the brain allocates expenditure to relatively watertight categories.
- *Loss aversion*: Losing a good has a higher mental value than acquiring it.

- *Denomination effect*: The customer tends to spend more easily when the amounts are small. For example, dividing an annual subscription into monthly or daily payments reduces the perception of costliness.
- *Framing effect*: The way in which the salesperson presents the cost has an influence on the perception of price.
- *Payment depreciation*: The brain tends to easily forget past expenditure.
- *Immediate discounting*: An immediate benefit is preferred to deferred possession of the product. Hence the importance of having a large stock of products in some sectors of activity, for example consumer goods.
- *Confirmation bias*: The brain tends to select and interpret pricing information that confirms its expectations. This effect is often complemented by the "post-purchase rationalization" effect, which involves justifying the cost of products after the event.
- *Bandwagon effect*: This means wanting to acquire the same goods that many other people are buying. In this situation, the reference to price is minimized.

This list of the effects on the brain's price-related decisions is not exhaustive. The interested reader will find more information in Trevisan and Jacquet's above-mentioned book.

The neuro-consumer confronted with pricing policies

In addition to the methods described above, the neuro-consumer is often confronted with a set of practices aimed at reducing his pain and increasing his immediate pleasure in relation to the price of a product or service. These practices have been described by several neuromarketing experts.[10]

Practices aimed at reducing the brain's pain when dealing with prices

Psychologists, neuroscientists, marketing experts, etc. attest to the use of practices that encourage purchases by reducing the pain felt by the neuro-consumer's brain when confronted with the prices of products and services. These practices enable commercial solutions to be found that partly respond to the effects felt in the brain's subconscious mental strategies when faced with a price. Among those that are frequently mentioned:

- Grouping products or services together contributes to reducing the purchaser's pain, which is renewed each time he has to pay for a separate product. This grouping usually takes the form of packs.

  For example, packs that group together the Internet and paper versions of a magazine, "luxury packs" containing several accessories to help sell a car or packs including different services linked to a banker's card or an insurance policy, etc. The grouping reduces the pain even further when it is accompanied by a special offer or bonus.

- Placing the same goods in bigger packs. This makes the customer's brain think that the pack contains more than the competitors' packs, which are smaller.
- Pricing the product monthly instead of hourly. AOL significantly increased its sales using this practice.
- Explaining to the consumer why the product is "premium" when its price is perceived as too high.
- Fixing the price a few cents below a given sum. Selling a product at $299 instead of $300 will often mean that the neuro-consumer's brain places it in the $200 price bracket.
- Placing products on a shelf whose packaging, design or simply their size makes them seem less interesting for an identical price alongside similar products of better quality. By offering the customer the opportunity to compare the perceived quality for the price offered, this technique highlights the product judged to be better and tends to increase its sales.

Two specific types of behavior: "tightwad" and "spendthrift"

Seeking to obtain a deeper understanding of the influence of price on the brain, Scott Rick, assistant professor of marketing at the Michigan Ross School of Business, Cynthia Cryder, assistant professor of marketing at Washington University, and George Loewenstein, professor of economics and psychology at Carnegie Mellon University, became interested in the behavior of two specific categories of neuro-consumers: "tightwads" on the one hand and "spendthrifts" on the other. Their reference study[11] is based on a survey of 13,000 individuals of whom 24% were "tightwads" and 15% were "spendthrifts," the remainder of the sample corresponding to those whom the researchers considered had a balanced feeling of pain in relation to pricing. "Tightwads" are defined as consumers feeling an excessive level of pain at the moment of paying when compared to a sample considered to be normal. On the contrary, "spendthrifts" feel a very low level of pain in a comparable situation. "Tightwad" neuro-consumers tend to reduce their pain and increase their purchases when confronted with certain pricing practices. They are attracted by discounts and price reductions. They are attracted by prices presented in a manner that appears cheaper. Reacting badly to a price of $120/year, they will suffer less with a price of $10/month or 33 cents/day. They are interested in packs with an all-in price. In restaurants, they prefer the fixed price to the à la carte menu. They feel that products or services that are grouped together will be less expensive than items sold separately. They need to feel that the item purchased is useful or sustainable.

"Spendthrifts" cause sales staff fewer problems. Sales staff can also improve their performance with "spendthrifts" by knowing how to respond to the characteristics of their brains. Appealing to their hedonistic tendencies, their interest in having an experience that is new, pleasant and interesting attracts them more than being told about the usefulness of the product or service in question. Offering a credit option can lead them to buy an item with a higher price but that is likely to provide them with immediate pleasure.

Notes

1. Enrico Trevisan and Florent Jacquet, *Psychologie des prix: Le Pricing comportemental*, De Boeck, 2015.
2. Brian Knutson, Scott Rick, G. Elliott Wimmer, Drazen Prelec and Georges Loewenstein, "Neural Predictors of Purchases," *Neuron*, vol. 53, no. 1 (2007), pp. 147–156.
3. Bernard Roullet and Olivier Droulers, *Neuromarketing: Le marketing revisité par les neurosciences du consommateur*, Dunod, 2010.
4. Knutson et al., "Neural Predictors of Purchases."
5. Dan Ariely, *Predictably Irrational: The Hidden Forces That Shape our Decisions*, Harper Perennial, 2010.
6. Roger Dooley, *Brainfluence: 100 Ways to Persuade and Convince Consumers with Neuromarketing*, John Wiley & Sons, 2012.
7. Hilke Plassmann, John O'Doherty, Baba Shiv and Antonio Rangel, "Marketing Actions Can Modulate Neural Representations of Experienced Pleasantness," *Proceedings of the National Academy of Sciences*, vol. 105, no. 3 (January 2008), pp. 1050–1054.
8. Benedict Carey, "More Expensive Placebos Bring More Relief," *New York Times*, March 5, 2008.
9. Trevisan and Jacquet, *Psychologie des prix*.
10. Roullet and Droulers, *Neuromarketing*; Dooley, *Brainfluence*.
11. Scott Rick, Cynthia Cryder and Georges Loewenstein, "Tightwads and Spendthrifts," *Journal of Consumer Research*, vol. 34, no. 6 (2008), pp. 767–782.

CHAPTER 19

The influence of experiential and sensory marketing

Neuroscience is used to improve product and service development and pricing policies, and to enhance the image of sales outlets and increase their revenue. It also serves to boost sales performances.

In the face of fierce competition from the Internet and smart phones, sales outlets are being forced to change tactics in order to survive. Alone, or in close collaboration with the brand's digital technology policy as part of multi-channel or cross-channel strategies, they aim to offer clients advantages that are missing from online shopping, such as the warmth of human contact and direct stimulation of certain senses, such as smell, taste and touch, which are difficult to convey via the Internet and mobile phones. To do so, they no longer limit themselves to just selling, but strive to provide spaces where the neuro-consumer enjoys coming and spending time because their senses have been elicited in such a way as to make them feel good. They aim to create a consumer experience for their clients by developing themes and dramatizing sales outlets. They offer visitors a series of additional services with no direct link to the products on display, such as childminding, cafes, restaurants, relaxation or therapy services and cultural or sporting events. To encourage the sensation of well-being in consumers' brains, they strive to make the experience perceived by the senses a pleasant one.

A new approach, sensory marketing, began to emerge tentatively in the 1950s. It was the subject of interesting experiments during the 1980s, then developed at a fast pace from the start of this millennium and has become widely used by companies today. "Sensory marketing" complements what experts call "experiential marketing." The aim is simple: to make the consumer feel good and stay in the sales outlet for as long as possible. Three university lecturers, Marie-Christine Lichtlé (University of Burgundy), Sylvie Llosa (IAE of Aix-Marseille) and Véronique Plichon-Guillou (University of Tours)[1] have shown that pleasing colors, smells and design of sales outlets can have an impact on customer satisfaction. The more time a customer spends in a sales outlet, the more products they will buy or the more they will consume.

> *Consumers tend to buy more if the presentation of products is better adapted to the human brain's modes of reception. This is the case, for example, with Ikea, Boulanger, Abercrombie & Fitch and Nature et Découvertes, etc. Sensory ergonomics concern a number of sectors besides mass retailing, such as catering, hospitality, vehicle dealerships, furniture and furnishings stores, travel agencies and bank branches.*

An increasing number of sales outlet managers are consulting in-house or external experts in order to give these spaces an "experiential atmosphere." The use of sensory marketing also contributes to the staging of sales outlets. It reinforces the perception of feelings incited to increase the sensation of enjoyment in neuro-consumers' brains during visits.

From experiential marketing to sensorial marketing

Patrick Hetzel, a politician and professor at the University of Paris II-Panthéon-Assas, stated that "experiential marketing involves five actions: surprising, proposing the extraordinary, creating connections, using the brand for the experiential policy and arousing the five senses,"[2] and Isabelle Frochot and Wided Batat[3] give several examples that use such marketing techniques, including Accor, Ikea, Caisses d'Épargne, Club Med, H&M, Relais et Châteaux, Le Louvre and French TV program *Nouvelle Star*.

In sales outlets, experiential marketing is mainly employed by defining themes, staging spaces and developing sensory marketing. The first two are fairly long established, although they continue to be used in modern approaches. The latter is much more recent. Experiential marketing mainly works through subconscious perceptions and reactions in consumers' brains. The advantages it offers retail chains have been bolstered by a number of emerging factors, such as:

- Fierce competition between retail outlets and selling online and via mobile phones. This forces shops to review their business models and place greater emphasis on their relational and sensory advantages.
- The growth in Europe of big international brands that have no hesitation in using the latest experiential marketing techniques, and in particular those derived from sensory marketing.
- The development of new technology managed by external experts, enabling colors, sounds or smells to be adjusted in any part of the shop and at any time to meet the expectations of the brains of neuro-consumers

browsing the shelves. The practice of "zoning" allows a different form of sensory marketing to be used for each area. The use of sprays and odor diffusers allows a specific smell to be given to each space. Specialist companies aim to create images, sounds and smells, etc. that correspond to the desired brand positioning in order to please the senses of their clientele.

- Changes in consumer behavior. Influenced by social media, consumers take an interest in brands that aim to create a community. The brain proves to be sensitive to new buyer experiences in a pleasant and inspiring atmosphere.

This is the case of Abercrombie & Fitch, which targets a community of young people. Shops that improve the ergonomics of their design usually see a significant increase in their revenue.

In order to meet these needs, three fields in particular are being developed by retail and services companies to procure enjoyment for neuro-consumers' brains: creating themes, staging and seeking to create sensory experiences for spaces. They are designed to create enjoyment for the brain and make it favorably disposed to buy.

Giving themes to sales outlets

Giving themes to sales outlets aims to attract communities of neuro-consumers with similar interests or passions, such as ecology, the love of nature, organic products, frugality, culture and sport. It is not a new thing.

For a long time, certain restaurants have been specializing in order to please a clientele defined according to a specific style of food, such as Indian, Thai, Japanese, Moroccan. The same is true of hotels that aim to differentiate themselves by adopting a unique character, for example the Relais & Château chain or Châteauforme.

The aim is to please a community of neuro-consumers with the same tastes and give the company a unique character that differentiates it from the competition.

To satisfy the desire for community, sales outlets are specializing by adopting a theme, like organic shops for example. They aim to appeal to consumers with a strong interest, such as in motorbikes in the case of Flash 76, a motorbike store in the French region of Rouen (whose territorial department number is 76) and one of the leading dealers for the Yamaha brand. They target citizens interested in nature, like Nature et Découvertes, or culture, like Fnac or the French Dialogue bookshop that opened in Brest in 1976 and whose concept has served as a model for several independent bookshops.[4]

Superstores are changing the layout of their products. They are adopting the principles of category management, which consist in grouping together products that are not similar but that correspond to groups of expectations in clients' brains in their purchasing experience, for example by grouping together products for the same sport or for baby care, etc.

The Auchan Val d'Europe hypermarket has adopted the concept of a "hypermarket for better living." It has reorganized its shelves to take account of the overall perception in the brains of clients: eating well, better housekeeping, taking an interest in leisure, etc. Carrefour Planet in Lyon has eight themed areas for its customers: food, organic products, beauty, frozen products, fashion, baby, the home, entertainment and multimedia.

Consumers are very sensitive to any discrepancies between the desired theme and the reality in the sales outlet.[5] The themed offering and its layout must give the brain an impression of authenticity and sincerity. The slightest discrepancy between the theme and its implementation may be perceived as a confidence trick and cause it to be rejected.

The neuro-consumer's brain would react negatively to the presence of plastic bags or trolleys in an organic food shop. It would not understand why sales staff in shops belonging to the 64 brand in the Basque Country, which advocates the Basque spirit, or in the 66 shops in Catalonia, which stand for Catalan uniqueness, do not know the region or are indifferent to local social or political problems.

The client must feel the coherence in the way the spaces are dramatized.

Sales outlet staging

Neuroscientists suggest that the brain is culturally wired to be attracted by art. For centuries drama has been an important component of art, and it interests the brain even more when it meets the need for interactivity. Well-known theater forms such as the Italian Commedia dell'arte (in which the actors adapt the scenario to the surroundings), street theater from the Middle Ages (which gets the audience involved), boulevard theater (like the sort shown by Marcel Carné in his famous film *Les Enfants du Paradis*) and more recently participatory theater, are all types of performances that may inspire the staging of sales outlets.

One of the pioneers in store staging was Aristide Boucicaut, who created Le Bon Marché in 1852. The concept of store staging was invented by Boucicault and was addressed at length by Émile Zola (1840–1902) in *Au Bonheur des dames*[6] (*The Ladies' Delight*), the 11th volume in the Rougon-Macquart series.

> *The aim of Le Bon Marché was to provide happiness for the modern woman by offering her most of the products she wanted under one roof. The fictional character of Mouret, invented by Zola and based on Aristide Boucicaut, staged his sales outlets. No space must remain empty. The founder demanded that noise, crowds and life be everywhere because "life," he said, "attracts life and generates and multiplies it."*[7] *The shop was filled with activity, with reading and writing rooms in which men could wait while the women were shopping, the provision of spaces for children to play in, a buffet where customers could enjoy free drinks and biscuits, a monumental gallery decorated in over-the-top luxury, temporary painting exhibitions, organized disorder to encourage spontaneous purchases, the laying out of items to give a feeling of discovery and the use of bright colors such as red, green and yellow.*

Certain ideas proposed by Aristide Boucicaut are still used in mass retail today.[8]

The staging of spaces is widely used in the services sector, including by holiday and leisure clubs and hotels that combine multiple services and entertainment for clients and children during holidays.

> *Michel Crouzet, a hotel manager in Aveyron in France, relaunched a campsite and family hotel in the small village of Saint-Geniez-d'Olt in Aveyron by staging them. For the campsite, he quickly followed the example of Club Med. He created a restaurant overlooking a large swimming pool called "Le Marmotel," offered games and a nightclub for young people and organized shows. Within a few years, his clientele included foreign guests from countries such as Holland, Belgium and Britain, and his turnover soon grew rapidly.*
>
> *At the same time, he inherited a small struggling family hotel in the middle of the village. He decided to target a new type of clientele consisting of older or retired people wanting to discover France. He created Hôtels Circuits France (HCF) in collaboration with hotel managers from other regions. The idea was to offer clients the opportunity to discover a region with its tourist sites and local specialties, through the intermediary of coach companies and tour operators. He organized the creation of themes and the staging of stays around the discovery of the local area through activities such as visits to typical regional sites with a local tour guide, discovery of life in a cowherd's hut in Aubrac and local cooking classes to learn how to make Aligot, stuffed boletus mushrooms, girolle mushroom fricassee or farçou rouergat.*

Evenings were occupied with visits from local entertainers, folk dances, etc. Thanks to staging, the hotel's occupancy rate during the low season was considerably improved and it became successful.

With the hotel and campsite, his enterprise became the second biggest employer in the region. The HCF concept is developing throughout France and may even be exported to Europe.

Michel Crouzet's experience shows that the staging of a space is not solely reserved for large brands. A shrewd small-and-medium-sized enterprise (SME) manager can adopt the concept and make a service company prosper by applying the idea with limited financial resources.

Staging of spaces is very relevant for retail and sales outlets for products and services.

Major international brands like Disney, McDonald's, Toys R Us, Nike Stores and others have been using it successfully for several years. In Europe, shopping centers such as Centre O in Oberhaussen, Germany, the Bluewater Center just south of London, Bercy Village in Paris and Ubbo on the outskirts of Lisbon combine a selection of additional services and entertainment with the large number of shops that constitute them. Some include amusement parks. For example, the Bluewater Center has adventure and animal parks, while Centre O has giant aquariums, exhibition rooms and concert halls.[9] *As for Ubbo, which recently went through a rebrand and refurbishing process, it now has two entertainment "parks," one space for collaborative work, workshops, exhibitions and street art.*

In France, Fnac has created and develops the theme of "culture for everyone." It employs culture enthusiasts to manage its different sections and provides additional training for them. It assigns large spaces in its shops to nonsales-related areas where it has forums, exhibitions, interviews with artists, relaxation areas, etc. The client does not just come to buy. He comes to spend time and read or listen to the latest musical hits free of charge. The concept is on the borderline between sales and culture. The closure of a Fnac shop is often experienced in the locality as a cultural disaster.

Decathlon provides a series of events that are in line with its image as a brand passionate about sport. It builds fitness parks outside stores with a terrace offering fast food. Clients can try out fitness training or African dance, learn roller-skate figures or take lessons in riding a bike in a city. The brand organizes events in connection with external activities such as "La Belle Rando," and its salespeople are trained and encouraged to only recommend and sell to clients the products that will suit them.

Experiments in sales outlet staging are increasingly common and judging by the success encountered in most cases where it is implemented, neuro-consumers' brains seem to find a certain satisfaction in them. Interior designers (individuals or teams in specialized enterprises such as Dragon Rouge, Carré Noir, Prinz Design) provide support for brands through their creative talent and experience. Collaboration with these experts helps create sales outlets that are as unusual as they are appealing. They have contributed to the creation of interesting shop concepts in different sectors, including Viadys (pharmacy), Histoire d'Or (jewelry), Joué Club (toys and games for children) and Marty (furniture store).

Sensory marketing is part of experiential marketing and makes a considerable contribution to store space staging.

Sensory marketing in sales outlets

The use of sensory marketing in sales outlets aims to procure a positive experience, most often subconscious, in the mind of visitors. It directly targets one or several of the senses. Each of the five senses can be aroused on its own or with others. The concept is described by Agnès Giboreau and Laurence Body in their book *Le Marketing sensoriel*,[10] and its applications are discussed in detail in the publication edited by Sophie Rieunier, a lecturer at the IAE of Paris (University of Paris I-Panthéon-Sorbonne), *Marketing sensoriel du point de vente*.[11] We chose some of the spaces described in the latter work to go and visit and, where possible, talk to the managers. Journalist Emmanuelle Ménage also gives examples of the cases of various brands including Ikea, Boulanger and Abercrombie & Fitch in a documentary.[12]

Sensory marketing in spaces and sight

Sight is one of our most-used senses: 80% of the information transmitted to our brains is visual. Sales outlets plan their design to give visitors a positive visual impression. Sensory marketing strategies linked with vision mainly concern the creation and layout of store paths, choices of colors, materials and lighting and the positioning of products in the displays.

Organization and layout of spaces Organization and layout of spaces are one of the first choices made in staging spaces visually.

> *Viadys pharmacies use a logical organization of space in order for the client to find their way around and avoid queuing needlessly. There is a clear divide between the space dedicated to prescriptions, an area for beauty products and a space in the form of a first aid cupboard for OTC products.*

Some brands create a clever impression of disorder, giving the brain the impression of searching and thinking it has found a piece of treasure like in Ali Baba's cave.

> *An attractive example of this layout can be seen in the Droguerie de la Marine shop in Saint-Malo in Brittany, or in Lingerie Coste, a traditional lingerie shop in Rodez in Midi-Pyrénées.*

In some supermarkets, the shapes of the stands are designed to control the flow of consumers, speeding it up when they are angular, and slowing it down when they are rounded.

Choice of colors This is particularly important in developing a visual policy for retail and service stores. It has been the subject of numerous studies, often carried out by researchers in American universities.[13] Colors influence the brain of the concerned neuro-consumer in several ways in terms of the positioning of the sales outlet, its attractiveness and the perception of the products or services on offer. Warm, bright colors like red, yellow and gold have a strong visual impact and power of attraction.

> *Garages of the Midas brand, which are gold and yellow, are known for being highly visible.*

According to the experts, bright colors like red inside sales outlets encourage impulse buying. In casinos, they lead to greater risk-taking by players. Some colors like white, black, gold and certain shades of gray are said to give shops a chic, high-quality feel. Cold colors are more often associated with the idea of a bargain and are used by some discount stores. Trends can change, however, with orange becoming a popular color for certain luxury brands, for example. Trend research institutes put together publications that predict which colors will be fashionable in the coming years. The most internationally famous institutes are American, such as Pantone Color Institute and Color Marketing Group. Their recommendations are studied and often adopted by retail and service brands before creating a visual strategy for their stores.

Unusual but successful chromatic choices are sometimes made with the aim of breaking established color codes within a profession.

> *Viadys pharmacies use red and blue, rather than the traditional green and white in order to make their spaces less daunting. The jewelry brand Histoire d'Or has abandoned the classic chic colors like black, white and gray in favor of stand presentations in bright red.*

Colors can help identify the specific nature of sales departments, as is the case in Tape à l'Œil shops. Different shades of color can be used to identify sales spaces according to the sex and age of consumers.

The use of color is all the more important because it can change the way things are perceived by the neuro-consumer's other senses. White makes a space feel bigger, while black tends to shrink it. Salespeople dressed in black seem more competent, employees in pharmacies are seen to provide better advice when they wear white coats, and staff in home improvement shops are considered more professional if dressed in blue overalls. Loud noises are attenuated in an environment of dark colors. Turquoise and white give an impression of coolness, whereas red and yellow create a feeling of warmth. Green increases the impression of freshness of the products on shelves in supermarkets.

Use of materials and lighting These also create cognitive impressions in the minds of visitors.

> *Shop chains in France such as DeliTraiteur (caterer) or Comptoir Richard (coffee, tea, herbal teas, hot chocolate, etc.) use traditional wooden materials to give their shops a "cozy" feel.*
>
> *Conversely, the Cyberdog shop (futuristic fashion clothes and accessories) in London has been designed using modern, innovative materials and made to look like the inside of a spaceship.*
>
> *Shops in the Joué Club chain (children's toys) use modern furnishings adapted to children's tastes. The walls are covered with festive colors and drawings tailored to suit the perceptions of the brand's target customers.*

Lighting is another element used to give neuro-consumers the impression of well-being in sales outlets. It is designed to encourage purchases. A number of studies have been published in scientific journals, mostly carried out by researchers in the United States.[14] They address the effects of lighting used in spaces on consumption and suggest that consumers look at more products in the shop when the lighting is brighter. When the shop has a darker ambiance, such as in Abercrombie & Fitch stores, products are brightly lit.

The brains of some sectors of the population, such as older people, tend to have less visual perspicacity and sensitivity. Sales spaces targeting this type of clientele often use brighter lighting for the items on sale.

> *Grand Optical, a brand for prescription glasses, uses strong lighting for its products, which are presented in a bright white environment.*

Some types of lighting, such as neon lighting, give an impression of inexpensiveness in the mind of neuro-consumers and are often used in discount stores.

Positioning of items on shelves Positioning is used to capture the visual attention of the brain. As shown in the documentary by Emmanuelle Ménage,[15] layout is skillfully arranged to attract the eye and give an impression that will encourage purchasing. The place in the shop, height, position on the display stands and the optimum number of products belonging to the same category or brand on a single shelf are often the subject of detailed studies carried out by experts.

Sensory marketing and hearing

The use of music to attract customers to a brand or sales offering is not a recent thing. Since the Middle Ages and even dating back to antiquity, traveling salesmen have sung about their goods in order to appeal to consumers. In the 18th century, philosopher Jean-Jacques Rousseau spoke of the incredible power of music.[16] He referred to a certain melody that had the unusual effect of causing Swiss people to burst into tears due to the nostalgic reminder of their country, way of life and childhood, etc.

Since the 1920s, the development of radio technology has led to more extensive use of sounds in sales outlets and reception areas. Cafes also often play music in order to attract and please customers.

> *In Europe, the Monoprix chain was a pioneer in the use of sound in its shops, having played background music since the late 1920s.*

In the United States, the Muzak company, created in the early 1920s, made a name for itself by developing "background music," which it proposed to shops, offices, workshops, etc. It was initially designed to increase workers' productivity and was later introduced into sales outlets to please customers and encourage purchases. This type of music, known as "muzak," has been very successful in America and is sometimes considered "overly present." It has been criticized by some American consumers who are irritated by its invasive character. Critics reproach it not only for its almost "totalitarian" omnipresence in the country's musical environment, but also for subliminally manipulating consumers through numerous repetitions. This movement of opposition has been nicknamed the "Pipedown Protest" in the United States.

There have been numerous research projects carried out in several countries, in particular in the United States and Europe,[17] on the perception of sound by neuro-consumers' brains. They all show that background music in sales outlets has a strong cognitive influence on neuro-consumers' brains. Sound allows sales outlets to connotate a universe and brand positioning, change the perception of time spent in a shop and direct the client towards certain departments or categories of products. Ambient music is considered to be preferred to silence, as when customers enjoy listening to the music, they tend to stay longer and therefore buy more things. However, while consumers usually expect background music or sounds, certain repetitive songs or sounds (such as at Christmas) can annoy them.

The chosen musical atmosphere directly influences the brain's perception in areas as varied as attentiveness to salespeople, the number of products bought or consumed, the perception of the level of quality in the shop and the perception of pricing.

Current techniques for broadcasting sound (Hi-Fi systems, CD/MP3, independent or ADSL jukeboxes, USB sticks, transmission by satellite, etc.) allow musical compositions to be adapted for distribution or service brands, as well as the incorporation of personalized messages. Sound can be broadcast simultaneously in all the brand's sales outlets, be adapted to different types of visitors (teenagers, older people, men, women, etc.), vary according to the day of the week or time of day or include messages for clients or staff members.

A number of companies propose to adapt the sound used in spaces to the specifications of a brand's musical strategy. This includes creating music designed to make customers remember the brand, as well as the use of sounds in order to please clients and make them more receptive to sales arguments or to increase sales. Mood Média is one of the best-known companies. Others, such as Médiavéa, Sixième Son, BETC-Euro-RSCG and AtooMédia, also provide a high level of support for service and retail companies that use effective ambient music.

Sensory marketing in spaces and smell

The sense of smell also plays a role in sales outlet staging. The mysterious power of aromas over human emotions has been acknowledged for centuries. Since the time of the Egyptian pharaohs, fragrances have been widely used in religious ceremonies and in medicine. In India, under the Maharajahs, some natural fragrances were more valuable than diamonds. Readers interested in the history of perfume and the power of smells can consult a large range of literary works available on the subject.[18] It wasn't until the work by Louis Pasteur that this power was demystified. The field of "aromachology" was later created in 1982 by the Olfactory Research Fund, defining it as the science devoted to analyzing effects of the inhalation of fragrances on the emotional condition.[19] Similarly, "aromatherapy" rediscovered the therapeutic power of aromas with the use of essential oils.

This mysterious aspect of smells is also reflected in language. For example, if someone seems innocent, it is often said that they "smell like a rose," while excessive piety or holiness can have an "odor of sanctity."

Sensory marketing in sales outlets uses the evocative power of smells to help improve revenue and several research projects have been carried out on the subject.[20] Mass retail and service companies are especially interested in certain aspects of the evocative power of smells, i.e., the power to create, like music, a relaxing or stimulating atmosphere by influencing the neuro-consumer's feelings. The diffusion of fragrances such as lavender, orange, sandalwood, rose and fennel in shops has a relaxing effect. On the other hand, jasmine, chamomile, lemon, musk or mint are designed to stimulate customers' senses.

At certain times of day, the smell of food emitted by the butcher's spit or the wafting aroma of hot croissants stimulates a gastric secretion sparked by the feeling of hunger in the neuro-consumer's brain. When well-managed, it can lead to an increase in food purchases.

The power for the emotional evocation of memories was cleverly described by German author and scriptwriter Patrick Süskind in his famous novel *Perfume*.[21] It helps improve the memorization of names and locations of retail and service brands.

Smells can also have the effect of reducing the neuro-consumer's perception of time spent in a shop. The more time they spend, the more they consume.

> *The primary use of fragrances in sales outlets is to eliminate bad smells, which create an unpleasant sensation in the brain and discourage the neuro-consumer. This is often the case in fitting rooms (the smell of sweat), casinos and hotel rooms (the smell of stale tobacco) or certain public places. In France, the RATP is making a big effort to avoid bad smells in the Parisian metro. In casinos, neutralizing fragrances are diffused to get rid of unwanted smells and in luxury shops, lingerie stores and restaurants, staff members are often specifically requested to wear deodorant.*

Smells can attract a customer to a display, confer a better assessment of the products on the shelves and increase revenue.

> *The documentary produced by journalist Emmanuelle Ménage[22] shows several experiments carried out in shops belonging to the Boulanger brand in France. The smell of clean laundry attracted customers and boosted washing machine sales.*
>
> *Bruno Daucé, French professor of marketing at University of Angers,[23] cites the following three examples. Hollywood Chewing-Gum saw an increase in the use of their vending machines when they diffused a mint fragrance when somebody draws near. Nestlé Waters Marketing Distribution, in collaboration with the company Audiadis that markets the Sensodis tool, is carrying out an interesting project in France involving the diffusion of fragrances relating to specific themes and calendar periods in the stores of various hypermarket chains: Leclerc, Carrefour and Auchan. When coordinated with sound, fragrances helped incite visitors to taste different flavored waters by Nestlé Waters: Vitalitos, Fraise Délire, Perrier Framboise and Vittel Pêche d'Enfer. For Bruno Daucé, "This*

experiment associates Nestlé Waters with an avant-gardist and dynamic presentation concept in sales outlets." The third experiment involves the creation of fragranced ATMs (automated teller machines) using marine aromas by the Banque Populaire (BPCE) in two branches in the Lot region in France. The fragrance emitted by these cash machines increased the number of visitors to the two branches concerned.

Fragrances are used to improve the memorization of enterprises and service brands.

The Accor group is looking into the use of fragrances. Along with the company Air Berger, the Novotel brand has developed a signature smell called "Cosy Lounge." This fragrance is used in all the company's hotel lobbies. The experiment seems conclusive. Other hotel chains are also choosing to create personalized brand fragrances. A number of brands such as Mercure (Accor group), Le Méridien (Starwood group) and the Park Hyatt hotels are going down the same route. They often call upon the experience of specialized partners such as Air Berger or perfumers such as Eddie Roschi and Fabrice Penot (founders of the perfume company Le Labo) for Le Méridien, and Blaise Mautin for Hyatt hotels.

A number of enterprises provide valuable advice to retail and service brands that wish to use olfactory marketing for their brand or spaces. Over the past decade, they have increased in number and size with, for example, Atmosphère Diffusion (a subsidiary of Mood Média), Air Berger, Aroma Company, Midis, Quinte & Sens, Audiadis and Olfactair.

Diffusing fragrances is a delicate task. They spread inside the shop and can seem inappropriate in certain sections. They can even irritate people sensitive to smell if they are too strong. They sometimes need to be personalized to avoid upsetting people. As a solution to the problem, the Crayon Hotel in Paris proposes a choice of ambient fragrances to its clients in their rooms. The International Fragrance Association (IFRA), which uses research work from laboratories such as the Research Institute Fragrance Materials (RIFM), makes recommendations on the use of aromatic raw materials in order to prevent high levels of discomfort. It can be useful to consult expert advisors who are either in-house or external to the brand.

As regards the other senses, the notion of congruence is essential in the implementation of olfactory marketing in sales outlets. The marketing policy must be established in harmony with the brand's chosen positioning or the theme of the products on sale. It must be adapted to certain elements such as the gender of visitors in a section or the seasons and hours of the day. It must be coordinated with the other senses if they are also stimulated. This is notably the case for seeing the decor, hearing ambient music and tasting foodstuffs.

Sensory marketing in spaces and taste

Although little research has been carried out with regard to the sense of taste in sales outlets, it is nevertheless an important part of human contact. Tayllerand (1754–1838), the famous French minister for foreign relations during the First Empire and then the Restoration, was reported to have said on the subject: "There can be no good diplomacy without good cuisine." For retail and service spaces, taste offers a real competitive edge in comparison to its very limited uses on the Internet and social networks. The ability to taste a product in a shop can be extremely useful in demonstrating its quality, for example with certain drinks such as wine, whisky, beer, coffee, etc.

Sales of numerous food products increase when they can be taste tested. This sense is best aroused at specific times when the customer is starting to feel hungry. Proposing food products for tasting is particularly efficient for advertising during these moments. When tasting a slice of foie gras, Parma ham or cheese, the brain automatically produces gastric secretions that increase the feeling of hunger. Polite neuro-consumers know that they cannot eat everything on the salesperson's tray and will compensate for the desire in their brain by buying a few products. According to Robert Cialdini, purchase is all the more probable given the brain's often instinctive tendency to want to return the favor to someone who gives them something.[24]

Taste becomes even more effective when it is congruent with the other senses, in particular sight, touch and smell.

Sensory marketing in spaces and touch

This sense has a uniquely important role in sales outlets. In referring to a study by IFM/MCA in 2003, Sophie Rieunier[25] reminds us that when a product is picked up, it is then purchased in 87% of cases. This fact is linked to the "endowment effect" described in Chapter 18. Like taste, the sense of touch cannot be easily stimulated on the Internet, something that confers an undeniable advantage to physical sales outlets that promote it successfully. Touch produces a hedonistic effect in the human brain that is sometimes referred to as "the need for touch." This sensation has been found to be more developed in women's brains, which Alka Citrin and colleagues suggest might explain why women often feel a greater need to touch products in shops before deciding whether to buy them.[26]

The use of touch and its effects on neuro-consumers in sales outlets has only been studied in a limited amount of scientific research.[27] Shops or service spaces tend first of all to get rid of negative influences that could disturb this sense, for example when the temperature is too high, goods are not easily accessible, customers' arms are already too full of items in festive periods, the product is too heavy or it does not have a pleasant feel. A negative influence can also be felt when clients feel frustrated at not being able to handle products in the same way, as with fabrics.

Ergonomics specialists have come up with clever propositions to alleviate the large number of disadvantages relating to touch that may perturb the perception in the brain of neuro-consumers in sales outlets. They first recommend eliminating negative stimulations, and then advise building a positive tactile environment. Touch is used as a source of information. It allows someone to feel the softness of a jumper or judge how easy it is to handle a product or if its weight corresponds to the idea of quality they have in mind, etc. It helps strengthen and gives credibility to the brand's image.

> *In restaurants, high-quality positioning can be sensed through the quality of the tablecloths, serviettes, tableware and furniture. The image of comfort targeted by certain brands in the Accor group is manifested through the pleasant feel of the sheets in the bedrooms. In Nature et Découvertes shops, handling products helps create a hedonistic feeling among customers.*

Like the other senses, touch is all the more effective if it remains congruent with other senses like taste, smell and sight.

Congruency in sales outlets

The implementation of a sensory policy in sales outlets is as efficient as it is congruent. It is guided by the brand's positioning and harmonized as regards the senses and the overall communication and relations policy chosen by the brand. It is a vital tool for creating themes and dramatizing spaces.

> *A variety of different brands place great importance on congruency of the senses, such as Nature et Découvertes, Sephora, Abercrombie & Fitch, Nespresso, Apple, Viadys, Ikea and the Accor group.*
>
> *In Belgium, the Colruyt brand, a leading discount store, coordinates a number of sensory perceptions to give neuro-consumers' brains the impression of inexpensiveness. The illumination from the neon lighting is diaphanous, the material on the floor is left in its raw untreated condition, products are displayed in cardboard transport packaging and refrigerators are replaced by a large cold room for displaying dairy products, cooked meats, etc. The ambient sounds and smells are adjusted to remain coherent with the positioning.*
>
> *In France and Europe, the Accor group has been experimenting for several years with the concept of "techno-sensory" rooms. An experiment was launched with Sofitel hotels, to be later extended to other chains such as Ibis. The group aims to afford clients a sensory experience that suggests both luxury and comfort, based on the congruency of the senses. Several can be stimulated at once, for example sight through the choice of colors and shape of the furnishings, through the ease with which the person can make the space*

their own using a touch screen, the materials in the bathroom, lighting, etc. Smell can be stimulated by offering visitors a selection of ambient fragrances for their room, hearing with the ability to listen to different radio stations or their own CDs. Touch and a feeling of well-being can be aroused through adjusting the room temperature, creating a new, very comfortable bed, improving the softness of the sheets and towels or fitting the bathroom with a "tropical shower head."

Notes

1. Marie-Christine Lichtlé, Sylvie Llosa and Véronique Plichon, "La contribution des différents éléments d'une surface alimentaire à la satisfaction des clients," *Recherche et applications en marketing*, vol. 17, no. 4 (2002), pp. 23–34.
2. Patrick Hetzel, *Planète Conso: Marketing expérientiel et nouveaux univers de consommation*, Éditions d'Organisation, 2002.
3. Wided Batat and Isabelle Frochot, *Le Marketing expérientiel: Comment concevoir et stimuler l'expérience client*, Dunod, 2014.
4. View the cultural spaces in the Dialogue bookshop in Brest on Google Images "librairie dialogue photos," www.unidivers.fr/dialogues-librairie-brest.
5. Bernard Cova and Véronique Cova, *Alternative Marketing*, Dunod, 2001.
6. Emile Zola, *Au Bonheur des Dames* [1883], Penguin Classics, 2001.
7. Ibid.
8. Marie-Louise Hélies-Hassid "Au Bonheur des dames ou la leçon de commerce de M. Zola," *Décisions Marketing*, vol. 20 (2000), pp. 35–46.
9. Photos relating to the examples given here and throughout this chapter can be consulted through various search engines online.
10. Agnès Giboreau and Laurence Body, *Le Marketing sensoriel: De la stratégie à la mise en œuvre*, Vuibert, 2007.
11. Sophie Rieunier (Ed.), *Marketing sensoriel du point de vente* (4th ed.), Dunod, 2013.
12. Emmanuelle Ménage, *Consommateurs pris au piège* [film], France 5, January 12, 2014.
13. Joseph A. Bellizzi, Ayn E. Crowley and Ronald W. Hasty, "The Effects of Colors in Store Design," *Journal of Retailing*, vol. 59, no. 1 (1983), pp. 21–44; Rie Mimura, "Color and POP: The Effective Use of Colors for Point of Purchase Display," *Journal of Undergraduate Research*, vol. 6 (2003), pp. 1–15.
14. Charles S. Areni and David Kim, "The Influence of Background Music on Shopping Behavior: Classical Versus Top-Forty Music in a Wine Store," *Advances in Consumer Research*, vol. 20 (1993), pp. 336–340; Teresa. A. Summers and Paulette R. Herbert, "Shedding Some Light on Store Atmospherics: Influence of Illumination on Consumer Behavior," *Journal of Business Research*, vol. 54, no. 2 (2001), pp. 145–150.
15. Ménage, *Consommateurs pris au piège.*
16. Jean-Jacques Rousseau, *The Complete Dictionary of Music* [1768], Gale Ecco, Print Editions, 2010.
17. Sophie Rieunier, *L'influence de la musique d'ambiance sur le comportement des consommateurs dans les points de vente*, PhD thesis, University of Paris IX

Dauphine, 2000; Rieunier, *Marketing sensoriel du point de vente*; Alain Goudey, *Une approche non verbale de l'identité musicale de la marque: influence du timbre et du tempo sur l'image de marque évoquée*, PhD thesis, University of Paris IV Dauphine, 2007.

18. Annick Le Guérer, *Le Pouvoir des odeurs*, Odile Jacob, 2002; *Le Parfum: des origines à nos jours*, Odile Jacob, 2005.
19. Maria Lis-Balchin, *Aromatherapy Science: A Guide for Healthcare Professionals*, Pharmaceutical Press, 2006.
20. Eric R. Spangenberg, Bianca Grohmann and David E. Sprott, "It's Beginning to Smell (and Sound) a Lot Like Christmas: The Interactive Effects of Ambient Scent and Music in a Retail Setting," *Journal of Business Research*, vol. 58, no. 11 (2005), pp. 1583–1589; Bruno Daucé, *La diffusion des senteurs d'ambiance dans un lieu commercial: intérêts et tests des effets sur le comportement*, 2000, PhD thesis, University of Rennes 1; Bruno Daucé, "Comment gérer les senteurs d'ambiance," in Rieunier, *Le Marketing sensoriel du point de vente*, pp. 103–143.
21. Patrick Süskind, *Perfume: The Story of a Murderer*, Penguin, 2010.
22. Ménage, *Consommateurs pris au piège.*
23. Rieunier, *Le Marketing sensoriel du point de vente.*
24. Robert Cialdini, *Influence: The Psychology of Persuasion*, Harper Business, 2006.
25. Ibid.
26. Alka V. Citrin, Daniel E. Stern, Eric R. Spangenberg and M.J. Clark, "Consumer Need for Tactile Input: An Internet Retailing Challenge," *Journal of Business Research*, vol. 56, no. 11 (2003), pp. 915–922.
27. Damien Erceau and Nicolas Gueguen, "Tactile Contact and Evaluation of the Toucher," *Journal of Social Psychology*, vol. 147, no. 4 (2007), pp. 441–444; Bianca Grohmann, Eric R. Spangenberg and David E. Sprott, "The Influence of Tactile Input on the Evaluation of Retails Product Offering," *Journal of Retailing*, vol. 83, no. 2 (2007), pp. 238–245; Heidi J. Hornik, "Tactile Simulation and Consumer Response," *Journal of Consumer Research*, vol. 19, no. 3 (1995), pp. 449–458; Boon Lay Ong (Ed.), *Beyond Environmental Comfort*, Routledge, 2013.

CHAPTER 20

The salesman faced with the neuro-consumer

Sales departments have long been interested in training their staff in customer psychology. They teach them to detect, analyze and understand consumers' subconscious attitudes and reactions. Experienced sales staff know how important this is in making a successful sale. They also know that they must make a good first impression in order to reassure customers and try to obtain their confidence. They have learned that a set of signs – that often have nothing to do with the item to be sold – contribute subconsciously to producing a good impression and encouraging a future purchase. Books on the psychology of selling are legion.[1] They show how the salesman can adapt his appearance and attitude to increase consumers' confidence. They present a set of detectable signs enabling the salesman to see whether the consumer has obtained a positive, satisfactory perception of the arguments he has used. They teach salesmen to adapt their sales pitch to the psychology of potential purchasers and to sell efficiently. They contain a lot of advice to improve commercial effectiveness for selling both consumer goods and industrial products in tender processes or business-to-business (B2B) sales. Neuroscientists and neuromarketers are interested in this question. In Europe, Patrick Georges offers companies advice and training courses aimed at improving the performance of their sales staff. In the United States, Patrick Renvoisé and Christophe Morin[2] use a method they call "Selling to the Old Brain" to improve the skills of sales staff. Both approaches have been designed using knowledge from neuroscience.

Sales-staff training programs

In Europe, after carrying out an ergonomic review of sales-staff working methods, Patrick Georges and his teams present a set of methods and techniques to decode how the brains of both the customers and sales staff function in a sales situation. After carrying out the ergonomic analysis, they identify "neuro-compatible" tools aimed at convincing consumers. Experts have developed interesting applications in purchasing ergonomics. They are used in both distribution centers and with major accounts. Georges and his teams have achieved interesting results in industries as varied as pharmacy, telecommunications, digital technology, telephony, etc. and they offer several programs of advice and training drawing on neuroscientific studies.[3]

"Sales Point"

The "Sales Point" method helps increase sales-point turnover through cognitive optimization of both design and sales-staff behavior. It improves the attitude of sales staff in their presentations, including their appearance, sales pitch, body language, etc. It also involves recognizing signs of stress, interest, desire and dissatisfaction in the customer. It enhances sales-staff management (instructions, sales process, motivation) using neuroscientific methods.

"Story" and "Top-Ten Sales"

The "Story" method tries to optimize commercial presentations using emotions and intrigue. It involves characters, decor, stress, joy and fear.

The "Top-Ten Sales" method aims to improve the performance of sales staff by advising them how to observe the customer and have an impact on his purchase decision, but also how to present themselves to him in an agreeable and pleasant manner. It starts by carrying out an ergonomic review of their abilities in different areas: resistance to stress, memorization of sales pitches and products, influence, working under difficult conditions, social intelligence, etc. It covers themes such as:

- What happens in the customer's brain during a sale?
- What happens in the salesperson's brain when he sells?
- How to affect the customer's and the salesperson's memory so that the latter remembers the sales pitch?
- How to dress?
- What mental exercises should the salesperson perform to prepare for the sale?
- How to influence the customer's decision?

The team of professionals uses cognitive techniques that can increase turnover by between 8% and 20% depending on the situation. These techniques include influence, physical contact with the interlocutor, the placement of the voice and the manner of speaking, the use of body language and how to adapt the sales pitch to the type of customer, etc.

"Selling to the Old Brain" (4 D's method)

In the United States, Patrick Renvoisé and Christophe Morin were among the first to use information from neuroscience, in particular studies of how the consumer's "reptilian brain" works, to improve sales. They have invented the "4 D's method," described in detail in their book and briefly outlined below.[4]

Diagnose the customer's pain Diagnosing the customer's pain relates to what neuroscientists refer to as the "self-centered" nature of the human brain, whereby the customer is often more interested in solutions that reduce or eliminate frustrations than in the characteristics of the product or service. Any good sale begins by diagnosing the frustrations (pain) felt by the customer, not only in relation to the product or service on offer, but also in terms of how it will be used. It is comparable to a medical diagnosis providing a clear understanding of the illness and its context before administering a remedy. Frustrations of this sort often reside in the potential buyer's subconscious.

> *In order to illustrate their ideas, the authors mention, among other examples, the case of Domino Pizza, a home-delivery pizza company. Research showed the company that its customers' main frustration was not about how the pizza tasted, nor about whether it would be delivered hot or cold, but rather about not knowing when it would be delivered. In response to this analysis, the company was able to deal with the frustration using an appropriate promise and slogan: "Delivered in 30 minutes or your money back."*

Frustrations have various origins. They may be financial, strategic or personal. They can be revealed by fear, desire, need or dissatisfaction. The 4 D's method shows sales staff how to use efficient dialog in order to analyze them. This analysis enables the salesperson to evaluate how aware the customer is of his own frustrations, their intensity and how urgent it is for them to be reduced or eliminated.

Differentiate the company's claims Differentiation makes use of the brain's attraction towards contrast and novelty. It aims to catch the attention of the primitive brain and reveal the originality of the product or service. Sales staff are advised to present these differences as veritable innovations. They must make the neuro-consumer feel that the product or service is unique, rare and exclusive. They help sales staff explain how their advantages deal as far as possible with the frustrations identified in the preceding analysis. The advantages for the potential customer may relate to technology, proximity, personalization of services, authenticity of products, exceptional after-sales service, etc. that competitors are unable to offer. Sales staff must highlight the singularity and novelty of what is being proposed.

Demonstrate the gain Demonstrating the gain appeals to the brain's liking for tangibility. The salesperson must not simply present the interest of the product or service but prove and demonstrate it in the most factual and concrete manner possible. Value is the difference – as perceived by the consumer's brain – between the advantage of the proposed solution and its cost. If this difference is not perceived as positive, the solution will not be accepted. Interest may be at several levels: financial, strategic, personal, etc. It is more effective when it is based on tangible proofs that can be presented in different ways: evidence or testimonials from credible customers shown in interviews, real or virtual visits of the company, presentation of concrete success achieved thanks to the use of the product or service, concrete demonstrations of the product's effectiveness using prototypes, performance figures based on irrefutable statistics, etc.

When selling turnkey in-house management training "universities" to company directors, Jean-François Guillon, managing director of HEC Exed-CRC, often offers potential customers the opportunity to meet satisfied clients who have bought and experienced this type of service in collaboration with HEC Paris.

Deliver the message to the old brain Delivering the message to the older brain relates to the brain's emotivity. The act of buying, as we have seen in the previous chapters, is largely based on the brain's irrationality. The neuro-consumer's decisions are triggered by emotions. The success of a sale, beyond the rational elements of a tangible demonstration of the interest, lies to a large extent in the ability of the salesperson to trigger an urge. In this last phase, the 4 D's method teaches the salesperson how to use the customer's brain's sensory systems and to involve its self-centeredness. He is taught how to use attention grabbers that will enable him to maintain the alertness of an attentive neuro-consumer's primitive brain and to encourage fundamental elements of the sales pitch to be memorized.

During a sales presentation, sales staff learn how to use Patrick Renvoisé and Christophe Morin's U-shaped information attention and retention curve as a function of time (Figure 20.1).

Renvoisé and Morin's U-shaped information attention and retention curve has been drawn using the observation that the brain remembers more from the beginning and end of a story and more about innovations. In order to make the best use of this curve, the salesperson is advised to talk about the most important elements likely to convince the potential customer in the introduction to his presentation and to restate them in his conclusion. He is trained to use original presentation methods to highlight the most significant suggestions in his pitch. They enable him to demonstrate that the product or service on offer deals with the potential customer's main frustrations. Intelligent use of repetition, humor and brief anecdotes all contribute to encouraging attention and memorizing.

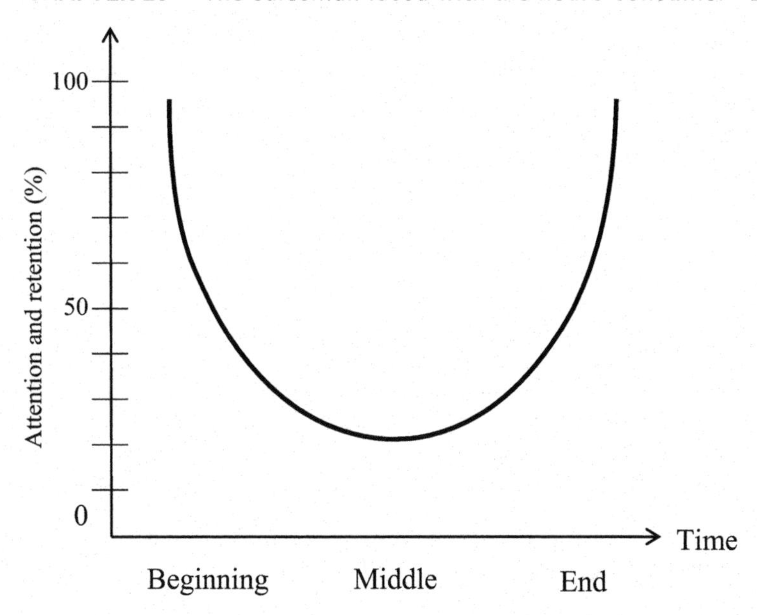

Figure 20.1 Attention and retention graph.
Source: Patrick Renvoisé and Christophe Morin, *Selling to the Old Brain* (2003).

Sales staff learn to use the subliminal effects of their voice and body language. The authors point out that the words in a message account for 5% of its impact on the brain, the voice for 38% and body language for 55%.

Certain methods are suggested for capturing the attention and arousing the interest of the potential customer during a sales interview. The professionals recommend using mini-dramas. Advertising specialist David Ogilvy suggests during his lectures: "If you sell fire extinguishers, start by lighting a fire in front of your prospective customers."[5] He also recommends using play on words, questions that get the customer involved, short anecdotes highlighting how the product or service can be used, etc. The presentation of objects, particularly those that recall childhood and symbolize the solution being provided, help the brain to memorize the pitch.

Involving the brain's self-centeredness makes for a successful sale. As noted above, the customer is often more interested in solutions that reduce or eliminate frustrations than in the characteristics of the product or service. The salesperson places the potential customer at the center of the debate. To achieve this, he includes him as a participant in the demonstration and in the discussion. As American advertising specialist Lisa Kirk puts it: "A gossip is someone who talks to you about others, a bore is someone who talks to you about himself; a brilliant interlocutor is someone who talks to you about yourself."[6] Several methods are proposed to involve the self-centeredness of the potential customer's brain:

- Use "you" instead of "we" and even less "I."
- Ask questions aimed at drawing the potential customer into the conversation and reformulating his objections.
- Ask for his opinion before introducing another argument.
- Solicit positive feedback and his opinion for the next stages of the sales pitch.

Notes

1. Gerhard Gschwandtner, *The Psychology of Sales Success*, Selling Power, 2007; Dennis M. Postema, *Psychology of sales*, Kindle edition, 2013; Andreas Edmüller, *L'Art d'argumenter, convaincre sans manipuler*, Ixelles, 2012; Brian Tracy, *The Psychology of Selling: Increase Your Sales Faster and Easier Than You Ever Thought Possible*, Thomas Nelson, 2006.
2. Patrick Renvoisé and Christophe Morin, *Selling to the Old Brain*, SalesBrain, 2003.
3. Patrick Georges, Anne-Sophie Bayle-Tourtoulou and Michel Badoc, *Neuromarketing in Action: How to Talk and Sell to the Brain*, Kogan Page, 2013.
4. Renvoisé and Morin, *Selling to the Old Brain.*
5. David Ogilvy, *Confessions of an Advertising Man*, Southbank Publishing, 2011.
6. Lisa Kirk, quoted in Renvoisé and Morin, *Selling to the Old Brain.*

Points to remember

- The brain is particularly interested in innovation, which has enabled the survival of the human race.
- Concepts of what is beautiful or ugly are perceived by the insular cortex (also known as the insula) area of the brain. They lead to direct feelings of pleasure or disgust with respect to innovations. Neuroscientific techniques provide information about the brain's positive or negative perception of innovations by observing the reactions of its insular cortex.
- Neuroscientific techniques contribute to improving the success rate in the launch of new products and services.
- The "neuro-aesthetics" of design represent a new line of research for neuroscientific analysis. Neuroscientific techniques enable the effects of design to be evaluated. Their applications are attracting increasing interest from a growing number of multinational companies. The creation of high-performance designs, agreeable to the brain, is an important element that makes a strong contribution to a product's success in the market.
- A number of neuroscientific researchers have noticed, when using MRI, significant variations in brain activity when the subject is shown different types of packaging.
- Certain specific elements of packaging directly attract the brain's attention and interest: images and iconography, colors, writing, size, shape and brand. When these elements are designed to be neuro-compatible, the brain finds them more attractive.
- The brain's perception of how products and services are priced has a significant impact on the amount spent. Pricing that is poorly matched to the brain's perception of value for money can kill a sale. The price perceived by the neuro-consumer's brain provides a fundamental basis for pricing policy. Neuroscientific analysis has made an important contribution to the understanding of this perception.
- Work carried out by behavioral economics researchers over the last few decades using neuroscience has contributed to a significant improvement in our knowledge of the effects of pricing. Various pieces of neuroscientific research attest to the fact that paying creates an impression of pain in the brain. Numerous methods can be used to reduce or eliminate this feeling: delayed payment, proposal of credit or deferred payment or use of a credit card to avoid the impression of actually handing over money, feeling of happiness at possessing the product quickly, arguments from the salesperson centered on possession, explaining that it is a "premium" product, offering grouped products or packs, complete menus in a restaurant, packaging or bottles made to look bigger, invoicing monthly rather than hourly, price fixed a few cents below a given price, etc.

- The brain has significant difficulty in forming an idea of the correct price of a product or service. Good knowledge of how it sets its pricing references helps in persuading potential customers and making sales.
- Pricing has an influence on the brain's perception of the quality of a product or service.
- Various effects coming from the mental strategies adopted by the brain when confronted with pricing strategies have been highlighted by professionals in the field. They particularly emphasize the anchoring effect, decoy effect, endowment effect, sunk cost effect, hyperchoice effect, payment decoupling, self-control effect, loss aversion, denomination effect, framing effect, payment depreciation, immediate discounting, confirmation bias, post-purchase rationalization and the bandwagon effect.
- Pricing must be adapted to the particular characteristics of the brains of specific categories of people, for example spendthrift or thrifty individuals.
- Sensory and experiential marketing at sales points for goods and services influences the perceptions of neuro-consumer's brains. Such places are designed by experts with themes and staging in order to make them more agreeable to the brain. This has the effect of prolonging the duration of customers' visits and increasing the amounts they spend.
- Sensory marketing in boutiques and shops is concerned with the subliminal (unconscious) effect produced in the brain when one of its five main senses is directly solicited.
- There are many examples of successful sales points in different fields when their design, carried out with experts specializing in one of the senses, creates places that are agreeable to sight, hearing, smell, taste or touch. Both personnel and customers say that they feel better in an environment where their senses are skillfully solicited. Personnel in direct contact with customers are more efficient and sell better. The customers feel good, stay longer, buy more, come back, recommend a chain, a brand and a place they feel to be agreeable.
- Sensory strategy is more effective when it is designed to be coherent with the positioning and attributes of the brand and is harmonized with other methods of communication and e-communication.
- Neuroscientific methods and techniques are increasingly used to improve the performance of sales staff. They are now included in training, motivation, work organization and commercial approaches, staff presentation, etc. Programs specifically designed to use knowledge drawn from neuroscience are offered by individuals and consultancy firms, mainly in Europe and the United States. They have evocative names such as: "Top-Ten Sales," "Sales Point," "Story" and "Selling to the Old Brain" and are generating an increasing amount of interest throughout the world from a growing number of companies from different industries.

PART V

The neuro-consumer's brain influenced by communication

Ever since early antiquity, under the influence of the sophists and rhetoricians, and until modern times with semiologists and marketing, communication has used numerous "subliminal" techniques to make itself ever more convincing. These techniques are designed to address the subconscious in the consumer's brain. The recent development of nudge theory and the use of neuroscience offer even more pertinent sensory artifices to convince neuro-consumers. They also endeavor to create "sensory brands" designed to entice customers over the long term.

Introduction

Communication, in its efforts to seduce the neuro-consumer, uses multiple subliminal techniques (beyond conscious perception) designed to address the subconscious directly. Some have been used since early antiquity, particularly in Greece and the Roman Empire. Their aim is to present a convincing discourse through a series of processes and positions used to make the elocution credible and seduce the audience. One of them, Aristotle's rhetoric, has been taught for centuries in schools in the West and is now used in advertising communication.

Advances in marketing research and more modern approaches, such as semiology, endeavor to provide a better understanding of consumers' subconscious behavior in relation to communication. Thanks to the use of neuroscientific techniques, in particular magnetic resonance imaging (MRI) and electroencephalography (EEG), major progress has been made in understanding the brain's reaction to the influence of subliminal advertising messages and their ability to influence it.

Communication experts and citizens are increasingly aware of the effectiveness of subliminal messages. Although prohibited in many countries, they are used in an indirect, legal form that is very effective. The most recent approaches include "nudges" (automatic decisions made by default in the brain and beyond all reasoning), as well as multiple artifices discovered by researchers using neuroscience. Their applications are growing in number with the aim of making the presentation of text, images, videos, websites and information on mobile phones, etc. more attractive for the human brain.

Brands play a fundamental role in influencing the neuro-consumer's brain in subconscious decision-making processes involved in purchasing. Neuroscientific analysis goes beyond marketing research to cast new light on their modes of influence. It highlights the importance of addressing all five senses, not just sight, to foster brand memorization and attractiveness. Neuromarketing is concerned with the creation of sensory brands to make traditional retail chains more enticing to the consumer.

CHAPTER 21

Advertising language

All forms of communication are constantly aiming to seduce the neuro-consumer. This is true both of speeches by politicians hoping to convince citizens of their ideas, and marketing that aims to persuade people of the advantages of the products, services or brands it champions.

To do so, communication has striven for thousands of years to find approaches or methods likely to make it more efficient in its persuasiveness. In early antiquity, Greek philosophers understood that the art of convincing was not limited to addressing the rational part of interlocutors' minds. Pre-Socratic philosophers and then rhetoricians believed that, to convince their audience, rational discourses were not enough. Listeners also had to be charmed and enticed. For this, they invented a new method: "rhetoric." It was taught for a long time in schools in the West and was especially prized in Jesuit education. Modernized by this century's authors, it once again guides reflection on improving the effectiveness of communication.

In addition to this ancient science, research teams develop approaches to provide a better understanding of the neuro-consumer's conscious and subconscious behavior when influenced by different modes of communication. The proposed methods concern writing, images, speeches, videos, etc. and all forms of advertisements used in mass media, events, on the Internet or mobile phones.

Professionals refer to concepts from semiology, innovations in advertising design and, increasingly, neuroscience, with the aim of acquiring a better understanding of what pleases or displeases the neuro-consumer's brain in a message.

Philosophers and communication

In the 5th century BC in Greece, a group of thinkers and professors of eloquence formed, claiming to teach the art of persuasion to make discourses convincing or even irrefutable. They were known as the sophists.[1] Criticized by Socrates and his disciples, their approach to the art of discourse was held in scorn for several decades, before being brought back to the forefront by modern philosophers in the 19th century. Opponents, in particular Aristotle, proposed a new method called rhetoric. It was taught in all European universities before being widely abandoned. It is once again attracting interest among advertising experts, who are constantly looking for ways to provide a deeper understanding of neuro-consumers' reactions to communication messages.

Sophists, true pioneers

Among the most famous sophists were Gorgias (485–380 BC), Protagoras (490–420 BC) and Prodicus of Ceos (470–399 BC). Their lessons, for which they charged a fee that could sometimes be quite high, were designed to increase the effectiveness of discourses presented mainly by thinkers, decision makers and governors, who were generally aristocrats. Sophists not only taught the art of language and dialectics, but also the psychological mechanisms that appeal to the audience's emotions. Some even claimed to be able to make two irrefutable speeches, taking one stance in the first and the opposite in the second. There are few works remaining today that come directly from the sophists. Their teaching is mainly known about through works by their critics, such as Plato reiterating Socrates' discourse in *Gorgias* and *Phaedrus*,[2] and Aristotle, Plato's disciple, in *On Sophistical Refutations*.[3]

The sophists came in for much criticism, mainly from Socrates and his disciples such as Plato and Xenophon, then later Aristotle. Besides the fact they commercialized their knowledge, they were accused of paying no heed to ethical or moral considerations linked to justice or truth in order to guarantee that their arguments were well received by an audience. Socrates and Plato even redefined the connotation of the word "sophism," making it a pejorative term synonymous with "manipulation." This denotation still remains today.

Protagoras and the sophists are sometimes considered the fathers of modern positivism and they affirmed that

> man is the measure of all things ... that nothing is true, all is relative ... that the laws of the city are not guided by what is good in itself, but by that which men are settled upon adopting.[4]

They are also mentioned as being the precursors of "postmodernity" by authors such as Jean Houssaye,[5] a professor and expert in learning sciences.

In a world in which the call for individualism, through the desire for personalized communication messages targeting clients, is much sought, the sophists' lessons are worth looking at again. They cast light on often irrational approaches and arguments that are likely to entice neuro-consumers.

Rhetoric in modern times

The word "rhetoric" is derived from the Greek *rhêtirikê technê* and the Latin *rhetorica*, which can be translated as "oratorical art, technique." For sophists, it is essentially an art designed to persuade. For Socrates and then Plato, who invented the "dialectic," it was to be used to reveal the truth. In the ancient Greek world, this method was largely promoted through discourses by the Athenian statesman Demosthenes (382–322 BC). In ancient Rome, rhetoric was practiced by well-known speakers such as Cicero and Quintilian. It became a science of "speaking well," or *bene dicendi sciencia*. Jean-Jacques Robrieux, a lecturer at the University of Paris Descartes, defines it as "the art of expressing oneself and persuading."[6] For Michel Meyer, a Belgian philosopher at the University of Brussels, "Rhetoric diminishes and alleviates difficulties, which simultaneously fade under the effect of eloquent discourse."[7] It relies on a series of techniques to correctly structure discourses with a view to persuading and convincing the audience. Rhetoric and its methods are described, explained and commented on in a large body of works.[8]

It was codified by Aristotle[9] (384–322 BC). He and his disciples held that discourse should be built on three central notions: "*logos*, *pathos* and *ethos*." *Logos* uses rational elements that allow the speaker to convince the audience through logic. *Pathos* aims to create an emotional connection with the audience in order to charm or seduce. Michel Meyer argues that *pathos* comprises three elements of passion: "The shock-question, the pleasure or displeasure it causes and the modality in the form of judgment it generates, such as love or hate."[10] *Ethos* is based on a dimension that is specific to the orator, namely their authority and competency to address a subject, as well as their morality and exemplary behavior, etc. It corresponds to the image of themselves they convey to the audience as a form of guarantee that makes their discourse credible.

Roman rhetoric is based on Greek principals, but at the same time opts for a more practical approach to discourse instead of prioritizing speculative and theoretical conjectures. It was widely promoted by the orator Cicero (106–43 BC) and the educator Quintilian (35–100 BC). *Rhetoric: for Herennius*,[11] often attributed to Cicero, gives a detailed, formal explanation of the rhetoric system. It provides a summary of Aristotle's arguments but with a more practical understanding that was better adapted to the expectations of Roman eloquence.

Quintilian presents rhetoric as a fundamental science.[12] He proposes five phases of training for rhetoric discourse, which were taught in schools in the West for centuries to follow. They are:

- *Inventio* (invention) consists in finding plausible or even irrefutable arguments to provide evidence to make the speech convincing.
- *Disposito* (or disposition) is linked to the text's structure and arrangement. It organizes the content of the invention and makes the discourse intelligible to convince the audience of the orator's point of view. It ensures that nothing is omitted, excessive repetition is avoided, and reiterations are organized.

- *Elocutio* (elocution) is specific to the drafting of the discourse. It aims to provide the invention with appropriate, powerful words and phrases, adapting the style to the perceptions of the audience in question.
- *Actio* (action) is the phase in which the discourse is pronounced. It enables the orator to adopt physical gestures in accordance with the words he speaks. Cicero spoke of the "elocution of the body." This stage involves teaching on how to use the voice, pertinent and codified gestural behavior, adaptation of tone, speed, breathing, etc. Quintilian believed that rhythm was essential, likening it to music. He speaks of "eurhythmy."
- *Memoria* (memory) teaches the art and methods that ensure the audience remembers the fundamental arguments of the discourse. It uses the structure of the speech itself alongside mnemotechnic methods and techniques to create emotion.

After long being considered and taught as a fundamental science and then neglected, rhetoric enjoyed renewed popularity among modern thinkers and philosophers. For Francis Bacon (1561–1626), it is "the art of applying reason to imagination for the better moving of the will."[13] In the 19th century, the Englishman John Stuart Mill[14] (1808–1873) studied the sophists and developed a classification of their different currents of thought. The German Arthur Schopenhauer[15] (1788–1860) defended the use of sophism for its dialectic effectiveness.

Roland Barthes used it in his research on semiology.[16] The philosopher Chaïm Perelman (1912–1984), a professor at the Free University of Brussels, founded the new school of contemporary rhetoric. In his works,[17] he returns to Greek rhetoric and proposes a "new rhetoric" that constitutes a theory of argumentation.

Advertising rhetoric or the art of persuasion for the neuro-consumer

In his recommendations on rhetoric, Cicero wrote that "it is necessary to prove the truth of what is affirmed (this is the art of *logos*), win the audience's goodwill (by using *ethos*) and awaken in them all emotions that are useful for the cause (through *pathos*)."[18] These arguments form the basis, whether deliberately or otherwise, of many advertising messages often put forward by advertisement creators today. Some make no secret of their keen interest in modern and ancient rhetoricians and study their respective contributions to the art of rhetoric. Thierry Herman and Gilles Lugrin,[19] professors of rhetoric and linguistics at the Swiss universities of Neuchâtel and Lausanne respectively, demonstrate the uses of the three rhetoric approaches – *logos*, *ethos* and *pathos* – in advertising campaigns and messages.

Logos corresponds to the choice of logical, rational and structured messages. These are largely based on success stories. They present irrefutable proof of the specific quality of a product or service or an innovative aspect that cannot be found among the competition. For Revlon, it is "the lipstick that doesn't smudge." This mode of communication offers up tests, illustrations and convincing demonstrations. This type of advertisement is used for multiple products such as washing powder, beauty creams and over-the-counter medication.

The role of *ethos* is to create trust and seduce the audience.

> *It is the guarantee conferred by the presence of the inventor, such as Dyson for bagless vacuum cleaners, Jean Louis David for hairdressing products and Alain Afflelou for glasses. Trust can also come from the presence of an expert, such as a professor in medicine for medication, a famous chef for a food product or a well-known sportsman for sports items. It can sometimes involve more modest figures such as a kindly grandmother for jam or a shepherd for cheese. In order to seduce, supermodels or movie actors are more often used, like Claudia Schiffer for L'Oréal beauty products, Eva Herzigova for Wonderbra bras, Brad Pitt and Alain Delon for Chanel and Dior perfume and George Clooney for Nespresso coffee.*

Ethos is also based on support from a known brand that guarantees quality and an image of luxury, which is liked or admired by a large number of people.

Pathos concerns messages based on emotions. Messages can be built on hedonistic designs targeting the search for inner well-being or pleasure such as luxury and sex. They can also target the desire to please a partner, children, family, friends and acquaintances or society. To provoke strong emotions and remain in the memory they sometimes use violent messages. This is often the case for road safety advertisements, which do not hesitate to use videos showing serious injuries and even dead bodies. The ability to provoke strong emotions is also used by some associations aiming to convince people of their cause.

> *In France, the Association for the Right to Die in Dignity (ADMR) did not hesitate to depict three of the most anti-euthanasia candidates in the 2012 French presidential elections, Nicolas Sarkozy, Marine Le Pen and François Bayrou, on their deathbeds with the words: "Dear candidate, must we put you in such a position to change the vote on euthanasia?" Benetton advertisements, inextricably associated with the photographer Oliviero Toscani, are known for combining strong and transgressive messages, without fear of creating a scandal in some countries.*

Jacques Durand describes rhetoric as the art of "feigned words."[20] Publicists can be considered versatile rhetors who, consciously or not, use rhetoric methods and techniques in artistic creation.

Inventio is sought through marketing studies and surveys on motivation, behavior, sociology, competing communication, etc. *Disposito* concerns the tagline, baseline or slogan, the rationale of the text or film. *Elocutio* aims to find original, attractive designs through visual choices, text, etc. *Actio* is the overall running of the campaign with harmonization of the different media and means used. *Memoria* uses techniques such as arousing emotions and various sorts of artifices to help the audience remember the communication.

After analyzing several thousands of advertisements, Jacques Durand identified all the classic figures of rhetoric as proposed by the major contributors to this science, such as addition, removal, substitution and exchange, and illustrates them using a number of concrete examples.[21] Recent neuroscientific studies confirm the appeal of certain rhetoric figures in advertising applications.

Semiology and decoding communication signs

First developed by linguistics in the 19th century, semiology is once again on the agenda. It has applications in fields that go beyond linguistics, medicine and sociology. It addresses the effect of signs used in advertisements on the perceptions, often subconscious, triggered in neuro-consumers' brains.

Semiology or the empire of signs

In his *Course in General Linguistics*,[22] Swiss linguist Ferdinand de Saussure (1857–1913) defined semiology as "the science that studies the life of signs within society." The term "semiology" was created for medicine by Émile Littré (1805–1888), a French lexicographer and philosopher. The word "semiotics," which is more commonly used outside of France, had been previously proposed in discourses by the American semiologist and philosopher Charles Sanders Pierce (1839–1914).

Although semiology appears to be a fairly recent discipline, signs had already been studied, especially in medicine, in Greece by Hippocrates (460–356 BC). Since ancient times, medical semiology has examined methods for observing and presenting signs to draw up a diagnosis of the patient's illness.

For Roland Barthes (1915–1980),[23] semiology considers everything as a signifier of power. It allows the interpretation of social phenomena, sign systems and the symbolic value of certain social facts. This French literary critic and semiologist considered that the semiology expert's role was to raise *mythos*, a silent and indistinct discourse, to the level of logical argument of *logos*. Modes of mass communication, such as clothing or advertising, are ignorant by themselves and must be deciphered.

The Belgian linguist Eric Buyssens (1910–2000) argued that semiology is "the science which studies the processes we use with a view to communicating our states of conscience and those by which we interpret what is communicated to us."[24]

Semiology and interpreting advertising communication

Semiology for communication is the study of the consciously and subconsciously perceived meaning of signs. These can be conventional, also known as "denotative messages" such as the highway code; railway, maritime and air traffic signals; Morse code; military bells; insignias; musical notation, etc. The elements of these messages have a constant meaning for an enlightened audience that is neither subjective nor open to personal interpretation. Their meaning does not change, regardless of the receiver.

Signs can also come from the "denotative messages," whose meaning largely depends on the receiver and context of their reception.

Olivier Burgelin[25] from the École des Hautes Études en Sciences Sociales in France, examines signs and, more specifically, sign systems in advertising communication. He looks for all the conscious and subconscious meanings a message may contain and studies message associations, which can vary from one reader to the next depending on their culture, social class, nationality, psychological characteristics, etc. A message can be broken down into the signifier, which is what we perceive in the form of text or an image in the message content, and the signified, which is what the message means. Anything incorporated into an advertising message can have a meaning. All elements are taken into consideration including words, text, images, verbal components, page layout, colors, configuration, etc. Semiology proposes grids for analyzing messages,[26] provides an understanding of the mechanisms of persuasion and helps prevent the communication from being ruined by an unwanted element. It sheds light on consumers' often irrational modes of perception in relation to advertising designs.

In the upstream process, it affords a better grasp of the attitudes of target consumers based on signs emitted when they are exposed to images of products or services. The way we stand, speak and move often translates a need that we are not necessarily aware of. Semiologists work with companies to interpret these needs and help propose and present the best offers.

Specialists in semiotics analyze how the meanings of the positioning and brand of a company or its competitors is perceived.

Semiology was largely used in advertising by the French semiologist Georges Peninou (1926–2001). In 1972 he applied it to the analysis of advertisements by advertising firm Publicis, then used this to develop the principles, methods and tools of his theory. He wrote:

> The specificity of the aim of semiology in the analysis of advertising designs is that it endeavors to verify that the designer has correctly fulfilled their obligation to express the signifier-signified function ... The semiologic examination of an advertisement concerns that which is explicitly shown: words and images and, within these images, objects, forms, situations, details, etc. In short, it concerns the emission of signs as well as the significant omission of signs, and within these signs, it is necessary to make the distinction between pertinent and non-pertinent features.[27]

Semiology is primarily a tool for analyzing advertisement designs. It relies on extensive skills used by experts who possess difficult-to-acquire concepts, methods and tools. Besides the reading mentioned in this section, the work by the French linguist Georges Mounin (1910–1993), *Semiotic Praxis: Studies in Pertinence and in the Means of Expression and Communication*,[28] provides an initial starting point for readers interested in finding out more about this science. Readers can also refer to *Semiotics, Marketing and Communication: Beneath the Signs, the Strategies* by Jean-Marie Floch.[29]

Research in communication and subconscious perceptions

At the end of the Second World War, publicity turned into advertising. A group of experts in the field, first in the United States and then in Europe, engaged in reflections and research to improve its efficiency by adapting messages to consumers' inner expectations.

Priority was first given to the use of marketing studies based on surveys, interviews, etc. and their applications to obtain a better understanding of consumers' attitudes when confronted with advertising. Some professionals soon began to realize that research based on the belief that clients responded and behaved rationally lacked foundation in the field of communication. In the United States, David Ogilvy[30] was one of the first to warn of the limits of marketing studies. In the same country, Rosser Reeves[31] (1910–1984), who was responsible for the success of Ted Bates, believed that the average consumer's brain could only remember one advertising argument for each advertising campaign. He developed the unique selling proposition (USP) theory, based on three principles:

- Each campaign must make a single proposition to consumers.
- The proposition must be specific, with no competitor able to do the same.
- The proposition must be strong enough to influence the behavior of millions of clients.

In Europe, Henri Joannis[32] examined the evaluation process for an advertisement using rational and irrational elements, Philippe Villémus[33] studied how to judge advertising designs, Jean-Noël Kapferer[34] considered paths of persuasion, Jean-Marie Dru[35] proposed a method to break with convention by creating communications that were "disruptive" for consumers' brains and Jacques Seguéla[36] emphasized the importance of the emotional quotient in advertising.

All these reflections, in addition to other concepts, aimed to provide a better understanding of the intimate behavior of consumers confronted with numerous advertising appeals. To avoid the failure of a communication campaign, which is always very costly, different tests are carried out to determine whether the messages they present effectively meet the brain's expectations among the clients they are destined for. These are called advertising "pre-tests" and "post-tests."

Pre-testing and post-testing

Pre-testing Pre-testing is carried out prior to the launch of the communication campaign. It serves to study consumers' behavior faced with agencies' designs and verifies that advertisements reflect the communication intent. It concerns various elements that convey messages such as television and media advertisements, images in reviews or on posters, the Internet and mobile phone advertisements. In particular, experts study the relevance of elements such as the perception of:

- the *scenario* (how the target clients remembered it, understanding, interest or approval, the presence of elements that hinder understanding or approval, etc.);
- the *message* (understanding, pertinence of the proposed benefit for the customer, brand differentiation with regard to the competition, brand legitimacy in relation to the proposed offer, etc.); and
- *communication patterns* (identification of the brand's codes, complicity between the consumer and the brand, etc.).

Pre-testing aims to evaluate the message's efficiency (benefit for the brand or product, appropriation by target consumers, intention to buy, etc.). It is carried out using qualitative and quantitative studies that are sometimes preceded by semiologic tests.

Qualitative tests are held in the very early phase of the campaign. They allow the creative direction to be validated and choices to be made between multiple propositions. They are largely based on "focus groups," which are often preceded by individual questionnaires filled out by the participants. One-on-one interviews can also be held to prepare questions or develop data derived from the meeting.

Quantitative tests verify that the design effectively reflects the communication intent. They usually involve a hundred or so people and are carried out close to where the advertisement is generally broadcast, such as on the street, in retail outlets or at home on a DVD. They alternate between using open and closed questions. These are often standardized and compared to scoring scales established by communication agencies for different sectors and countries.

Post-testing Post-testing takes place after the creation of the communication. It has no impact on the success or failure of the campaign, but provides a wealth of useful information for the design of future advertisements. It aims to understand the way target consumers perceive and experience messages. The main elements studied concern spontaneous memory (the person questioned is asked which advertisements they remember), assisted memory (the interviewee is presented with several advertisements and asked if they remember them), the advertisement's impact, attribution or link to the brand, approval, incitement to buy, etc. Interviews are the typical method used, carried out at intervals. This is also known as "tracking," which consists of multiple surveys over several months. It establishes comparisons between competing brands in the same sector or identical products or services categories, as well as the mix of perception and efficiency criteria. Post-testing also attempts to establish a link between the communication campaign and the sales reality using "test markets."

Over the past few years, publicists and advertisers have endeavored to understand what takes place in consumers' brains when confronted with messages in communication, e-communication and m-communication. Major progress has been made thanks to the use of marketing surveys on consumer behavior, analyses in the fields of sociology and psychology, modernized concepts of rhetoric and semiology and multiple research projects on the efficiency of communication. Answers are not easy to find due to the complexity of the brain's responses in this field. John Wanamaker (1838–1922), founder of the famous Wanamaker's shops in the United States, declared more than a century ago: "I know that half the money I spend on advertising is wasted, but I don't know which half."[37] This statement is still aired today by various publicists, who sometimes even claim credit for it. It has to be acknowledged that in spite of the significant progress made in marketing and communication in terms of understanding consumer behavior, Wanamaker's remark remains partially true.

Communication and classical marketing surveys are limited by the traditional methods and techniques used in this discipline. Bernard Roullet and Olivier Droulers highlight that, in the 1980s, researchers were already beginning to rebel against the rational or informative advertisements often produced for shampoo, washing powder and other cleaning products. They underline the particular importance of feelings and emotions in communication. The authors believe that "we are constantly immersed in an emotional pool; the 'non-emotional' state does not exist." Techniques based on questionnaires "only allow us to explore the implicit part. Part of the elements remembered by the individual is not accessible through questionnaires." It is therefore not surprising to observe that most neuromarketing survey firms are interested in this topic. The neuroscientific approach "allows us to 'objectively' address the role of emotions in both the appreciation of advertisements and their memorization."[38]

Neuroscience for understanding the consumer confronted with advertising

Studying the effects of advertising on the neuro-consumer's brain is one of the main focuses of neuromarketing consultancy firms. The demand is particularly high among big international advertisers who devote sizeable budgets to communication with sometimes disappointing results. They demand that their agencies improve methods to allow them to maximize the efficiency of advertisements among consumers. Nielsen, the top international marketing survey firm, seems to have perceived this need back in 2010 when they acquired NeuroFocus, a leader in the use of neuroscience to evaluate communication. In Europe, Ipsos, an international leader in market surveys, in partnership with Neurohm, proposes the use of neuroscience as part of their implementation reaction time (IRT) program. Neuroscience-based research will soon begin to complement the system of measures already used among communication consultancy agencies and market survey firms and to attempt to understand the effects of advertising on neuro-consumers' behavior.

The first experiments by researchers using electroencephalography (EEG) or electromyography (EMG) for the evaluation of campaigns date from the 1970s–1990s.[39] The range of techniques employed to analyze the pertinence of messages has increased more recently. Most of the neuroscientific tools described above are now used alone or in pairs. They provide better knowledge of the effects of communication on target neuro-consumers' brains.

Professor A.K. Pradeep from NeuroFocus posits that thanks to neuroscientific analysis it is becoming possible to scientifically predict whether a communication campaign is likely to succeed or fail, and to explain the causes. Using EEG, sometimes combined with other neuroscientific techniques, he analyzes a series of elements that are essential to a campaign's success.[40] These include the following:

- Attention, emotions and level of memorization perceived directly in the brain during each second in which the advertising message is displayed. Particular attention is paid to the first five seconds, to observe whether it

attracts attention and interest, as well as the final seconds to see whether it correctly gets across the value proposition or brand attribution.
- Elements that become strongly ingrained in the subconscious and those that are only a little or not at all.
- The degree of effectiveness on the brain linked to the repetition of messages.
- The effect of compressing messages (shorter duration of the same advertisement) on the different criteria of a communication's efficiency (attention, interest, impression of newness, memorization, intention to buy, etc.).
- Sensory stimulation of the brain. Does an advertisement stimulate the senses such as taste, touch and smell?
- Activity of the mirror neurons implying, for example, a desire to purchase or consume a product.
- Elements in the message that generate the highest level of engagement in the brain leading to a stronger intention to buy.
- Perception of the coherence between sound (music, voice, etc.) and images (people, visualization of the storyline, etc.) presented in the communication. The auditory cortex and the visual cortex in the brain are constantly interacting. The perception of disharmony between the two is very quickly perceived as a negative discordant element.
- Legitimacy and credibility of the storyline and actors.

For Pradeep,

> The brain has specific preferences. There are things it likes and things it doesn't like. We have identified these factors for brands, products, packaging and distribution, and we have found them for advertising. As we explain to advertising practitioners, neurological tests provide knowledge and suggestions which can help develop more effective messages. In our view, science can serve both art and commerce in this field. As consumers, we can all benefit from better advertising, if we define "better" as more relevant, reliable, interesting, informative, distracting, memorable and, consequently, motivating.[41]

Many elements relating to communication have been predicted through reflections produced by rhetoric and semiology studies, as well as through the extensive traditional research on communication efficiency. Neuroscience provides a direct view of reactions in the neuro-consumer's brain. It allows us to verify, shed light on and sometimes complement hypotheses and suggestions produced by previous approaches. Neuroscientific experiments are used in particular to provide further knowledge on the effects of advertising communication in certain domains. They notably concern the role played by mirror neurons, emotions, memory, creative context, etc.

Mirror neurons and communication

Mirror neurons play an important role in the creation of messages. In her research using neuroscientific tools, Sophie Lacoste-Badie,[42] a lecturer at the University of Rennes I, shows that advertisements in which the product is handled by the main character are better retained in terms of memory and recognition. She also notes that the observed consumers were more favorable toward this type of communication.

Advertising efficiency linked to the imitation of choices made by a famous person featured in the communication is another source of interest. Benjamin Neumann,[43] head of communication at the Agence Française de Développement (AFD), indicates that certain actors or elite athletes reap the benefits of the value of this mirror-neuron effect among consumers. Actors such as Nicole Kidman and Brad Pitt or golfer Tiger Woods have contracts for sums nearing $5 million. According to the same author, footballer Zinédine Zidane receives between €1 and 1.5 million per year to represent brands such as Danone, Adidas, Orange, Ford and Generali. Various research projects using MRI, such as the one carried out by Vasily Klucharev and colleagues at the University of Basel in Switzerland,[44] show that the legitimacy of the celebrity plays an important role in the evaluation and memorization of the product promoted. The connection between a product and a person such as the tennis player Roger Federer is better remembered when there is a coherent link between the product and the player's image. The same goes for film stars, supermodels, etc.

The role of emotions and memory

Neuroscience-based advertising research confirms the importance of emotions in the effectiveness of advertising communication. In the 1970s, Albert Mehrabian and James Russel,[45] professors at the University of California, Los Angeles (UCLA), examined the role of surrounding stimuli on the buyer's emotions. In the late 1990s, neurologists Larry Cahill and James McGauch,[46] also at UCLA, studied neuronal mechanisms by examining the activation of the right amygdala in the brain, which is essential for remembering emotional events, using positron emission tomography (PET). In an experiment they showed that extracts from emotional films were better remembered and consolidated in the observed individuals' memory than extracts from neutral films. Tim Ambler, professor at London Business School (LBS) in the United Kingdom, and Tom Burne, professor and neurobiologist at the University of Queensland Centre for Mental Health Research in Australia,[47] carried out research in which they exposed subjects to predominately informative or emotional advertisements. The messages were incorporated in a documentary. They observed that the emotional advertisements were better remembered than the informative ones. During a conference in Toronto in Canada, Olivier Droulers and Bernard Roullet[48] explained that advertisements were better remembered when inserted into violent programs than when they were incorporated into neutral programs. All these studies attest to the important role of emotions in the memorization of advertising messages.

The creative context and its influence

The context in which an advertisement is presented plays an important role in ensuring its credibility. More specifically, this means the image of the media or material it is included in. This observation has already been made by rhetoricians, and has also been observed in neuroscientific research. In Europe, neurology professor Michael Deppe[49] and colleagues at the University of Münster in Germany call this phenomenon the "framing effect." After studying advertising messages in four well-known magazines in Germany and asking whether the same ideas were true or false, they observed an increasing difference in credibility depending on each magazine's image of reliability among the German public. The experiment reveals the subconscious influence that a medium or material can have on a message's perceived reliability. The credibility of a message is drawn from consideration of surrounding elements, which can significantly influence the perception of a rational message in the communication. After being analyzed for the press, this fact has also been observed for radio and television channels and social media. It provides food for thought for advertising on the importance of the consistency that must be established between the choice of media, the material and type of message proposed. Communication and marketing experts are already aware of this need, which has been confirmed by observations using neuroscientific techniques of direct responses in neuro-consumers' brains.

Notes

1. John Dillon and Tania Gergel, *The Greek Sophists*, Penguin London, 2003; Jean Houssaye (Ed.), *Premiers pédagogues: de l'antiquité à la renaissance*, ESF, 2002; Jean Dumont, Daniel Delattre and Jean Louis Poirier, *Les Présocratiques*, Gallimard, coll. "Bibliothèque de la Pléiade," 1988.
2. Plato, *Complete Works*, Hackett, 1997.
3. Aristotle, *On Sophistical Refutations*, Kessinger, LLC, 2010.
4. Ibid.
5. Houssaye, *Premiers pédagogues.*
6. Jean Jacques Robrieux, *Rhétorique et argumentation*, Armand Colin, 2010.
7. Michel Meyer, *What Is Rhetoric?* Oxford University Press, 2017 and *Histoire de la rhétorique des Grecs à nos jours*, Le livre de Poche, coll. "Biblio Essais," 1999.
8. Laurent Pernot, *Rhetoric in Antiquity*, Catholic University of America Press, 2005; Anthelme Édouard Chaignet, *La Rhétorique et son histoire* [1888], Hachette Livre BNF, 2012.

9. Aristotle, *Rhetoric*, Dover Publications, 2004.
10. Michel Meyer, *Rhétorique*, PUF, coll. "Que sais-je," 2011.
11. Cicero, *Rhetorica Ad Herennium*, Loeb, 1989.
12. Quintilian, *The Institutio Oratoria*, Nabu Press, 2010.
13. Bacon quoted in Meyer, *Rhétorique*.
14. John Stuart Mill, *A System of Logic: Ratiocinative and Inductive*, University Press of the Pacific, 2002.
15. Arthur Schopenhauer, *The Art of Always Being Right: The 38 Subtle Ways to Win an Argument*, Gibson Square Books, 2009.
16. Roland Barthes, *Elements of Semiology*, Atlantic Books, 1997.
17. Chaim Perelman, *The New Rhetoric and the Humanities: Essays on Rhetoric and Its Applications*, Springer, 1979.
18. Cicero, *Rhetorica Ad Herennium*.
19. Thierry Herman and Gilles Lugrin, "La Rhétorique publicitaire ou l'art de la persuasion," *Commin-Com Analysis*, Lausanne, 2001.
20. Jacques Durand, "Rhétorique et image publicitaire," *Communications*, vol. 15 (1970), pp. 70–95.
21. Ibid.
22. Ferdinand de Saussure, *Course in General Linguistics*, Columbia University Press, 2011.
23. Roland Barthes, "Rhétorique de l'image," *Communications*, vol. 4 (1964), pp. 40–51, "Le message publicitaire," *Cahiers de la publicité*, vol. 7 (1963), pp. 91–96 and *The Empire of Signs, Anchor Books*, 1983.
24. Eric Buyssens, *Speaking and Thinking from the Linguistic Standpoint*, North-Holland Publishing, 1954.
25. Olivier Burgelin, "Sémiologie et publicité," *Cahiers de la publicité*, vol. 15 (1965), pp. 98–104.
26. Laurent Gervereau, *Voir, comprendre et analyser les messages* (4th ed.), La Découverte, 2004; Dominique Bounie, *Sémiologie de l'image*, Politech' Lille, 2006.
27. Georges Peninou, *Intelligence de la publicité: Étude sémiotique*, Robert Laffont, 1972.
28. Georges Mounin, *Semiotic Praxis: Studies in Pertinence and in the Means of Expression and Communication*, Springer, 1985.
29. Jean-Marie Floch, *Semiotics, Marketing and Communication: Beneath the Signs, the Strategies*, Palgrave Macmillan, 2001.
30. David Ogilvy, *Confession of an Advertising Man*, Southbank Publishing 2011 and *Ogilvy on Advertising*, Crown Publishing, 1983.
31. Rosser Reeves, *Reality in Advertising*, Random House, 1988.
32. Henri Joannis, *Processus de création publicitaire* (3rd ed.), Dunod, 1991.
33. Philippe Villémus, *Comment juger la création publicitaire*, Éditions d'Organisation, 1997.
34. Jean-Noël Kapferer, *Les Chemins de la persuasion*, Dunod, 1991.
35. Jean-Marie Dru, *Disruption: Overturning Conventions and Shaking Up the Marketplace*, Wiley, 2008.
36. Jacques Séguéla and Christophe Haag, *Génération QE*, Pearson Editions, 2009.
37. Jacques Lendrevie and Julien Levy, *Mercator*, Dunod, 2012.
38. Bernard Roullet and Olivier Droulers, *Neuromarketing: Le marketing revisité par les neurosciences du consommateur*, Dunod, 2010.

39. Sydney Weinstein, Ronald Drozdenko and Curt Weinstein, "Brain Wave Analysis in Advertising Research," *Psychology & Marketing*, vol. 1, no. 3–4 (1984), pp. 83–95.
40. A.K. Pradeep, *The Buying Brain: Secrets for Selling to the Subconscious Mind*, Wiley, 2010.
41. Ibid.
42. Sophie Lacoste-Badie, *La présentation du packaging dans les annonces télévisées: étude des réponses mémorielles et attitudinales des consommateurs*, PhD thesis, University of Rennes, 2009.
43. Benjamin Neumann, "Leur image, c'est leur capital," *L'Expansion*, January 1, 2006.
44. Vasily Klucharev, A. Smidts and G. Fernandez "Brain Mechanism of Persuasion: How Expert Power Modulates Memory and Attitudes," *Social Cognitive and Affective Neuroscience*, vol. 3, no. 4 (2008), pp. 353–366.
45. James Russel and Albert Mehrabian, "Evidence for a Three-Factor Theory of Emotions," *Journal of Research in Personality*, vol. 11, no. 3 (1977), pp. 273–294.
46. Larry Cahill and James Mc Caugh, "Modulation of Memory Storage," *Current Opinion in Neurobiology*, vol. 6, no. 2 (1996), pp. 237–242.
47. Tim Ambler and Tom Burne, "The Impact of Affect on Memory of Advertising," *Journal of Advertising Research*, vol. 39, no. 2 (1999), pp. 25–34.
48. Olivier Droulers and Bernard Roullet "Does Crime Pay for Violent Program-Embedded Ads?" *Advances in Consumer Research*, vol. 31 (2004), pp. 646–651.
49. Michael Deppe, W. Schwindt, J. Krämer, H. Kugel, H. Plassman, P. Kenning and E.B. Ringelstein, "Evidence for a Neural Correlate of a Framing Effect: Bias-Specific in the Ventromedial Prefrontal Cortex During Credibility Judgments," *Brain Research Bulletin*, vol. 67, no. 5 (2005), pp. 413–421.

CHAPTER 22

Subliminal influences of communication on the brain and the nudge concept

Subliminal communication provides direct access to the subconscious brain. Deprived of the conscious barriers installed by the neocortex, the neuro-consumer is subject to subconscious and often undesired influences from advertising communication. For a long time, the use of elements such as entertainment, art and eroticism has shown the subliminal effect obtained on the brain. More recently, "nudges" have been presented as a noncoercive means of leading people to involuntarily make decisions they have not chosen. Neuroscience has shed light on how the tricks of written, oral or video communication can guide neuro-consumers' subconscious choices.

Significant precautions are necessary to ensure that the use of these methods does not overstep moral and ethical rules. Beyond the legal aspects, such precautions are essential to limit their application to what is appropriate.

Subliminal communication to address the subconscious brain

The word "subliminal" comes from the Latin *sub* (under) and *limen* (threshold), and applies to communication that is below the conscious brain's perception threshold. At the cinema, on television and in other media, a subliminal image is one that the neuro-consumer does not have the time to perceive visually, but that he remembers because it is imprinted in his brain. The word subliminal qualifies a visual or auditory stimulus deliberately designed to be directly stored in the brain without going through the checks provided by the subject's intelligence. There is a risk that it will influence the subject without his knowing it.

The effectiveness of subliminal communication

In 1957, James MacDonald Vicary (1915–1977), a marketing manager from New Jersey in the United States, started a big debate about the effectiveness of subliminal communication. Vicary maintained that, by inserting subliminal spots in a film shown at the cinema, such as "drink Coca-Cola" and "eat pop-corn," sales of these products increased by 18% and more than 50% respectively. The story created a big sensation in America, provoking many publications and debates. The Central Intelligence Agency (CIA) was ordered to investigate. In 1958 subliminal messages were banned in television advertisements in the United States, Australia and the United Kingdom. This ban has since been applied in most countries in the world. After investigation by scientific experts, the author admitted a few years later that he had invented the data. The debate on subliminal communication has since been the source of numerous controversies.

Dutch researchers, such as Johan C. Karremans,[1] professor at the University of Nijmegen, and his colleagues, have conducted experiments showing that subliminal presentation of messages only has an impact under very limited conditions. Drew Western,[2] professor in the Department of Psychology and Psychiatry at Emory University in Atlanta, shows that subliminally presented stimuli influence thought and emotions. They are also likely to attract attention, as described by psychologist Stanislas Dehaene, a professor at the Collège de France, and Olivier Naccache,[3] a doctor at La Pitié-Salpêtrière hospital.

Despite the ban, the debate continues about the continued use of subliminal messages in certain cinema films to reinforce levels of anxiety or attention. They are also said to be present in a few television and radio series. The Conseil supérieur de l'audiovisuel (CSA), the French audiovisual regulatory body, having observed the phenomenon several times in 2002 and 2003, carried out investigations and made recommendations.

Communications advisers to politicians are often accused of using subliminal communication tactics. In the United States, during Georges W. Bush's campaign in September 2000, a television spot is said to have been broadcast in which the word "rats" was inserted just after a photo of his political opponent, Al Gore.[4] In 2008, Internet users spotted a photograph of the Republican candidate John McCain and his wife Cindy in the opening titles of the Fox 5 News channel.[5] In France, the journalist Jean Montaldo, quoted by Vladimir Volkoff,[6] affirmed that, in 1988, not long before the presidential campaign, a photo of François Mitterrand was slipped discreetly into the opening titles of the news broadcast on what is now France 2.

The influence of subliminal messages has been the subject of many publications. Among the best known is *The Hidden Persuaders* by economist, sociologist and writer Vance Packard (1914–1996).[7] In his book, the author denounces the growing influence in the United States of techniques developed by research institutes, advertising agencies and advertisers to plumb consumers' subconscious in order to persuade them to buy what they have not chosen or to win them over to a political cause. The author carried out an in-depth investigation into the methods and gives examples that attest to the surprising efficiency of this manipulation. When the consumer's or citizen's mind has become conditioned, it is questionable whether he is acting under his own free will. Originally intended as an analysis of American society, Packard's book makes a big contribution to critical understanding of people's behavior in consumer society, as well as that of citizens in a modern democracy.

Apart from concealed messages, subliminal communication reaches directly into the consumer's subconscious brain by other means. While remaining legal, this is no less effective. We have already mentioned the subliminal effect that can be produced in the neuro-consumer by a simple, natural smile from a shop assistant or salesperson. Neuroscientific analysis confirms their influence on the brain. Among those frequently mentioned are entertainment, art and eroticism.

Entertainment in advertising

Entertainment comes from the Latin *inter* meaning "among," and *tenere* "hold," thus to keep in a certain frame of mind. It describes activities that enable people to take their minds off their essential preoccupations. It takes different forms depending on the place, period and culture. Among the most ancient forms of entertainment are games, music, dance and painting. Among the most recent are photography, cinema and Internet gaming on tablets and mobile phones.

The importance of entertainment was brought to light by the philosopher Blaise Pascal (1623–1662) in his *Pensées*. For Pascal, entertainment is essential to allow people to escape from their everyday lives. He wrote:

> If we leave a king alone without any satisfaction of the senses, without any leisure of the mind, without company and without entertainment, think of him at his leisure, and we will see that a king without entertainment is a man full of misery.[8]

In 1947, the writer Jean Giono (1895–1970) used Pascal's ideas in his novel *A King Without Distraction*.[9]

Neurologists agree that entertainment, like sleep, plays an important role, allowing the brain to rest while it sorts the information that is important and useful to it. Advertisers have understood its importance in producing their campaigns and some think that it can increase the impact of messages by associating them with a fun moment or entertaining content, often using buzzwords such as "advertainment" or "entertising."

One of the proponents of entertainment in advertising is Jean-Marie Dru, chairman of the TBWA Worldwide agency. He is the author of several nonconformist books about communication.[10] Dru argues that in order to interest consumers in advertising messages it is necessary to first entertain them. He writes: "The only way to interest them is to come up with an ad that enchants and amuses them."[11]

"Advertainment" is attracting increasing interest in many forms of communication, mainly cinema, television and the Internet. Humor, which triggers pleasure through laughter, attracts the brain's interest. The use of stories and plots in campaigns linked to the promotion of brands, products or services tries to render the publicity more attractive than purely demonstrative or rational advertisements. Advertainment tries to create an agreeable experience, capable of entertaining and informing at the same time. It is frequently used in communication on the Internet and enables websites to generate traffic, recruit an audience and keep it loyal. It is present on social media with the creation of humorous films, often essential for messages to be dispersed widely among the online community.

To our knowledge, little neuroscientific research has been carried out into the level of interest generated by an entertaining advertisement compared to an informative one, thus representing scope for future neuroscientific studies.

Art in its different forms is often considered to be an important type of entertainment, as is eroticism. These two themes are the subject of numerous debates among communication experts. Neuroscientific research is interested in the subliminal role that they could play.

The role of art in subliminal communication

For a long time, philosophers have believed, contrary to Plato in *Ion*[12] and *Hippias Major*,[13] that art is not a simple representation of nature. This was already the case for his disciple Aristotle (384–322 BC), who wrote: "Art not only imitates nature, but also completes its deficiencies."[14]

For many philosophers and thinkers, art enables us to speak directly to the soul, to attain the absolute. For example, Immanuel Kant (1724–1804) states: "Art enables us to attain the absolute in ourselves," something he calls the "noumenon."[15] Georges W.F. Hegel[16] (1770–1831), meanwhile, claims that art, particularly old styles like Greek and Flemish, can reveal the absolute. Art addresses itself to the "spiritualized sublime." It manifests the spirit. For Hegel, Greek art is a real union between the tangible and the spiritual. In 1935, Martin Heidegger[17] suggested that only art could show us the truth of the being. Sigmund Freud substituted the idea of the subconscious for that of the soul. For the painter Vassily Kandinski (1866–1944), "art, and in particular abstract art, is a language of the soul, and is alone in having that capacity."[18]

As we have already seen in Chapter 17 regarding design, 21st-century neuroscientists are showing greater interest in research into the relationship between art and the brain's perceptions. Professor Sémir Zeki[19] is studying the correlation between the idea of beauty and the activation of areas of the brain. Neuroscientist Jean-Pierre Changeux has carried out studies in conjunction with musician Pierre Boulez[20] (1925–2015). They consider that the brain is able to recognize the harmony that exists between the whole and one part of a work of art.

Neuroscience professors Cinzia Di Dio, Emiliano Macaluso and Giacomo Rizzolati[21] claim, from research using cerebral imagery, that harmonious proportions exist in works of art and are likely to produce neuronal activation, creating a sensation of beauty in the observer's brain. Their research focuses in particular on the harmony of proportions that follow the rule of the "golden number."

The "golden number" and "Vitruvian proportions"

The "golden number" is a ratio that has been known since ancient times. It was defined initially in geometry and then in algebra. It is: 1.6180339887 to 1.

$$\frac{1+\sqrt{5}}{2} = \textit{golden number}$$

The harmony of the great pyramid of Gizeh in Egypt and the Parthenon in Athens, for example, is based on an architectural calculation that respects this ratio. It is first mentioned by the Greek mathematician Euclid (around 300 BC) in his writing about architecture.[22]

An Italian Franciscan monk, Luca Bartolomes Pacioli (1445–1517), brought it to prominence during the Renaissance.[23] He called it the "divine ratio," associating it with an ideal communicated by God. The mysticism attributed to the golden number developed over the centuries, at the same time being the subject of much argument. Certain artists elevate it to the level of an esthetic theory speaking directly to the subconscious. We find it in many art forms including architecture and sculpture, but also in poetry, painting and music, etc.

In *The Golden Number: Pythagorean Rites and Rhythms in the Development of Western Civilization*, frequently considered to be the origin of the modern myth of the golden number, the Romanian prince and artist Matila Costiesco Ghyka (1881–1965) wrote: "Archaeology offers us proof that the golden number's epitomizing of beauty is universal."[24] Many works of architecture, sculpture and painting, such as the theater of Epidaurus, the Temple of Solomon, the façade of the Parthenon, the Sistine chapel, Polykleitos' Athlete and Botticelli's *The Birth of Venus* are all considered to have been conceived by respecting the divine proportion that comes from the golden number.

> *Leonardo da Vinci, in his famous Vitruvian Man (1600) – incidentally used as a logo by the human resources company Manpower – like Albrecht Dürer and other artists later, refers more to the esthetic rules that use the proportions decreed by Vitruvius (from 90–20 BC), an architect living in Rome, in his treatise De Architerctura.*[25]

Many 20th-century artists, motivated by the work of Matila Ghyka, claim to have been inspired by the "golden number." The ratio is found in the poetry of Frenchman Paul Valéry (1871–1995), the music of the Greek Xenakis (1922–2001), the painting of Spaniard Salvador Dali (1904–1989) and the architecture of Le Corbusier (1887–1965).

In order to find out whether the brain judges the harmony of the golden number as being more esthetically pleasing, Italian neuroscientists Cinzia Di Dio, Emiliano Macaluso and Giacomo Rizzolati[26] carried out an experiment. They used cerebral imagery to observe the reactions of subjects' brains when they were shown three versions of Polykleitos' famous *Doryphoros* sculpture.

In two of the versions, the proportions were slightly modified, no longer corresponding to the golden number that is the basis of the masterpiece's creation. The researchers observed that the insula, an area of the brain relating to rejection and disgust, activated in the great majority of subjects when they looked at the images of the sculpture with altered proportions. They concluded that the brain does perceive an objective beauty. They suggest that the harmony of the golden number provokes a specific neural activation pattern underpinning the notion of beauty in the observer's brain.

Neuroscientific studies show that the brain perceives ideas of harmony and beauty directly. This confirms the thinking of many philosophers concerning the subliminal role of works of art. Communication experts have been aware of this phenomenon for several years. They often call upon well-known painters, directors and composers to help create visual or auditory logos for brands, improve the music in radio advertisements, illustrate posters or direct advertising films. If new neuroscientific studies confirm the existence of a correlation between art and the subconscious, advertising agencies will be even more ready to bring in creative artists to improve the quality of their campaigns by striking a chord with deep-seated subconscious expectations in neuro-consumers' brains.

The subliminal role of eroticism in the perception of messages

Food and sex are two important priorities for the human brain. This goes back to the dawn of time and involves two fundamental needs: ensuring the survival of the individual in the first case, and that of the species in the second.

As early as 1905, Sigmund Freud, the founder of psychoanalysis,[27] established a close relationship between many aspects of human behavior and strong subconscious sexual tendencies. Freud considered that life is built around tension and pleasure. The tension is due to the "libido" or sexual energy; all pleasure comes from its release. The tendency towards all kinds of perversion exists as a subconscious force.

Very early on, advertising specialists understood empirically how important eroticism could be in communication. Advertisements including people or objects evoking sexuality or passion directly, but also symbolically or subliminally, appeared in the 19th century.

> *For instance, the advertisement of the Robette absinthe liquor dating from 1896 shows a naked woman in a transparent garment offering a glass full of the drink. The seductive clothing has a strong sexual connotation although there is no obvious link between the woman, the product and the brand.*

In 17th-century Europe, wooden sculptures and images of attractive, naked women were used in drinking establishments to attract customers. In the 19th century, the walls of big cities such as New York, London and Paris displayed advertising posters of an erotic nature. They showed seductive women offering a variety of products and services, such as drinks, medicines, books and shows. The artists of the period such as Jules Chéret, Henri de Toulouse Lautrec and Alfons Mucha were much in demand to produce them.

> *In 1910, sales of a women's beauty soap, Woodbury Facial Soap, were declining. New publicity showed romantic couples associated with a promise of love and intimacy for the women that used the soap. Sales quickly took off again.*[28]
>
> *The erotic aspect of the male is also used, for example a half-naked man to promote Barnum's circus. Later, he is dressed, but in such a way as to express his virility or sensuality. Other examples include the famous Marlboro cowboy, actors such as George Clooney for Nespresso coffee and Brad Pitt for Chanel No. 5 perfume.*
>
> *Some brands regularly use eroticism in their advertising. Calvin Klein produced one of its first controversial advertisements showing actress Brooke Shields, then aged 15, wearing jeans and saying: "What gets between me and my Calvins? Nothing."*[29] *In recent years, Sisley (Benetton group) has used the codes of chic porno and "trash" in its advertising campaigns. Diesel has used provocation to show that wearing jeans made the wearer at ease with their sexuality. Its competitor, American Apparel, bares its models to extol the merits of its clothes.*

The use of eroticism in advertising is particularly widespread. One-fifth of all the advertising in the world has a link with eroticism or sexuality. A great number of the viral films forwarded within online communities are directly linked to sex. Eroticism is used for a very large category of products and services that have, nevertheless, no direct relationship to it.

> *A few years ago, in the United States, it was even used to sell coffins. A very scantily clad young woman is lying on the coffin with the slogan: "You supply the body and we supply the rest." Organizations also use sex to draw attention to their causes. In France, Gérard Andureau, chairman of the association DNF (rights for nonsmokers), defends the production of a noteworthy advertisement showing that tobacco, like fellatio, represents a symbol of submission. In the Catalonia region of Spain, the Young Socialists, whose objective was to eliminate voter abstention, ran a campaign with a spot entitled "Voting Is a Pleasure," showing a woman placing her ballot paper in the box while in the middle of having an orgasm. Both campaigns were the subject of protests.*[30]

Does sex really sell?

Neuroscientists agree that eroticism has a subliminal effect on the brain, while advertising specialists empirically observe its power of attraction. However, one question arises more and more frequently: Does the use of sex in advertising actually increase sales? Opinions are divided.

The forms used by eroticism have been found to have different effects on male and female brains. Whereas explicit or suggested female nudity seems to attract and please the male brain, studies suggest that the image of a completely naked man does not seem to have the same appeal for the female brain. A study carried out in the United States in 2005 by MediaAnalyzer[31] involving 200 men and 200 women, performed with the help of eye tracking, shows a difference in eye movement and concentration of the two sexes as they look at the same posters and newspaper advertisements. Men look first at the model's bust or face, then at the product or the catchphrase. Women look at the model as a whole, then the product and the catchphrase. In terms of sensual perception, 48% of the male subjects found the advertisements sexy and interesting, compared to only 8% of the females.

An awareness of the apparent distinctions between male and female brains is essential to adapt the forms of eroticism used in advertising to each gender's subconscious perceptions. Erotic messages seem to have a positive effect when there is a relationship between the products or services on offer and a desire for seduction – feminine or masculine – during their use. This includes advertisements for underwear, jeans, perfumes, cosmetics, alcohol, cigarettes, holiday clubs for single people, etc.

When there is no link between the erotic message and the offer, the effect is less obvious and the message may even have a detrimental effect on the product or brand. Two professors from University College of London, Elie Parker and Adrian Furnham,[32] show that too much sex in an advertisement can disturb the viewer's concentration and change the process of memorizing the brand. They also note that these phenomena are accentuated when such spots are broadcast during a program with sexual connotations. In 2009, researchers in Canada and the United States found that eroticism is not always effective in selling products.[33]

Neuroscience specialist Martin Lindstrom also questions the effectiveness of sex in today's advertisements. Comparing the current period with the 1980s, he explains that sex has lost a lot of its influence because it is now everywhere. He writes: "It suffices to simply type the word 'sex' into a search engine to find images that are much more shocking than the suggestive campaigns that we can see every day."[34]

"Nudges": a soft method for influencing decisions

By observing and studying how the human brain works, certain communication experts think that the use of soft methods, adapted to their subconscious perceptions, will enable citizens' decisions and choices to be orientated. Among these methods are "nudges." They were first used to propose solutions to problems concerning public health, savings and the environment and are now widely used in politics, in relation to major social causes and marketing.

Nudges to improve citizens' behavior subconsciously and without constraint

"Nudge" – literally meaning "to push someone with the elbow" – expresses the action of leading someone to do something with a little unconscious help. The nudge method was designed to guide people towards taking decisions for their own good – and of their own free will – in terms of health and education.

The concept of nudges – and the method – was developed by two American professors from the University of Chicago: economist Richard H. Thaler, awarded the 2017 Economics Nobel Prize, and lawyer Cass R. Sunstein. In 2010, David Cameron's government invited Thaler to help set up the UK's "Behavioral Insights Team" or "Nudge Unit." Sunstein became an informal advisor during President Obama's presidential campaign. He then took on important administrative responsibilities in the US Federal Regulation apparatus. He was made an administrator of the Office of Information and

Regulatory Affairs (OIRA). The two professors develop their theory in a book that has become a worldwide best-seller: *Nudge: Improving Decisions About Health, Wealth and Happiness.*[35] They call their new doctrine "libertarian paternalism."

Thaler and Sunstein's approach draws inspiration from the work of David Kahneman and Amos Tversky. Kahneman[36] applies psychological insights to economic theory, particularly in the areas of judgment and decision-making amid uncertainty. He studies human irrationality in the field of economics, clearly questioning the assumption of human rationality that has prevailed in modern economic theory for years and the importance of his contribution was recognized when he was awarded the Nobel Prize for Economics in 2002. Tversky is a renowned psychologist and has worked with Kahneman for many years to establish a typology of the cognitive errors systematically made by humans. Kahneman and Tversky are still generally considered to be the founding fathers of behavioral economics, although their ideas in this field were preceded by several theories from pioneers such as Herbert Simon, Maurice Allais, Reinhart Selten, Paul Slovic, Robert Zajonc and Robert Cialdini.

Richard Thaler and Cass Sunstein begin by identifying what it is that leads people unthinkingly into error. Examples include tending to overestimate a risk that in reality has little likelihood of occurring just because it is very much present in the mind, having an excess of optimism and of self-confidence, being afraid of losing and tending to favor the status quo, etc. For example, a study by Thaler and Sunstein found that 90% of the drivers they questioned considered themselves to be above average and 94% of professors at a large university believed themselves to be better than the average professor.[37] Nudges are introduced to compensate for these tendencies. Public policies can make use of these forms of inertia by suggesting default solutions that subconsciously push citizens to take more advantageous decisions. For example, the automatic opening of a savings account leads to a very significant increase in the number of employee savings plans.

Another idea is to lead people to deliberately avoid waste, anticipate mistakes and simplify complex options.

> *To reduce domestic electricity consumption by 40% at peak periods, the nudge method suggests installing an "Ambient Orb" into each household. This is a small lighting ball that turns red when energy consumption is too high. It turns green when consumption is moderate. When the South Californian Edison Company introduced this system, it proved to be conclusive. Nobody obliged the consumers to reduce their energy consumption. Faced with a system that induced guilt, the great majority did it automatically.*
>
> *In London, at each pedestrian crossing, the words "look right" have been painted on the ground so that foreign tourists, who are used to people driving on the right in their countries, are not knocked over.*

An example often quoted by Richard Thaler is that of the fly engraved into the urinals at Amsterdam airport to encourage men to aim better. Thanks to this simple initiative, the cleanliness of the toilets has been significantly improved.

Thaler and Sunstein's book and the nudges it suggests are enjoying real success in the United States and the United Kingdom where the researchers' ideas are making political programs more effective. Their approach, with its links to behavioral economics, has inspired many others, mainly in the United States. Some researchers, including George Loewenstein and Dan Ariely, have written numerous publications and gained major reputations in this "anti-Cartesian" field of thought. Behavioral economists are now developing their approach, inspired by neuroscientific research and the neuro-economy. For example, the 2019 winners of the Nobel Prize in Economics, Abhijit Banerjee (MIT), Esther Duflo (MIT) and Michael Kremer (Harvard University), have studied programs that successfully nudge farmers in the developing world to use more fertilizer.

Nudge marketing and subconscious purchasing decisions

The fields of marketing and communication are using nudges to better orientate consumers' purchasing decisions. Eric Singler,[38] managing director of BVA Group, pioneer of the nudge approach in France and founder of the Nudge France think tank, carries out research in experimental shops that he has designed himself, reconstituting the life of a convenience store. Using specific tools, some of them taken from neuroscientific approaches, the experimental shops enable him to study the reactions of consumers confronted with a variety of marketing and communication stimuli. Having presented a major synthesis of the components of behavioral economics and its value, he proposes several methods including those of the BVA Nudge Unit. They are based for the most part on teachings from behavioral economics that study consumer behavior in real life, rather than being based on purely theoretical approaches. Behavioral economics emphasizes the largely irrational and subconscious factors that govern our decisions as consumers and make us, as Dan Ariely suggests, so "predictable."[39] It has applications in the public domain, but can also improve companies' marketing while remaining compatible with customers' real interests. As deputy CEO of the BVA Nudge Unit Etienne Bressoud pointed out at the 2019 Nudge Conference at HEC Paris, nudges can indeed be used to help improve consumers' experience.

Beyond nudges, the increasing depth of knowledge provided by neuroscientific tools about how the brain works supplies complementary ideas on ways in which it may be influenced.

Ways of subconsciously influencing the brain brought to light by neuroscience

Neuroscientists are interested in what subconsciously influences the brain during advertising. The different tricks and their influence are the subject of research published in scientific reviews and widely quoted by experts such as Roger Dooley, A.K. Pradeep and Patrick Georges.[40] These kinds of influences, discovered from experiments into how the subconscious brain works in humans, have been applied to improving the effectiveness of advertisements. They may be considered to be nudges for publicity, although neuroscientists rarely use this term. Their application can be seen in the writing of messages, the design and placing of images in relation to text and the production of videos. Certain procedures inspired by direct observation have been used pragmatically in advertising for a long time. Neuroscientific analysis has the advantage of confirming the interest of these procedures in relation to neuro-consumers' brains and complementing them with scientific research.

Tricks for written communication

A study carried out by the Millward Brown agency in 2009[41] using magnetic resonance imaging (MRI) shows that writing in newspapers and magazines has a stronger emotional impact than digital messages and it is also better memorized. Created in 1972 in the United Kingdom by Gordon Brown and Maurice Millward, the agency figures among the biggest market study companies in world markets. It is recognized as an authority in the field of brands. Despite the development of video communication and the spectacular progression of e-communication and m-communication, writing remains an efficient method of reaching the neuro-consumer. It is all the more efficient when it uses cognitive rules developed from neuroscientific studies on how the brain perceives information.

In several articles and a book written in collaboration with Anne Sophie Bayle-Tourtoulou and Michel Badoc,[42] Patrick Georges gives advice on how to make written messages as "neuro-compatible" as possible:

- *Turn the titles into questions*: The brain is programmed to find solutions when it meets a question. This process helps to memorize them.
- *Write in columns*: Columns six to eight words wide are better captured by the eyes.
- *Write short sentences and use words and characters that are easy to read*: This facilitates quicker reading by the brain.
- *Eliminate asymmetry*: For the ancestral brain, anything that is asymmetrical is not good and is therefore best avoided.

- *Use contrast to highlight what is important – bigger, bolder, brighter*: To be adequately perceived, such contrast should not be used on more than 10% of the text.
- *Optimize the placement of important information*: Whatever is at the top left or center of the page is seen best.
- *Repetition*: Do not hesitate to repeat ideas that need to be memorized.

Other researchers using neuroscientific tools suggest methods that make written communication more effective because it is more compatible with the brain. On the basis of their experiments, Hyunjin Song, assistant professor at Arizona State University, and Norbert Schwarz,[43] a professor at the University of Michigan, show that the brain's perception of information is considerably affected by the simplicity of the characters, words and sentences used. Simplicity is an important rule for making communication more effective. The exception to this rule is when the advertiser wishes to promote an expensive or sophisticated product, as for example with technological products or the menus of top-end restaurants. In this situation, using writing that is more challenging or scientific terms can reinforce the image of advanced technology or experienced chefs.

The use of difficult characters is sometimes recommended to help the brain retain certain information, for example a telephone number or website address. The presentation makes it more of an effort for the brain to read the characters, which can reinforce the memorizing process.

Three American researchers and marketing lecturers working on dietetics – Cargar Irmak, associate professor at the University of Miami, Beth Vallen, assistant professor at the University of Villanova and Stephanie Rosen Robinson, assistant professor at North Carolina State University[44] – demonstrate the importance of using appropriate wording in communication related to dietary products.

> *For an assortment of pasta, the use of the word "salad" makes the dish appear lighter. The expression "fruit chewing gum" suggests a more natural and less sugar-rich product. Creative advertising copy-writers try to change expressions in order to make them more attractive to the brain.*
>
> *The expression "shower gel" is more agreeable than "liquid soap." On an airplane, asking customers, "Would you like our scallops in soubise sauce or the chef's pork tenderloin?" is more mouth-watering than, "Would you like the fish or the meat?"*

Brian Wansink,[45] professor of marketing at Cornell University, presents various tricks, in particular those used in restaurants, which push the consumer to eat more. He shows that the way in which the food is described on the menu can increase turnover by 27%. He explains various wordings that favor the choice of a dish. We have adapted his ideas to apply them to a European restaurant environment.

> *The description can prove to be more effective if it evokes a context: regional (for example, "fillet of beef from the Aubrac," "Scotch salmon"); sensorial ("prime rib of beef cooked over a wood fire," "fillet of trout smoked at the fishery by artisan-fishermen"); nostalgic ("sheep's-milk cheese produced by small farmers on the Larzac" or "in the Carpathian mountains"); supported by a brand name ("rum baba made with Saint James' agricultural rum from Martinique"); freshness ("dish of the day prepared by the chef"). Some words such as "handmade," "natural," "traditional," "home-made," "chef's specialty" or, for an industrial item, "innovation," "at the forefront of progress" trigger a positive response from the brain.*

Roger Dooley[46] also evokes the importance of certain words such as "free" and "new."

> *According to Dooley, Amazon has increased its sales in several countries by offering free shipping for the purchase of a second item.*

Results are also perceived differently by the brain depending on whether they are presented as numbers or percentages. Neuro-economist Jason Zweig[47] gives interesting examples obtained with the help of neuroscientific studies. He explains that speaking about "one person in ten," an expression that implies real people, is more evocative for the brain than "10%." It appears to be more effective to mention that nine consumers out of ten judge a service to be excellent rather than talking about 90% who think this way. It is preferable to use real numbers rather than percentages to express a positive message. On the other hand, it is better to use percentages to present figures with a negative connotation.

Using these tricks in written communication can make it more effective. Good knowledge of how they work enables advertisers to improve their messages and neuro-consumers to avoid being caught out by certain artifices likely to orientate their decisions without them realizing it.

Tricks for images

For centuries in the West, images succeeded oral transmission to comment on stories, in particular religious history. They enabled the population, largely illiterate during those periods, to understand the writings of the Old and New Testaments, but also the lives of the saints and martyrs. They have been used abundantly to illustrate the stained-glass windows and walls of churches and are still admired today. It is not surprising that the brain's reptilian or limbic region has conserved the idea of their importance through transmission from generation to generation.

Images accentuate the sensory impact of products on the brain. Coordinated with writing, they reinforce emotional responses using shapes and colors. They make it possible to highlight the luxurious or natural aspect of a product, or to include it in a beautiful landscape. Research suggests that images in advertisements attract the brain's attention more than writing and that it is more effective to place the image on the left or in the center of the advertisement and the text on the right or above.[48]

According to Morten L. Kringelbach,[49] professor at the Faculty of Medicine at Aarhus University in Denmark, the brain is wired to recognize the human face and is more particularly attracted by a baby's face. Its attention is also captured by an adult face or a product whose shape reminds it of a baby's face. The success of cars like the Volkswagen Beetle and the Austin Mini is sometimes attributed to this resemblance. Using eye tracking, Australian psychologist and photographer James Breeze[50] has studied how consumers look at advertisements that include babies. When the baby is photographed from the front, they look at its face directly. On the other hand, if the baby is presented sideways, looking towards a product, the viewer tends to move his eyes in the same direction.

A.K. Pradeep[51] makes suggestions for increasing the attention paid to an image by the brain. He contends that the brain likes things that are new, that surprise it, make it question itself, etc. He recommends the use of original images that "flash" in the middle of the advertisement, or disruptive images, for example a bird with a dog's head. He suggests that the enigmatic smile on Leonardo da Vinci's famous painting of the Mona Lisa attracts the attention of the consumer's brain, which tries to understand what the smile is hiding.

Roger Dooley[52] argues that the fact of showing photographs of people in an advertisement strengthens the observer's empathy, for example it might be people helped by a charitable organization, the advisors or sales staff of a service company, the students in a training center or the clients of an industrial company.

Tricks for videos

Videos are used in four main types of advertising media: cinema, television, the Internet and mobile phones. They have been found to provoke the greatest degree of solicitation of mirror neurons and, as a result, the viewer is better integrated into the presentation. A number of cognitive rules can make them more effective.

One of the most important is to ensure good harmony between the verbal language, the physical attitude and the music used. The brain's visual and auditory cortices are constantly interacting. Dissonance between the information that they receive can lead to rejection of the video.

American political consultant Frantz Luntz[53] explains that to convince an audience it is important to choose the right words, prepare the listeners to receive the message and select the order of the elements in the presentation. For example, it is necessary to establish the credibility of the person who speaks about the product or the brand before letting him communicate a message.

The brain is wired to remember stories. From the dawn of time, before the existence of writing, humans used them to transmit knowledge. Using MRI, American scientist Wray Herbert[54] shows that the brain's neurons are activated when people read or listen to a story. A story is all the more efficient in promoting products or brands when they play an active role in what is happening. Some advertising uses sagas based on a common theme illustrating a coherent series of stories.

Finally, the brain retains emotional stimuli better than rational ones. Two eminent American publicists, Hamish Pringle and Peter Field,[55] have tried to demonstrate this.

> *People retain information even better when it is both emotional and surprising. For example, in France the Crédit Mutuel advertisements show a familiar animal, a dog, talking to and even advising its master on the choice of a good bank.*

The subliminal effect on the brain plays a very important role in communication. Since early antiquity, philosophers, particularly sophists and rhetoricians, have used tricks to make speeches more convincing. Today, the use of nudges and neuroscientific techniques are considered to make all modes of communication – written, visual and oral – more effective.

It is, however, essential to remember that, when he has time to think, the neuro-consumer's neocortex is activated and can rapidly alert him to subliminal tricks used in advertising. There is a risk of the neuro-consumer reacting negatively and rejecting advertising messages that break ethical rules or the simple principles of good taste.

Notes

1. Johan C. Karremans, Wolfgang Stroebe and Jasper Claus, "Beyond Vicary's Fantasies: The Impact of Subliminal Priming and Brand Choice," *Journal of Experimental Social Psychology*, vol. 42, no. 6 (November 2006), pp. 792–798.
2. Drew Westen, *Psychology: Mind, Brain and Culture*, John Wiley & Sons, 1999.
3. S. Dehaene, L. Naccache, G. Le Clec'h, E. Koechlin, M. Mueller, G. Dehaene-Lambertz, P. F. van de Moortele and D. Le Bihan, "Imagine Unconscious Semantic Priming," *Nature*, vol. 395, no. 6702 (1998), pp. 597–600.

4. "Bush Says 'RATS' Ad Not Meant as Subliminal Message," CNN, September 12, 2000.
5. "Fox News Caught Flashing McCain TV Subliminal," Infowars.com, May 13, 2008; "Une image subliminale de McCain sur Fox News," *Le Nouvel Observateur*, September 12, 2000.
6. Vladimir Volkoff, *Petites Histoires de la désinformation*, Éditions du Rocher, 1999.
7. Vance Packard, *The Hidden Persuaders* [1957], Penguin, 1961 and IG Publishing 2007.
8. Blaise Pascal, *Pensées* [English ed.], Pinnacle Press, 2017.
9. Jean Giono, *A King Alone* [1947], NYRB Classics, 2019.
10. Jean-Marie Dru, *Beyond Disruption*, John Wiley & Sons, 2002; *How Disruption Brought Order: The Story of a Winning Strategy in the World of Advertising*, St. Martin's Press, 2007.
11. Dru, *Disruption.*
12. Plato, *Ion*, Kessinger, 2010.
13. Plato, *Hippias Major*, Hackett, 1982.
14. Aristotle, *Poetics*, Penguin Classics, 1996.
15. Immanuel Kant, *Critique of Pure Reason*, Penguin Classics, 2007.
16. Georges Wilhelm Friedrich Hegel, *Hegel's Aesthetics: Lectures on Fine Art*, Oxford University Press, 1998.
17. Martin Heidegger, *De l'origine de l'œuvre d'art*, Rivage Poche, 2014.
18. Vassily Kandinski, *Concerning the Spiritual in Art*, Dover Publications, 1977.
19. Semir Zeki, *Inner Vision: An Exploration of Art and the Brain*, Oxford University Press, 1999 and *Splendors and Misery of the Brain: Love, Creativity and the Quest of Human Happiness*, Wiley Blackwell, 2008.
20. Pierre Boulez, Jean-Pierre Changeux and Philippe Manoury, *Les Neurones enchantés*, Odile Jacob, 2014.
21. Cinzia Di Dio, Emiliano Macaluso and Giacomo Rizzolatti, "The Golden Beauty: Brain Response to Classical and Renaissance Sculpture," *PLoS ONE*, vol. 2 (2007), pp. 1201–1209.
22. Euclid, *Elements of Geometry of Euclid*, Rarebooksclub.com, 2012.
23. Luca Bartolomes Pacioli, *On the Divine Proportion*, CreateSpace Independent Publishing Platform, 2014.
24. Matila Costiescu Ghyka, *Le Nombre d'or: Rites et rythmes pythagoriciens dans le développement de la civilisation occidentale*, Gallimard, 1976.
25. Vitruvius, *Les Dix Livres d'architecture de Vitruve* [1673], Editor Pierre Mardaga, 1995.
26. Di Dio et al., "The Golden Beauty."
27. Sigmund Freud, *Three Essays on the Theory of Sexuality* [1905], Martino Fine Books, 2011.
28. Tom Reichert and Jacqueline Lambiase, *Sex and Advertising: Perspectives on the Erotic Appeal*, Routledge, 2002.
29. Ingrid Sichy, "Calvin to the Core," *Vanity Fair*, April 2008; Joanne Eglash, "Actress Brook Shields celebrates 47th Birthday: Get Her Slimming Secrets," *The Examiner*, May 31, 2012.
30. Damien Grosset, "Le sexe envahit la pub," *eMarketing*, February 1, 2011.
31. *Adweek*, October 17, 2005.

32. Ellie Parker and Adrian Furnham, "Does Sex Sell? The Effect of Sexual Program Contents on the Recall of Sexual and Non-Sexual Advertising," *Applied Cognitive Psychology*, vol. 21, no. 9 (2007), pp. 1217–1228.
33. Anemone Cerridwer and Dean Keith Simonton, "Sex Doesn't Sell – Nor Impress! Content, Box-Office, Critics and Awards in Mainstream Cinema," *Psychology of Aesthetic Creativity and the Arts*, vol. 3, no. 4 (2009), pp. 200–210.
34. Martin Lindstrom, *Brandwashed: Tricks Companies Use to Manipulate Our Minds and Persuade Us to Buy*, Kogan Page, 2012.
35. Richard H. Thaler and Cass R. Sunstein, *Nudge: Improving Decisions About Health, Wealth and Happiness*, Yale University Press, 2008.
36. David Kahneman, *Thinking, Fast and Slow*, Penguin, 2012.
37. Thaler and Sunstein, *Nudge.*
38. Eric Singler, *Nudge Marketing: Comment changer efficacement les comportements*, Pearson Education, "Village Mondial" collection, 2015 and *Nudge Management: Applying Behavioural Science to Boost Well-Being, Engagement and Performance at Work*, Pearson, 2018.
39. Dan Ariely, *Predictably Irrational: The Hidden Forces That Shape Our Decisions*, Harper Perennial, 2010.
40. Roger Dooley, *Brainfluence: 100 Ways to Persuade and Convince Consumers with Neuromarketing*, John Wiley & Sons, 2012; AK. Pradeep, *The Buying Brain: Secrets for Selling to the Subconscious Mind*, Wiley, 2010; Patrick Georges, "Rédiger des rapports plus intelligibles," *Trends, Be*, February 23, 2012.
41. Millward Brown, "Using Neurosciences to Understand the Role of Direct Mail," www.millwardbrown.com/
42. Georges, "Rédiger des rapports plus intelligibles"; Patrick Georges, Anne-Sophie Bayle-Tourtoulou and Michel Badoc, *Neuromarketing in Action: How to Talk and Sell to the Brain*, Kogan Page, 2013.
43. Hyunjin Song and Norbert Schwartz, "If It's Hard to Read, It's Hard to Do: Processing Fluency Affects Effort Prediction and Motivation," *Psychological Science*, vol. 19, no. 10 (October 2008), pp. 986–988.
44. Caglar Irmak, Beth Vallen and Stefanie Rosen Robinson "The Impact of Product Name on 'Dieters and Non-Dieters' Food Evaluations and Consumption," *Journal of Consumer Research*, vol. 38, no. 2 (August 2011), pp. 390–405.
45. Brian Wansink, *Mindless Eating: Why We Eat More Than We Think*, Bantam, 2006.
46. Dooley, *Brainfluence.*
47. Jason Zweig, *Your Money and Your Brain: Become a Smarter, More Successful Investor – the Neuroscience Way*, Souvenir Press, 2007.
48. Pradeep, *The Buying Brain.*
49. M.L. Kringelbach, A. Lehtonen, S. Squire, A.G. Harvey, M.G. Craske, I.E. Holiday, A. L. Green, T.Z. Aziz, P. C. Hansen, P.L. Cornelissen and A. Stein, "A Specific and Rapid Neural Signature for Parental Instinct," *Plos One*, vol. 3, no. 2 (2008), p. e1664.
50. James Breeze, "You Look Where They Look," Usable World Blog of James Breeze, March 16, 2009.
51. Pradeep, *The Buying Brain.*
52. Dooley, *Brainfluence.*

53. Frantz Luntz, *Words That Work: It's Not What You Say, It's What People Hear*, Hachette, 2007.
54. Wray Herbert, *On Second Thought: Outsmarting Your Mind's Hard-Wired Habits*, Broadway Books, 2011.
55. Hamish Pringle and Peter Field, *Brand Immortality: How Brands Can Live Long and Prosper*, Kogan Page, 2008.

CHAPTER 23

The subliminal influence of brands

Corporate interest in the perception of brands in neuro-consumers' brains is a highly topical matter. Brands, although intangible and based on emotions, nevertheless command unique financial value. Their worth, calculated by specialist firms such as Interbrand and Millward Brown, and published yearly in American financial reviews like *Business Week* and the *Financial Times*, can reach extraordinary sums. Top international brands such as Google or Apple are some years rated at more than $100 billion.

> *Microsoft, Coca-Cola and IBM are nearing this value. Major collaborative economy brands like Uber and Airbnb are also starting to reach vertiginous ratings. The value of European luxury brands such as Louis Vuitton, L'Oréal, Chanel and Hermès is more modest, nearing or exceeding $20 billion. The size of such sums explains executive committees' interest in any information allowing them to add value to their brands and logos.*

As Jean-Noël Kapferer, an international specialist in the field, rightly indicates, "a brand is a name with power":[1] power to positively influence the client's brain in order to make them choose it, as well as influencing staff members' brains to encourage them to promote it with pride and sometimes passion. It also has power to create a unique place for itself in the environment it is to be positioned in.

A brand draws this power from the emotional impact it has on a significant number of consumers. In various conferences, Amazon founder Jeff Bezos has stated that it is not what the brand says to the consumer that is important, but rather what the consumer says of the brand: "Your brand is what people say about you when you're not in the room."

Numerous professionals have striven to propose methods and approaches designed to procure the success of companies' brands by adapting them to the emotional expectations in neuro-consumers' brains.

Among the large number of researchers and consultants, Georges Lewi believes that establishing a comparison between brands' missions and narrative and the foundational mythology of humanity or nations contributes to their success. Meanwhile, Jean-Noël Kapferer and Jean-François Variot[2] advance "the brand identity prism" that assimilates brands' principal characteristics with those of a person.

Neuroscientific analysis is enjoying increased interest for its ability to help companies identify the brand ingredients that attract and please neuro-consumers' brains, often subconsciously, and numerous researchers all over the world use neuroimaging, electroencephalography (EEG) and other neuroscientific techniques to study the matter. Their studies are directed by three main concerns:

- identification of brand characteristics likely to make the brain react positively or negatively;
- neuro-consumers' perception of the legitimacy of brand extension to new offerings beyond the original range;
- creating brands that address all neuro-consumers' senses, known as "sensory brands."

The neuro-consumer and brand mythology

Neuro-consumers' brains seem to be particularly attracted to brands whose history reminds them of the great myths embedded in their subconscious and transposed by somatic markers. This is the theory defended by brand expert Georges Lewi,[3] who believes that brands have more significance and are more memorable among consumers when their story recalls the "great myths of humanity."

Myths and brands

Lewi suggests that there is a close connection between the mission and rivalry of several long-standing major brands and the passion or struggles attributed to the gods and heroes of Greek mythology. The link allows brands to penetrate the depths of consumers' collective subconscious. It bestows upon these brands a particularly strong appeal that will lastingly seduce a sizeable audience that is sensitive to myths. Establishing this similarity helps shed light on their success.

Readers interested in Greek mythology can refer to the fascinating work by Denis Lindon called *Les dieux s'amusent. La mythologie.*[4] This book describes with accuracy, humor and simplicity the saga of the principal gods and heroes of ancient Greece.

Lewi refers to the quests of certain Greek gods whose objectives and stories he believes have inspired those of major brands.

Poseidon (Neptune) Poseidon (Neptune) was the brother of Zeus and husband of the daughter of Oceanus. As sovereign of the sea, his reign spanned a vast area. He gave the first horse as a gift to the Athenians to help them conquer space and distance.

> *This mission could be seen to be reflected in the brand of indigo-colored trousers, or "Genoa sailors' trousers," Levi's. The brand has been universal for several generations. Just as the sea god reigned over the Mediterranean region, the Levi's brand has spread its global industrial and commercial presence to all the major regions of the world. It brings its production into proximity with its end markets. "It enters the mythical era by progressively abolishing physical global distances and consumption distances."*[5] *It offers men and women a second skin, a feeling of liberty.*

Zeus (Jupiter) Zeus (Jupiter) ruled as king of Olympus. By killing his father Chronos, he took over and abolished time as it was, and established his own reign and control of time.

> *Like Zeus, the brand Breitling "fought Chronos and declared a new time to man, that of a legend to recover."*[6] *Breitling supplied watches to the heroic pilots of the Royal Air Force and designed a specific watch for American cosmonauts: the Navitimer. Scott Carpenter wore this watch on his wrist in space during Project Mercury. Where space meets time, "the Breitling man" embodies the fundamental values of adventure and freedom in aviation, which is able to abolish the concept of time. Only by plane can a man leave New York at 8 o'clock in the morning and arrive in Paris at the same local time.*

Aphrodite (Venus) Aphrodite (Venus) abolished ugliness and embodied beauty, love and irresistible seduction.

> *Like Aphrodite, the brand Dior demonstrates a permanent mission to protect beauty. Its products are made to enhance it. "The challenges and promises of this brand are to banish ugliness and push back the visible boundaries of aging. Dior wrote its own myth by establishing a classic vision of beauty and standardizing criteria."*[7]

Dionysus (Bacchus) Dionysus (Bacchus) was a dual-role god, who evokes ecstasy and renewed liberty as well as wild brutality and hysterical orgies. He breaks all chains. He brings men together around the vine. He granted them drunkenness and showed them the path to freedom from the limits and rules laid down by society.

> *The brand Jean Paul Gautier reflects the characteristics of Dionysian mythology with its founder's often ambivalent behavior and publicity: "Its perfumes' success swings in a constant balance between the destructive and the beneficial and transparency and business."*[8]

Other Greek gods or heroes Georges Lewi points to other similarities between major Greek gods or heroes and the success of famous brands.

> *For example, Hephaestus (Vulcan) symbolizes technical ingenuity with the brand Bouygues; Apollo, the Sun god, gives man civilizing "logos" and awareness of their power with Microsoft. The same can be said of Hercules: the brand Ariane Espace constantly searches for solutions to almost insurmountable challenges. It has confronted space since the start of its history in a legendary, herculean challenge where strength, bolstered by courage in the face of trials, faces up to major risk.*
>
> *The brand Michelin, similar to Achilles, is the archetype of a brand that continuously strives to surpass customer expectations with its tireless search for safety. Just like the hero in* The Iliad, *its ambassador the Michelin Man conveys a sense of friendship and generosity.*
>
> *The brand IBM, which resembles Odysseus, suffered and erred after a great victory and a strong position. Although it has experienced failure and sometimes succumbed to doubt, it eventually picked itself up again all the stronger and triumphed.*

Other myths Apart from the work by Georges Lewi at the High Co Institute, a European center for studies, research and exchange on brands, little research has been carried out into the reactions in neuro-consumers' brains to potential links between brands and mythology.

The author has focused his approach on Greek mythology, but other national or regional legends can also inspire the conception of brand stories, such as the founding legends of the United States from the arrival of the first colonizers to the conquering of the West, Texas and gold-bearing regions, the Indian Wars and the Industrial Revolution. Historian Élise Marienstras,[9] Emeritus professor at the University of Paris VII-Diderot, provides an interesting perspective on the subject.

> *For example, Marlboro takes advantage of the cowboy image deeply rooted in the American mindset.*

Stories from tales and legends embedded in the individual and collective memory can also bring success and longevity and help build brands that please the brain and stimulate emotions that help make the brand memorable. There will likely be more research in this field in the future.

The influence of the perception of myths on brand positioning

Brands' success largely depends on the specificity and quality of their positioning. It creates a high remembrance rate linked with a feeling of affection in the brains of the neuro-consumers concerned. The influence of criteria that determine the development of humanity's greatest legends provides an interesting insight into the development of relevant positioning designed to promote a lasting brand. Comparisons can be drawn by answering three fundamental questions on brand positioning and choices made by the principal heroes of mythology.

Why? A legendary hero cannot exist without a fight for an important cause, such as a transgressive and bold choice in favor of humanity that often leads to a tragic destiny for the hero himself. Prometheus takes on the mission of improving the well-being of humanity by giving them fire. He steals from the gods at the risk of a tragic fate. Just like heroes, true brands are usually built with a mission that manifests itself as a quest to champion a cause to improve human well-being in some way.

> *Apple aims to make personal computing accessible to everyone so as to help change the way we think, work, learn and communicate. Facebook wants to give people the power to share and make the world more open and connected. L'Oréal wishes to preserve beauty and youth among women. Danone not only wants its food products to be flavorsome, but also to contribute to improving consumers' health. Mauboussin and its CEO, Alain Nemarq, aim to make luxury and high-quality jewelry accessible to all women, whatever their social or economic status.*

For whom? The mission of mythological heroes was generally to improve a people's well-being. Big brands' missions frequently represent a similar project with a global or limited vision corresponding to the offering.

> *For Facebook, Microsoft and Nike, etc., it is the whole world. For other brands, the chosen audience is more restricted but still shares common interests in their attributes. Chanel wishes to make the free and independent woman seductive. Breiz-Cola aims to take the Coca-Cola flavor to Bretons. Belle-Iloise offers quality canned fish with products made according to traditional recipes and authentic knowledge for consumers who prioritize taste over other criteria.*

Against whom? Like the heroes of mythology brands must have one or more valiant enemies to fight in order to seduce. Their battle against strong adversaries makes their destiny uncertain and deepens the value of their acts. Without enemies, their ongoing struggles are not as seductive. Zeus' glory comes from his fight against the Titans and the Giants. The Trojan War can only be won if Achilles manages to defeat the bold Hector.

> *Coca-Cola must always be stronger than Pepsi, Nestlé fights against Danone, Nike must constantly defeat competition from Reebok and Adidas, as well as new competitors such as Converse.*

Brands draw strength from this constant battle with other strong corporate names. Like a football team in the eyes of its supporters, it must always be "at the top." It has to regularly win battles and attract admirers.

> *Losing a battle creates a tragic destiny that brings it closer to the human condition. This was the case with IBM when it was defeated by Microsoft, but the "Big Blue" became even stronger when it came back from its defeat and renewed its brand.*

Creating a story Like myths, brands need to tell a story, a unique saga that can be recounted within a community and passed down through the generations. Stories have a specific effect on the brain, which has used them since the origins of humanity to pass on methods and secrets allowing the human race to survive and progress.

> *The saga can be linked to the exceptional destiny of the founder of the brand, such as Gabrielle Chasnel (Chanel), Steve Jobs (Apple), Mark Zuckerberg (Facebook), Bill Gates (Microsoft) or Larry Page and Sergey Brin (Google).*
>
> *It can refer to heroic events the brand is associated with, like the Second World War for Chevignon, Mac Douglas and Breitling, the conquest of space for Breitling, Bridgeston and Coca-Cola, or sports victories for Lacoste, Fred Perry and Nike.*

Having a secret The quest for a holy grail, the pursuit of material or spiritual treasure or the existence of a secret to discover are the prerogative of great myths. Human intelligence is hardwired for discovery. Some brands skillfully exploit this aspiration in the brains of neuro-consumers.

> *This is the creation of a secret, such as the Coca-Cola recipe, the Nike air bubble, the famous recipe invented by Colonel Harland Sanders (1930–1980), founder of Kentucky Fried Chicken, or the complexity of the criteria and rules in Google's search algorithm.*

Tirelessly repeating a strong idea

Repetition of a strong idea linked to a brand anchors it deeply in neuro-consumers' brains. Orators in ancient Rome boasting perfect mastery of the art of discourse and persuasion knew how to use this technique. Cato the Elder (234–149 BC), a Roman censor, developed the habit of ending his speeches before the Senate with a phrase that eventually became famous: "*Carthago delenda est*" ("Carthage is to be destroyed"). With this incessant repetition, the idea gradually took root in the minds of the Romans, and a third war was declared by Rome against her major Mediterranean rival, ending in the total destruction of Carthage.

American psychologist Robert Zajonc[10] shows, from multiple experiments on animals and humans, that familiarity with a stimulus leads to a positive assessment of it. Simple contact with stimuli that have become familiar through repeated exposure is enough to produce a more positive mood in a consumer. The repetition of slogans linked to a brand contributes significantly to favorably boosting its memorization and renown.

> *Examples include the famous "Nespresso, what else!" for the well-known brand Nestlé, "Just do it" by Nike and "Yes we can," now attributed to the former US president, Barack Obama.*

The six facets of the "brand-identity prism"

How does the brand Perrier distinguish itself from other sparkling water labels, or the Italian coffee brand Lavazza from the numerous other brands selling this beverage, or French banks BNP Paribas from Société Générale and LCL from Crédit Agricole? And what about European insurance companies? What makes Axa, Allianz or Generali unique? The notion of positioning has its limits. It provides a basic distinction between organizations addressing different clients with different products, but offers little help to neuro-consumers in the same target market and looking for the same type of product or service in their choice of partner. This situation is characteristic of the competition in many sectors. To get around the limits of positioning, Jean-Noël Kapferer,[11] professor at HEC Paris, and publicist Jean-François Variot[12] propose an original operational methodology to identify where a brand's uniqueness lies. Their approach aims to compare characteristics comprising a brand's reflected identity to what is intimately perceived by consumers, as though it were a real person. Their concept allows the creation of a "brand-identity prism," which can be linked to what Antonio Damasio calls the "somatic markers of the brain." These markers are linked to the brain's recording of all the major happy or unhappy experiences every customer has had in the past with the brand.

According to Kapferer and Variot, a brand's identity can be broken down into six facets similar to a human profile: physical appearance, personality, culture, relationships, reflection and inner mentalization. These aspects form the "identity prism" of every emitter, whether it is a company or a brand. All companies can find their distinctive identity and longevity in one or several of these aspects.

The physical facet

The physical facet represents the characteristics of the company, its resources, products or services. It is the traditional realm of communication, of "product strengths." It is also a facet that can be fairly standardized as benefits/services are often very similar.

> *For Perrier, it is sparkling water, the glass bottle and the color green. For Lavazza coffee, it is the typology, design and Italian-sounding name.*

The personality or character facet

The personality or character facet corresponds to an anthropomorphic vision of the company, the personification of the advertiser.

> *If L'Oréal, Ikéa and BNP Paribas bank were people, how would they be described in terms of gender, age, style, social status, personality traits, etc.?*

Some companies are perceived as "serious" while others as "smiling" or "dynamic."

> *Perrier has a sporty, dynamic, young and oneiric personality. Lavazza coffee is authentic, traditional, open.*

The identity aspect is integrated in the working methods of many communication firms. The firm Ted Bates, which created the notion of the unique selling proposition (USP), also talks of the "unique selling personality." Jacques Séguéla,[13] the cofounder of RSCG communication agency, organizes his firm's working methods around two notions: physical appearance and character. A company's identity, however, cannot be limited to these two aspects. They only specify the "built advertiser," the speaker. Two other aspects concern the "built recipient": the reflection facet and the mentalization facet.

The reflection facet

The reflection facet is based on an identification mechanism. All companies implicitly specify through their communication what type of individual they are addressing. This is not the target (the objective recipient), but a recipient built from a reflected image (even if it does not correspond to an objective portrait of the client).

> *Saying "the bank for successful people" creates a specific recipient by offering clients a certain image of themselves. For Perrier, it is a positive person who is proactive in life, refined and cultivated. For Lavazza, it is an intergenerational consumer who is modern and yet respects tradition.*

The mentalization facet

The mentalization facet corresponds to the "inner" aspect of the built recipient. While reflection corresponds to the external mirror, mentalization refers to the internal mirror.

> *Deep down, what relationship does the client have with himself through his association with a certain bank or insurance company? Is he offered the opportunity to see himself as clever, thrifty, a forerunner, pioneer, shrewd manager, a thoughtful and considerate father, etc.?*
>
> *Drinking Perrier means belonging to an elite, a group of superior people. For Lavazza, it is an authentic connoisseur, someone who likes drinking real Italian coffee in an international context.*

The relationship facet

The relationship facet corresponds to an observation: every communication offers recipients a certain form of connection with the advertiser, a certain type of relationship. The nature of this relationship is an essential aspect of identity.

> *It can be an informative relationship such as "the bank that teaches you"; a partnership such as "the bank where the client is an associate"; or a relationship of full management, control, mothering, admiration, complicity, etc.*

The relationship facet puts into perspective the notion of "star strategy" developed by Jacques Séguéla: "Considering the company as a star gives the recipient a relationship of admiration or even adulation. However, this type of relationship can't be the norm."[14]

Communication can offer other types of relationships between the company and the recipient.

> *For Perrier, the relationship is closely linked with social distinction, personal and professional success and an interest in culture and art. For Lavazza, it is the return to the origins of real coffee.*

Complicity is just one type of relationship proposed. Each brand must identify which one corresponds best to their core identity and their consumers.

The cultural facet

The cultural facet refers to the cultural roots of a company. Through the company, the client gets access to its universe and myths.

> *Having a checkbook from the bank Rothschild means symbolically appropriating part of the saga. Being a client with Barclays means adopting the symbols of the British Empire, the City. For Perrier, it means adhering to a bourgeois culture, the "American Dream," a culture linked with sporting values that is positive, young and dynamic. For Lavazza, it means adopting a culture linked to authentic and hedonistic values, respect for expertise and the pleasure of consuming in the traditional way.*

A brand's identity is projected onto the cultural facet: What kind of world does the company open doors to? Manhattan, Europe, London, Paris? What myths are revealed?

Before launching a communication campaign, all brands benefit when they establish their identity based on each of these six facets. By comparing the brand to a person, the identity prism allows a more intimate relationship to be built between it and the neuro-consumers. It helps bestow a unique character on it and allows it to adopt a more accessible and human image among consumers.

Neuroscience researchers such as Yoon and colleagues[15] show that brands are processed in the brain as different entities to people. These authors believe that brands are more closely linked to mental representations associated with objects than living beings.

The notion of brand personality is nevertheless a useful analogy in reflections on the creation of an identity through marketing.

The perception of brands as studied by neuroscientific analysis

A brand's influence on the neuro-consumer's behavior is currently a hot topic in neuroscientific research. According to Martin Lindstrom, the evocative power of brands, which are "just emotion,"[16] their persuasive power among consumers, the marketing expenses required to create them, and their intrinsic value – nearing $100 billion for the strongest – stimulate keen interest. Neuroscientific studies today provide a better understanding of brands' influence on purchasing decision processes among neuro-consumers. They show that regions of the brain such as the ventromedial prefrontal cortex, involved in managing emotions or the reflection of the self, are activated when solicited by the presence of certain known brands.

Neuroscientific research and brands

In 2006, a group of German researchers[17] used magnetic resonance imaging (MRI) in a study by the Radiological Society of North America to show that strong brands activate regions associated with positive emotions, reward systems and self-identification in the brain. Conversely, weak brands activate regions of the brain connected with memory and negative emotions.

> *As shown by Read Montague, Samuel McClure[18] and their colleagues in an experiment carried out with the brands Pepsi and Coca-Cola, a strong brand such as Coca-Cola has the power to modify the brain's perception of taste in its favor.*

Martin Lindstrom[19] suggests that a consumer who is satisfied with a product or service becomes attached to the brand. The opposite is just as true in the case of negative experiences. Once these associations have been established in the brain, it is difficult to change them, except in cases of strong disappointment among clients in the offering, distribution, after-sales service, communication, etc. To remain strong in the eyes of clients, a brand must constantly strive for quality. It must also always ensure consistency between its actions and the positive image it wants to convey.

Besides satisfaction and pleasure, certain authors conducting neuroscientific research believe that strong brands can produce similar effects in the brain to those produced by adhering to great social, humanitarian or religious causes.

Kate Newlin,[20] founder of the eponymous consultancy firm, suggests that big brands go as far as creating a sense of evangelization among their followers. One essential condition for creating "brand passion" is whether consumers detect genuine passion for the brand among managers, staff, distributors and all other personnel. This passion is passed onto clients and stimulates their attachment to the brand. This factor leads some brands to set up unique, exclusive shops managed directly by the brand or to create franchises[21] specifying its attributes.

During his research using MRI, Martin Lindstrom[22] studied brain responses in specific citizen groups. While studying a group of nuns, he noticed that a very specific part of their brain, a region generally devoted to joy, serenity and love, was activated when names such as God, Jesus and the Virgin Mary were mentioned. As he continued his research, he was surprised to see that the same region was activated among aficionados of major brands such as Apple, Coca-Cola, Google, Microsoft, Nespresso or Starbucks when their name was mentioned or when certain specific characteristics unique to them were presented.

Strong brands do not simply meet the needs or expectations of neuro-consumers, but can also produce emotion in their brain. Emotion can turn into affection, in certain cases bordering on the idea of great distinction and reaching almost mystical heights.

> *Some people with "the bug" are willing to spend long hours waiting to get the latest Harry Potter book or be the first to buy the new Apple product. Through its strong presence in the brain, a powerful brand has a driving effect on clients' purchasing decisions.*

Creating a brand that is considered attractive in the neuro-consumer's brain is far from being easy. To succeed, it must produce the feeling that it not only does something, but also makes consumers experience something. It must produce enough affection so that the consumer buys its products not only because they like them, but also because they like the brand, because they want to belong to a community of fans who also love the brand.[23]

When correctly targeted, the senses, through their direct connection to the brain, contribute to creating an experience of familiarity and affection for a brand. A successful brand policy consists in creating and maintaining what Maurice Levy, CEO of Publicis, calls in his conferences "a familiar relationship with it."

In the brand world, luxury products seem to occupy a specific place in the brain. Two German professors from the Neurology Department at the University of Otto von Guericke in Magdeburg, Michael Schaefer and Michael Rotte,[24] show that luxury brands possess the unique ability to activate the region of the prefrontal medial cortex, which is known to be involved in the self-perception process. For Bernard Roullet and Olivier Droulers, "Luxury brands seem to reflect a desire to incorporate the prestigious brand into the consumer's self, in a sort of transfer of the prestigious brand's status to the client's projected status."[25]

From neuroscientific evaluation of brand essence to its legitimacy of extension

Researchers and neuromarketing consultancy firms use different neuroscientific techniques to understand the behavior of neuro-consumers' brains when confronted with brands. The main ones are MRI and EEG. Some, used by the firm Neurofocus, are described by A.K. Pradeep[26] and are often implemented alongside traditional marketing surveys, which are already numerous (pre-tests, post-tests, brand loyalty, etc.) and often based on surveys using questionnaires, observations, opinion polls and focus groups. Neuroscientific research is mainly interested in two areas:

- perception of the brand attributes that permeate the brain; and
- understanding the legitimacy of brand extensions as experienced by the subconscious part of human intelligence.

Attributes linked to brand essence that permeate the neuro-consumer's brain

Using functional magnetic resonance imaging (fMRI) or EEG, neuroscientific studies can test whether specific elements in the brand "identity prism" or other attributes interest the brain, generate emotion or penetrate deep into the memory.

Neurometric tests examine positive and negative perceptions in the brain regarding elements such as:

- physical and tangible aspects of the brand (name, logo, signature, iconography, design, music or jingle, etc.);
- values supported by the brand on a social, societal, moral, human, humanitarian or environmental level, etc.;
- Benefits proposed by the brand for consumers: seduction, life improvement (personal, professional, family or other), hedonistic or intellectual pleasure, etc.;
- coherence between the brand's ambitions and its actions on the ground;
- originality of the features proposed by the brand's products in its sector compared with competing offerings;
- emotions produced by the brand among neuro-consumers.

The perception of these brand attributes and their influence on the neuro-consumer's brain are an important source of information in the field of marketing and help to improve the management of brand policies.

Perception of the legitimacy of brand extension in the brain

The advantage of a known, appreciated or popular brand is that it can become an umbrella label for other products that can then benefit from its assets. This is the case for many luxury brands that take advantage of their seductive, selective, high-end image to expand their offering.

> *Hermès extended its brand of luxury horse saddles to scarfs, clothes, perfumes, etc. Chanel shifted from producing haute couture women's hats to clothes, perfumes, etc.*
>
> *The creation of umbrella brands, which is common in the luxury market, can also be seen among less exclusive companies. Some*

European banks such as Société Générale use their company's trustworthy image to offer clients insurance products and debit and credit cards; BNP Paribas, LCL, Crédit Mutuel and La Banque Postale also offer property insurance, home security products and telephone services, while CIC sells cars.

The ability to diversify from an existing brand is a bonus for many companies.

It becomes dangerous when the products presented are not perceived by the clients as being legitimately associated with the brand and can sometimes lead to the failure of the new product. Besides product launch failure, offerings not considered legitimate by clients' brains can also be damaging to the brand's image.

Companies take a keen interest in the extent to which their brand is considered legitimate by clients when they want to extend their product range. Neuroscientific analysis helps them better understand reactions in the neuro-consumer's brain in relation to this change. Researchers and neuromarketing consultancy firms can shed further light on the perceived legitimacy of brand extension and are particularly valuable for companies seeking to limit failures in the launch of new associated products or services.

In China, a group of professors in the neuro-management laboratory at the University of Zhejiang in Hangzhou are conducting major research on the subject.[27] In 2008, they used EEG to study the level of acceptance of brand extension in the neuro-consumer's brain. The subjects were first shown the name of a drink that was well known in the country, such as Pepsi, Coca-Cola, Wahaha and NongfuSpring. Next, they were presented with a generic drinks category (soda, tea, milk, fruit juice, etc.) as well as a category of food or an everyday object (sweets, cake, bread, biscuits, a shoe, television, T-shirt, etc.) In all experiment conditions, the research team observed a positive wave that appeared approximately 300 milliseconds after the stimulus, called "P300." When a brand extension in a given category of products is accepted by the consumer, the P300 wave has a greater amplitude. It is present in a diffuse way in all the parietal and occipital regions of the brain. Conversely, when the brain considers a brand extension is not very credible, such as a drinks brand extended to shoes, the P300 has a smaller amplitude and is only present in the right occipital region. In these cases, the researchers also detected the appearance of a negative wave in the frontal region of the brain, called "N400." According to the study's authors, this second negative wave signals the brain's rejection of the brand's extension to the product in question. The researchers believe that the amplitude of the P300 wave is an interesting neuronal indicator of the acceptance of brand extension in client's brains.[28]

Research carried out on brand extension in university laboratories and by major neuromarketing consultancy firms such as NeuroFocus and NeuroSense is likely to see real progress in the future and provide a response in line with the scale of the economic stakes for companies in terms of their success or failure.

Sensory brands

The neuro-consumer's brain perceives brand essences though all its senses. A name, smell, piece of music or taste can reawaken the "somatic markers" embedded in the depths of the subconscious.

Brands and the senses

Sight The primary sense involved in brand perception is sight. It is stimulated by the name, logo, signature, design and colors chosen to express the brand's core concept.

The name of a brand can bring back connotations integrated in the "somatic markers" in the brain. A French name could evoke the idea of a certain refinement in sectors such as clothing, perfume, gastronomy, etc. An Italian name could reflect a certain chic style, creativity, etc. An English name could evoke quality and traditional sophistication. Brands with a German name are often considered to evoke efficiency and reliability, while for an American name, it could be competence, modernity, etc. The simple fact of choosing a name that reflects a certain country can bestow image attributes upon the brand. This explains certain states' growing desire to adopt and communicate their own positive brand image.

A logo is emblazoned on business cards and every document used for communication. It portrays the brand's values in the most faithful, general and creative way possible. The graphic presents the chosen colors, forms and symbols of the brand and carries a specific meaning that it hopes to make clients' brains perceive.

> *The logo used by BNP Paribas, as Sylvie Dubois from the bank's communication and publicity department remarks, was designed according to the following rationale:*
>
> *The aim of this logo is to promote a new strategic impetus from the merger with Paribas. We see the BNP Paribas stars turning into birds in a dynamic trajectory. The star is both a guide and a reminder that BNP Paribas belongs to the European Union. The bird is ready to travel the world and settle, just as our customer advisors are ready to accompany clients in their future projects. The green color of the star turning into a bird symbolizes hope and creates notions of transparency and concern for environmental issues to the curved flight trajectory. BNP Paribas deliberately stands out through a clear manifestation of its desire for innovation, ability to anticipate change and ambition of international openness.*[29]

The signature highlights the message connected to the image that the company seeks to communicate to its clients' and prospective clients' brains. Signatures evolve over time: they have shifted from a more company-focused message at the end of the 1980s to becoming closer to consumers and placing them at the heart of the relationship.[30] The signature can strengthen the association between a strong idea and a product brand, company or even a person.

The design and colors allow the brand to strengthen its visibility and image in the brain.

> *The design gives the product its uniqueness. For example, the knives by the French brand Laguiole have an original, practical and comfortable design, giving them a feeling of modernity and creativeness.*

Colors like red, black and white attract clients' attention. Others are used to produce a general or specific positive feeling appreciated in certain cultures, countries or regions of the world.

Sound Sound generally complements sight to encourage brand memorization. Most major brands adopt a "sound logo" designed to complement their visual signature. Sounds that recall a nostalgic or pleasant song are sometimes used to give the brand this attribute.

> *This is the case in France, for example, with the insurance company MAAF, which links its image to the song "La ouate."*

Smell Smell is detected in the cerebral regions in a similar way to memory. It is used to stimulate memory and facilitate the recognition of certain clothing brands or the identification of hotel chains, points of sale, etc.

Touch Touch is another sense used to help identify a brand. In the hospitality sector, identical textures for bedding, carpets and wood furnishing are used to produce recognition in the client's brain. Touch is also frequently used in the clothing industry. The tactile impression is carefully designed to make certain luxury or everyday objects identifiable with a brand.

Taste Except in certain industries like the food industry, taste is less commonly used to make the brain remember a brand.

> *Spanish Serrano, Italian San Daniele and French Bayonne hams make a concerted effort to give their products brand recognition or recognition of origin through their taste identity.*

Research into brand identification through taste will likely develop in other sectors of the food industry, such as dairy products, preserves, cooked meats, etc.

Developing a brand's sensory policy is all the more effective if it is consistent with the brand positioning across all distribution channels and communication methods. The client must also be able to sense harmony in the combination of sensory ingredients used. The brain negatively perceives incoherence between the different physical, relational and sensory attributes proposed by a brand.

Sensory congruence for brands

A brand's sensory policy more effectively transforms the customer experience in terms of emotional and affective habituation if it creates a sense of harmony between the different senses stimulated. Experts call this "sensory congruence." Brands must construct a coherent relationship between the different sensory logos through all means of contact and communication with neuro-consumers. Moreover, harmony between the stimulated senses must correspond to its positioning and identity.

For example, to convey the image of a calm, reassuring and prudent corporate brand, pastel colors (pink, blue, green, etc.) are generally used alongside soft, slow music and the diffusion of soothing fragrances (lavender, rose, orange, sandalwood, etc.).

An increasing number of brands talk about the need to create a sensory experience for their clients.

> *The luxury cosmetics brand Decléor, by L'Oréal, which specializes in aromatherapy using essential oils, shows a real desire to promote sensory experiences among its customers with its products.*
>
> *Nespresso wants to confer an image of rarity, exclusivity and status to its brand. To do so, it opts for high-end shops, exclusive addresses reserved for club members, a luxury limited-edition review and communication that strengthens the idea of selectivity.*
>
> *The famous New York chocolate maker, Hershey's, aims to offer chocolate lovers a true sensory experience with chocolate and its brand. Its store, Hershey's Chocolate World, is located in several cities in the United States and offers visitors a full-on chocolate experience during which they can even make their own chocolate bar.*
>
> *Abercrombie & Fitch, for its branded T-shirts and other products displayed in its stores, builds a specific atmosphere based on a combination of sensory stimuli including lighting, music, olfactory signature and social representation through the sales assistants' dress code.*

Sensory marketing and neuromarketing are also becoming more popular among retail brands in sales outlets, as discussed in Chapter 19 on sensory marketing.

In an increasing number of sectors, growing corporate interest in building emotional brands is turning sensory marketing and neuromarketing into fundamental components of their creation and growth. It will certainly become an even more significant subject in the future.

Notes

1. Jean-Noël Kapferer, *The New Strategic Brand Management: Advanced Insights and Strategic Thinking*, Kogan Page, 2012.
2. Ibid.
3. Geoges Lewi, *Les Marques mythologie du quotidien: Comprendre le succès des grandes marques* (2nd ed.), Village Mondial, 2009.
4. Denis Lindon, *Les dieux s'amusent: La Mythologie*, Flammarion, coll. "Jeunesse," 2010.
5. Lewi, *Les Marques mythologie du quotidien.*
6. Ibid.
7. Ibid.
8. Ibid.
9. Elise Marienstras, *Les Mythes fondateurs de la Nation américaine*. François Maspero, 1976.
10. Robert Zajonc, *The Selected Works of R. B. Zajonc*, Wiley, 2004.
11. Kapferer, *The New Strategic Brand Management.*
12. Jean-François Variot, *La Marque postpublicitaire: Internet Acte II*, Village Mondial, 2011.
13. Jacques Séguéla, *Hollywood lave plus blanc*, Flammarion, 1992.
14. Ibid.
15. C. Yoon, A. Gutchess, F. Feinberg and T.A. Polk, "A Functional Magnetic Resonance Study of Neural Dissociations Between Brand and Person Judgments," *Journal of Consumer Research*, vol. 33, no. 1 (2006), pp. 31–40.
16. Michael Lindstrom, *Brand Sense: Sensory Secrets Behind the Stuff We Buy*, Free Press, 2010.
17. C. Born, S. Schoenberg, M. Reiser, T. Meindl and E. Poeppel, "MRI Shows Brains Respond Better to Name Brand," *Proceedings of RSNA*, November 28, 2006.
18. Read Montague, *Your Brian Is (Almost) Perfect: How We Make Decisions*, Plume Books, 2006.
19. Martin Lindstrom, *Brandwashed: Tricks Companies Use to Manipulate Our Minds and Persuade Us to Buy*, Kogan Page, 2012 and *Buy.Ology: How Everything We Believe About What We Buy Is Wrong*, Random House Business, 2009.
20. Kate Newlin, *Passion Brands: Why Some Brands Are Just Gotta Have, Drive All Night For, Tell All Your Friends About*, Prometheus Books, 2009.
21. Lindstrom, *Brandwashed.*
22. Patrick Georges, Anne-Sophie Bayle-Tourtoulou, Michel Badoc, *Neuromarketing in Action: How to Talk and Sell to the Brain*, Kogan Page, 2013.
23. Laurence Body and Christophe Tallec, *L'Expérience client*, Eyrolles, 2016.
24. Michael Schaefer and Michael Rotte, "Favorite Brands as Cultural Objects Modulate Reward Circuit," *NeuroReport*, vol. 18, no. 2 (2007), pp. 141–145.

25. Bernard Roullet and Olivier Droulers, *Neuromarketing: Le marketing revisité par les neurosciences du consommateur*, Dunod, 2010.
26. A.K. Pradeep, *The Buying Brain: Secrets for Selling to the Subconscious Mind*, Wiley, 2010.
27. Q. Ma, X. Wang, L. Shu and S. Dai, "P 300 and Categorization in Brand Extension," *Neuroscience Letters*, vol. 431 (2008), pp. 57–61.
28. Ibid.
29. Sylvie Dubois cited in Christelle Rancev, *Quels sont les enjeux de la marque*, master's thesis, University of Paris X Nanterre, 2002.
30. Rancev, *Quels sont les enjeux de la marque*.

Points to remember

- Since early antiquity, philosophers have taught us that we should not limit ourselves to the rational aspects of a message in order to convince an audience. The sophists in Greece were the forerunners of this idea. Aristotle, by inventing rhetoric, proposed an art of speech that went beyond simple language to seduce listeners. His method was perfected by Roman orators such as Quintilian and Cicero. It places emphasis on the physical and emotional aspects that allow an orator to make his message credible. After being taught in European schools for centuries and largely forgotten for a number of decades, it is being brought up to date by philosophers today. In Brussels, David Perelman, in collaboration with Lucie Olbrechts-Tyteca, has returned to Greek rhetoric and proposed a "new rhetoric" that establishes a theory of argument.
- Today, publicists are endeavoring to benefit from the theories and methods of rhetoric to improve the credibility of their communication. Professors Thierry Herman and Gilles Lugrin reveal how rhetoric (*logos*, *ethos* and *pathos*) is used in publicity and advertising campaigns.
- Semiology (or the study of signs) is another technique used to decode certain subjective elements in an advertisement and evaluate their role in consumers' subconscious in order to improve effectiveness.
- Many advances have been made by marketing and communication experts to better understand and perceive the importance of the phenomena, often subconscious and irrational, that lead to a campaign's success or failure. A series of tests can be used to see whether the messages proposed are coherent with the internal expectations of neuro-consumers' brains.
- The use of neuroscientific techniques, mainly MRI and EEG, provides further knowledge to make communication more pertinent through observation of the direct responses in neuro-consumers' brains. These techniques provide an analysis of various aspects including attention, emotion, memorization, the "mirror-neuron" effect and the level of interest generated by messages during each second of their diffusion.
- Subliminal communication circumnavigates the barrier of reasoning by directly addressing the subconscious. Although it is, in theory, forbidden in many countries, the reality is quite different. In his internationally bestselling book, *The Hidden Persuader*, Vance Packard presents the many methods used by advertisers to sound out and reach consumers' subconscious. He demonstrates their effectiveness and warns against the danger of their use.
- Some perfectly legal methods can have a subliminal effect. They have been successfully used in advertising for decades, and include entertainment in advertising, the use of art in communication, appealing to the

rules of harmony directly perceived by the brain such as the "golden ratio" and the "Vitruvian proportions" and the use of eroticism.

- The use of "nudges" (automatic decisions made subconsciously by default) is a gentle method among more recent subliminal techniques used to influence citizens' and consumers' decisions. First developed by Richard Thaler and Cass R. Sunstein during the 2000s, nudge theory has enjoyed growing success in political and marketing applications.
- Neuroscientific research, based on more in-depth knowledge of the neuro-consumer's brain, proposes different artifices to make advertising design more effective. This concerns written, image and video-based communication.
- Brands play a fundamental role in influencing neuro-consumers' brains in their purchasing decisions.
- Georges Lewi has established a positive subconscious efficiency ratio for brands that draw from famous myths, particularly Greek mythology. Jean-Noël Kapferer and Jean François Variot aim to improve brand popularity by giving them a human personality using the "brand identity prism."
- Neuroscientific analysis provides new elements that allow researchers to see how the brain reacts to brands as well as their influence on the neuro-consumer's subconscious. It provides a better understanding of the phenomena linked with their essence and legitimacy for brand extension.
- The neuro-consumer perceives a brand's essence through all five senses and not just sight. A name, smell, melody, taste or feel can awaken the "somatic markers." Neuromarketing makes recommendations for implementing a brand sensory strategy, ensuring coherence between the strategy and the sensory elements used and creating sensory "congruence" between them.

PART VI

The neuro-consumer's brain influenced by the digital revolution

The spectacular development of digital technology has caused a profound transformation in neuro-consumers' behavior. The interactivity associated with the advent of Web 2.0 and 3.0 is changing them into "neuro-consumer-actors." With the increasing influence of online communities and social networks, there is a gradual shift from the individual to a collective consciousness. New types of community are appearing, united around senses such as music or taste and marketing must adapt quickly to these profound behavioral changes in the brain.

Introduction

The digital revolution and the emergence of online social networks has fundamentally changed the behavior of neuro-consumers. Neuro-consumers now have access to a wide range of information, obtained in real time and on a global scale. Before making a purchase, neuro-consumers can compare and get information via online communities and social networks, reflecting a gradual shift from an individual to an increasingly collective consciousness.

This media revolution compels the brands to immediately adapt and respond to the demands of the neuro-consumer, who, with a significant desire for interactivity, becomes a "neuro-consumer-actor." Together with the arrival of "big data," traditional marketing and communications are forced to transform themselves and to use new techniques of e-marketing, m-marketing, e-communications, "one to one," mass customization, permission marketing, desire-based marketing and inbound marketing. New professions such as web masters, community managers and social network officers have also arrived on the scene and are used by companies to devise digital and community policies and actions adapted to new behaviors, perceptions and expectations of the neuro-consumer-actor's brain.

CHAPTER 24

When the digital revolution changes the brain's perception

The digital revolution is on the move. Digital marketing can be defined as "an offer on a market whose information products or related services are digitized, in other words, transformed into data which may be processed and transmitted by means of a computer network."[1] Among these "information products" are printed, sound, visual, multimedia or computer documents, or the access or exchange of media.

The digital revolution is an inexorable and fast march generated by the spectacular development of techniques originating from information and communications technology (ICT). The World Wide Web (WWW) was introduced by the British inventor Timothy John Berners-Lee on August 6, 1991. Its development was meteoric and in only few years, the four major companies of the digital world, Google, Apple, Microsoft and Facebook, have driven major societal change via their dominance in online activities.

Many Americans take 1995 as the year that marked the move toward the interactive transformation. This was the year when, for the first time in the United States, there were more PCs sold than televisions and where the number of e-mails sent exceeded the number of handwritten letters.

Initially, the Internet, also called "Web 1.0," was a simple single-dimension communications channel for companies to communicate with consumers. As a measure of its development and customers' interest, the brands found it useful to encourage "feedback" from Internet users. Interest in it increased exponentially and it is now the preferred tool for direct and relationship marketing.

The arrival of online social networks when Facebook was set up in 2004 by Marc Zuckerberg, an American student at Harvard University, brought about changes in the Internet's applications. It became a new collective worldwide collaboration platform with networks using easily accessible data in the "open-source" environment. This functional transformation is known by the name of "Web 2.0."

A new digital development is already on the way, referred to as "Web 3.0," the third generation of Internet services, which aims to focus on using a machine-based understanding of data to provide a data-driven and semantic Internet, creating more intelligent, connected and open websites. Web 3.0, often called the "semantic web," is thus also commonly used to refer to the internet of things or IOT.

As a result of these developments, there has been a profound change in consumers' behaviors. This is particularly the case for generations Y and Z (described in Chapter 9), who were born into the digital world. For neuro-consumers, the entry into the virtual era changed the perception of the environment and relations with others. The possibility of permanently having a large quantity of communications tools at their disposal, summarized by the acronym ATAWAD (anytime, anywhere, any device), has been found to have a multiprogramming and sometimes addictive effect on consumers. How their brain behaves in this new context is therefore modified, creating an increasing shift from an individual to a collective awareness.

The Internet revolution and the neuro-consumer's behavior

Before the Second World War, the British mathematician Alan Turing (1912–1954), who became a hero of the war by deciphering codes used by the enemy, established the basis of what became modern information technology. He theorized the idea of the computer in his "Turing Universal Machine" as a different tool from the calculator through its ability to carry out operations using algorithms. In 1950, he established the groundwork for artificial intelligence and waged "that within 50 years there would be no means of making a distinction between replies from a man or from a computer on any subject."[2] In 1997, at an unforgettable competition, "Deep Blue," the computer designed by IBM, won a chess game against the then world champion, Garry Kasparov. In 2016, Lee Sedol, the South Korean world champion of one of the most complex games – the Go game – was defeated by a machine developed by Google, named Alpha Go.

In 1957 German engineer, Karl Steinbuch (1917–2005) coined the term "informatik," a neologism for the words "information" and "automatic." In 1965, American engineer Gordon Earl Moore, co-founder of the company Intel, noted that, since 1959, when semiconductors were invented, their complexity had doubled every year at a constant rate. In an article published at the time in *Electronic Magazine*, the "Moore's Law" appeared that anticipated the exponential development of their sophistication at a constant size and at a reduced price in the future.[3] This prediction has now been confirmed while ensuring the fortune of Intel that Gordon Moore helped to found.

In 1969, the US Defense Department created ARPA, an acronym for ARPANET, the forerunner of the Internet. The network soon made it possible for research centers in major universities all over the world to communicate with each other by e-mail.

The inexorable development of the Internet environment

In 1991 the World Wide Web became publicly available. Since that time, all new information and communications technologies (NICTs) linked by the Internet experienced astonishing progress on the basis of techniques and tools that, every year, became more numerous, smaller, intelligent, autonomous and faster. They were devised to function in a network, with or without the intervention of humans. They improved the users' possibilities while increasing their power.

In 1975, Bill Gates and Paul Allen initialized a revolution in the systems and software industry in creating the company Microsoft, in Albuquerque, New Mexico.

Apple was set up in 1976 and in 1977 the company developed the personal computer that conquered the world at the beginning of the 1980s. Innovations followed one another: the iPod was launched in 2001, the iPhone in 2007, the iPad in 2010, the Apple Watch in 2014.

So that Internet users could make the most of NICT applications and the Internet, the number of inventions greatly increased, mainly in the United States, but also all over the world.

> *IBM designed the first smartphone in 1982 and its use became general from 2000. Amazon opened up e-commerce in 1995, the same year that eBay was set up. Google proposed a worldwide search engine in 1998 and 2001 saw the start of Wikipedia, the first major collaborative encyclopedia on the Internet, as well as the dating site Meetic. In 2004, the first major social network, Facebook, was born and Twitter appeared in 2006.*

Despite the collapse of the Internet bubble in 2000, known as the "dot-com crash" in which investors lost trillions of dollars, the start-ups grew faster, particularly in Silicon Valley (San Francisco Bay) and Silicon Alley (New York). Experts believe that revenues from the Internet will grow spectacularly in the coming years. Some of them humorously suggest that revenues from electronic commerce will increase from 0 million dollars in 2005 to 0 billion in 2020.

The worldwide empire of the Internet slowly built itself around big leaders such as Google, Apple, Facebook, Amazon and Microsoft, known as the GAFAM companies. China is also now a big player in the online world with giants such as the social media specialist Tencent and its WeChat messenger, the e-commerce leader Alibaba and its social platform Weibo, the smartphone and internet-of-things giant Xiaomi and Baidu, the most-used search engine in China. Together, they are known by the acronym BATX.

The multiple uses provided by the Internet, such as access to information, knowledge, education, entertainment, communications, e-commerce, services, remote working, global communication, etc., are described in many works.[4]

However, despite its advantages, the Internet is not without criticism. In 2013, Edward Snowden created a worldwide scandal by revealing on the Internet the scale of the espionage programs concerning citizens all over the world. Some of the main flaws attributed to the Internet are frauds, scams, the dissemination of sites or messages contrary to the principles of good morality or ethics, risk of addiction to certain sites, difficulties in addressing sensory messages, lack of human warmth in communications, etc.

Changes in behaviors noted in neuro-Internet users

The Internet penetration rate exceeds 80% in the majority of European countries and in the United States. The smartphone rate is more than 60%. Predominantly used more by young people initially, smartphones are now widely used by consumers of all generations.

The development of the Internet has brought about changes in the behavior of the neuro-consumer, with easy access to information with searches on Google, Bing and other search engines; the participative encyclopedia Wikipedia; contact with extended online social communities with Facebook or increased professional relations with LinkedIn, Viadeo, etc.; the possibility of at-home online learning via MOOCs; sharing opinions on any subject globally with Twitter; online dating networks such as Meetic and e-darling, etc.; online shopping; buying or reselling on eBay, Amazon, etc.

Owing to the many possibilities offered by the digital space, neuro-consumers with interactivity become "neuro-consumer-actors." The ease of use and automatic reflexes of the many Internet users make their brains more sensitive to forming rapid and instinctive responses, which may also give rise to manipulation. They gradually enter a virtual world where the classic reasoning standards are often called into question.

Close to hand, 24 hours a day, mobile phones can become an inseparable "security blanket" for many people, particularly teenagers, and much research suggests that the increasing use of mobile phones may lead to problematic behaviors, though the qualification of addiction doesn't seem to be scientifically appropriate.[5]

A new behavioral generation of "multiprogrammed neuro-consumers" is appearing. At the same time, they can have breakfast as a family while watching television and replying to their messages. A study carried out on 1,000 smartphone users in 2013 by Google, called *Our Mobile Planet: France – Understanding the Mobile Consumer*, shows that 78% of them consult their smartphone at the same time as using other media: 52% while watching television, 47% while listening to music, 36% while consulting the Internet, 32% while watching a movie, 19% while playing a video game, 18% while reading magazines or newspapers, 9% while reading a book.[6]

On the other hand, it seems that "brain multiprogramming" is not the same as "multi-attention." The speed required for replies to messages on mobiles and to e-mails relies primarily on the reptilian and limbic part of the brain. Gradually we are moving into a world where "zapping" replaces logic and thoughts. Where individuals favor sequential information based on emotion rather than what is linear and based on conceptual reasoning.

Other practices developed by Internet network players lead to significant changes in purchasing behaviors observed among Internet users. Various trends are directly affecting companies' marketing strategies.

The increasingly frequent use of mobile phones in stores A study made by GFK states that already,

> 24% of buyers in stores use their mobile to compare prices, 32% to contact a friend or the family for advice, 26% to take photos of products, 17% to take a picture of the description, 15% to scan the bar codes or the QR codes, 10% to buy a product using an application, 9% to order on a website.[7]

The gratuity economy Above all, neuro-Internet users are drawn to companies proposing special or free offers. This practice compels companies to learn how to split up their approach in order to remain profitable. The first free offer is designed to attract, then to generate advertising or lead the consumer towards a purchase. LinkedIn is a model of this type: The initial services are free, hoping that they will rapidly lead the customer to use a billed "LinkedIn premium." The MOOCs offer free online teaching hoping to sell a diploma course afterwards.

"Uberization" or the participatory economics of purchasing Thanks to new sites that appear every day, "internauts" can buy services at knock-down prices compared to those charged by professionals by appealing directly to other individuals. Uber and BlaBlaCar have revolutionized the world of transport. Airbnb and GuestToGuest have done the same for holiday accommodation and apartment rental. Sites for loans between private individuals and the advent of "crowdfunding" have revolutionized the world of investment. These companies are growing exponentially to achieve huge market capitalization. Despite strong opposition from some industries, apps of these companies are downloaded on a growing number of mobile phones and the neuro-consumer's brain is progressively adopting these new ways of buying.

A return to auctions Specialized firms such as eBay have initiated a return to auctions. A large number of sites offer Internet users the possibility to hold direct or reversed auctions.

Expectations of last-minute purchase proposals The aim is to get a bargain by getting early or last-minute purchases.

Choosing from multiple references Amazon has more than 70 million references while, in a European hypermarket, they vary from 30,000 to 100,000. iTunes presents more than 26 million titles, Google Play Store, a million applications. The references on websites have become almost unlimited.

Geo-location Companies are increasingly using the geo-location application thanks to iBeacons. It enables them to attract consumers to a point of sales, an exhibition or a restaurant by presenting a tempting offer when they pass nearby. It may offer a free guided tour of a museum, as is the case, for example, with the Rubens House in Antwerp.

The "cross-channel" or "omni-channel" It is no longer enough for the neuro-consumer just to use the various sales channels as an alternative. He uses them in parallel. He sees an offer on a social network, researches it on the Internet, he can then buy it online or choose to go to a boutique on the high street to choose it in person, he can order by telephone, etc. The customer constantly moves from one channel to another. To increase their opportunities to sell, firms must adapt and be capable of following and assisting the consumer throughout his visits to multiple channels.

New practices and proposals appear every day New practices and proposals appear constantly that profoundly change the neuro-consumer's behavioral habits. Marketing must continuously adapt to such changes. As François Cazals, an expert in the field, remarked:

> The new paradigm in digital strategies is not that of a confrontation between the strong and the weak, but that of the agile against the slow. The important thing is to rapidly pre-empt the new "Blue

Ocean" growth territories to reap "a first mover advantage." Obviously, this strategic approach is particularly appropriate for the SMEs, mid-market companies, and high-growth companies.[8]

To adapt the Internet to the perceptions of the neuro-consumer-actors' brain

The neuro-consumer's brain is bombarded by Internet communications in addition to the increase in the number of other means of obtaining information. In an interview with AFP, former CEO of the television channel TF1, Patrick Le Lay, expressed concern about finding an available space in the brain where an attractive advertisement could be placed, stating: "What we're selling to Coca-Cola is available human brain time ... nothing is more difficult than obtaining this availability."[9]

The increase in the quantity of information, which regularly appears on smartphones, available 24 hours a day, is often accompanied by a loss of meaning. The consumer ends up by favoring whatever comes from the loudest, most repeated sources. As the message is not checked, there is the risk of mystification and deceit. The concern for speed and the need to respond to the numerous appeals, for a brain that can only process one-fifth of the information received at a time, becomes the significant cause of errors of judgment. The primitive (to respond rapidly), or limbic (to deal with emotional messages) parts of the brain are called on and go into action as a priority. When the neocortex (the "intelligent" part) takes control, the Internet user who is not sufficiently informed runs the risk of being manipulated, resulting in a feeling of inevitable frustration towards communications on the Internet – contrary to the policy of companies whose main aim is to maintain loyal customers. The traditional "marketers" can no longer continue to use the methods that made marketing such a success in the past – they must refer to new rules and adapt speedily to the invasion of information picked up by consumers via the multiplicity of receivers and shared with communities on social networks. The opening up to new marketing concepts aimed at permission, desire and virality, are capable of better meeting the behaviors of the neuro-consumers who are committed to the digital world.

Permission e-marketing

In the United States, Seth Godin,[10] a pioneer in online marketing, was one of the first experts in this discipline to rebel against what he calls "aggressive marketing" or "interruption marketing" in his now famous work, *Permission Marketing: The Internet Lessons in Marketing*. He particularly targets consumers' relations with the Internet. He is amazed that, for many years, we have been seeing the paradox of communications budgets being exponentially increased while their impact on consumers continues to weaken. One of the main reasons comes from brains being overloaded with so much information in all areas, from a large number of receivers such as television, mobile telephones, tablets and so on. Godin argues that advertisers must end the costly waste of uncontrolled messages inundating consumers without any discernment. To put a stop to this waste, Godin proposed "permission marketing." This consists of communicating only with those who give their prior consent to receive messages from a particular brand. All the initial messages, in particular on the Internet, must not try to sell something to a future customer but to obtain his permission to continue to send him messages. To achieve this, the author designs a method based on the use of desirable "bait" that is of interest to the consumer and makes him willing to open the e-mails intended for him. The "bait" consists of a reward designed to interest the consumer within the targets the company has chosen. It is presented by e-mails or through mobile applications (apps).[11]

> *Microsoft gives priority to feeding its customers, whom it considers to be leaders of opinions, trendsetters with new information, to increase their knowledge in the fields that interest them. This preferential information, apart from improving their knowledge, reinforces their image as a leader with the members of their community. To obtain it, the interested web user clicks on the web pages. He then increases his relations with the company's experts who continue the conversation by meeting his expectations.*
>
> *British Airways provided the members of its "Executive Club" with regular information in the form of newsletters. They deal with the company's programs such as the opening and closing of routes, changes of frequencies and times, etc.*
>
> *In its program "Danone and you," Danone sends a newsletter twice a month to more than 250,000 registered people who agreed to receive it. All the media used lead towards the website "www.Danoneetvous.com." Several million people visit the site every year. It offers information in the form of articles and videos and gives advice. The main topics discussed deal with subjects such as health, slimming, energy, personal development, young mums, etc.*
>
> *The bait may be presented by means of discount e-coupons offered for a future purchase. They are sent by e-mail to be printed or again in the form of bar codes on a smartphone. The "Tesco ClubCard" uses them abundantly in the United Kingdom by means of mobile phone apps.*

To retain their appeal, the baits constantly change to maintain consumers' interest and to adjust to changes in their expectations. The use of internal data bases within the customer relationship management (CRM) system plays its part in providing the information for renewing them, taking into account what interests the consumers.

"Permission marketing," apart from avoiding overloading customers' brains by increasing the number of "spams" or superfluous e-mails, enables companies to avoid useless spending on e-mailing. The simple fact of asking customers if they would like to receive certain information before sending it systematically to everyone on the file, results in significant savings. This new practice avoids irritating consumers and filling their inbox or wastepaper basket with unwanted communication.

Desire-based e-marketing

Desire-based marketing is a topic already discussed in our Chapter 11 and Seth Godin highlighted it in a second book entitled *Purple Cow: Make Your Products and Your Company Remarkable!*[12]

Brian Halligan and Dharmesh Shah[13] propose to attract customers via the Internet by creating value-added content about the firm, its sector, the products and services it offers. They develop a methodology that helps to convert a random visitor into a profitable customer through a four-step process ("attract/convert/close/delight").

In his book *Inbound Marketing*: *According to the Sherpa strategy*,[14] Gabriel Szapiro develops the concept of "permission and desire-based marketing." The author suggests increasing companies' digital effectiveness and attracting consumers' brains by using four values in their message to increase desire: "humor, plots, the unexpected and seduction."

Many experts agree that, in a world inundated by marketing, communications and e-communications, it is not enough just to respond to consumers' tastes, needs and expectations. To attract the attention then the empathy of the neuro-consumer, it becomes necessary to take an interest in his desires, sometimes even his fantasies. This is particularly true on websites where it takes less than a second to decide to click on a page then whether or not to jump to the second one.

Surprising the Internet user with a new and original offer and communication policy has become necessary in order to make companies more appealing to consumers and to stand out from their competitors. It is crucial to position the brands of products, services or retailers in a "blue ocean," making competition irrelevant thanks to the "value innovation" proposal.

The neuro-consumer-actor must then become a neuro-partner linked to the brand by a genuine affective relationship.

> *Brands such as Apple, Nespresso, Sephora, Nature et Découverte, Abercrombie, Ikea, Décathlon and many others endeavor to create this relationship in order to establish a desirable, permanent and affective partnership link with their customers.*

The communications effect is reinforced by the search for surprises that the brain finds of interest. Frequently, they rely on the use of plots and the unexpected. The presentation of surprising events often wins acclaim and interest.

> *The publishers of* Harry Potter *promoted new books in the series late at night. Among experts in the field in Europe, Didier Reynaud, a former manager in the Generali Group, set up the Affiliance communications agency to promote original and surprising events for its customers. After having used Zinedine Zidane's collaboration to successfully improve the reputation of the insurance group, he created Generali-On-Ice, a show based on figure skating for the customers and insurance agents. More recently, he launched Melomania around the nostalgic theme linked to the legend of the musical comedies.*

The surprise effect on the Internet linked with the offer's originality and the desirability of the bait is an important factor in getting Internet users to start a relationship with a company and then to continue it through the company's website or blog. Yet, however important the communications that are trying to reach neuro-consumers' desires may be, the previously cited authors agree that they must always favor "the customer's permission" prior to any new attempts to enter into relationships.

The necessity for a personalized or "one-to-one" relationship with neuro-consumers

Neuroscientists agree that "self-interest" or "self-centeredness" is one of the brain's characteristics. ICTs and the Internet can take advantage of this to enable them to create a personalized relationship with each neuro-consumer-actor.

Owing to the development of microprocessors, companies can now store and manage large quantities of data in reduced spaces, at a very low cost. A PlayStation or an Xbox console costing only a few hundred dollars can now store as much information as an impressive computer in the 1970s, which, at that time, cost several million dollars. Computers' storage capacity is of less importance now that data may be archived outside of the hardware. Storage facilities, located in the "clouds" with "cloud computing," contain almost unlimited arrangements at attractive prices.

Thanks to these facilities, combined with those of the Internet, companies are gradually moving into an era of "big data," with "data driven-marketing" or "social CRM." Its aim is to build a database that is reliable and able to transform information coming from a variety of sources to allow them better knowledge of their clients. It enables personalized marketing, sales and communications policies to be set up and optimized on various relational contact points.

A vast quantity of data on customers may now be recorded, processed and returned in an operational way to the CRM systems, allowing companies to access information about customers' behavior and habits in order to set up a personalized and interactive relationship with them.

This new relationship approach is called "one-to-one" and is explained by Don Peppers and Martha Rogers[15] in their books and on their website: 1to1.com. The customer is identified within the CRM by crossing various knowledge elements that are specific to him: needs identified in the marketing databases supplemented by those detected through interactive relationships, profitability and risk criteria specific to him, assessment of his individual or family value for the company over time ("lifetime value"), the change in his anticipated needs from models of projected behavioral analyses, etc. Managing these data enables an appropriate service to be proposed or a personalized dialogue to be set up in real time. "One-to-one" proposals are made possible by optimization, thanks to the use of intelligent search agents, of the neuro-consumer's detected or expressed expectations linked with profitability or risk indications specific to him. The messages and the service may be accessed in real time by the Internet from a personal computer, mobile phone, tablet or television. "One-to-one" makes it possible to meet the personalization and interactivity expectations of the neuro-consumers' brain.

Making the website neuro-compatible to improve its effectiveness

The neuro-consumer's decision to open a website is made in a matter of milliseconds. An article published by several professors at Carlton University in Ottawa, Canada, Gitte Lindgaard, Garry Fernandez, Cathy Dubeck and John Brown,[16] showed that an Internet user only takes 50 milliseconds to decide whether or not to open a page. The bounce rate, which checks that the Internet user opens a second web page after having consulted the first, is also an important criterion enabling the quality and interest of an advertisement online to be assessed. To encourage "clicks," the neuro-consumer's brain needs to feel attracted almost immediately. Difficulties arise when websites tries to convey certain elements such as taste, smell and touch, which are important to the brain's perception. Experts try to compensate for these shortcomings in various ways.

Responding to the difficulties in addressing certain senses such as smell, taste and touch on a website

The difficulty in addressing certain senses is a disadvantage for websites, particularly in comparison with other points of sale. Retailers who seek to provide a sensorial experience have undeniable advantages for the neuro-consumer's brain. The "multichannel" policy combining the Internet with physical trading, to which a number of activity sectors have given priority, provides a response to this difficulty. Nevertheless, professionals endeavor to find ways of compensating for the Internet's sensorial shortcomings in order to appeal to the neuro-consumers' brain.

> *In 2004, Yvan Régeard set up Exhalia. It uses a patent that France Télécom created and then sold. Its aim is to transmit smells via the Internet. The company proposes synchronizing pictures, sounds and smells on the Internet, linking information sent to professional or personal disseminators who make it possible to experience the essences of the products presented. The world leader in the field, this company has more than one million smells. It offers an olfactory USB key called iSample. Exhalia produces olfactory visits to the Bureau Interprofessionnel des Vins de Bourgogne (BIVB; Interprofessional Bureau of Burgundy Wines) vineyard and the cellars on the Internet. It works with a Japanese hotel school that experiments with cooking courses online.*[17]

Among the main criticisms of online shopping raised by neuro-consumers is that of transmitting a tactile experience through the Internet. Retailers are aware that if a customer can feel a product, the chances are greater that it will be bought rather than being put back on the shelf. Known as the "endowment effect" or the "possession effect," put forward by Daniel Kahneman, Jack Knetsch and Richard Thaler[18] in 1991, it shows that the consumer places a higher value on objects they "own" in relation to objects they do not and explains why a growing number of brands such as Apple allow their products to be touched on the shelves.

Margaux Limoges[19] proposes a set of solutions to compensate for the lack of tactile experience with products online:

- The inclusion of pictures in 2-D or 3-D on the website to allow the qualities of the products to be shown.
- Staging the product to include the consumer in the action, such as stirring a pot of yoghurt on the side of the consumer's dominant hand, picking up a hamburger with the fingers. The aim is to help the neuro-consumer's brain to imagine using the product through stimulating his mirror neurons. The effect of sending an imaginary handling sensation is also known as the "motor fluency effect."[20]
- A description or sensorial comments to bring to mind the tactile properties of the product (texture, feel of the materials, dimension, size, weight, etc.).

The softness of cotton on the skin for a bath towel. The pleasure of feeling a moisturizing cream produced by aromatherapy on the face. The ease of picking up, of handling, the lightness of a drill or an electric screwdriver.

According to Limoges, "Women tend more to touch an object for the pleasure while men touch them more to obtain information on the product." In her study, she also notes that "85% of the women interviewed said that what they missed most when buying 'material' products online is the fact that they cannot touch them, compared to only 53% of the men interviewed." She also commented that "a product's experts give more importance than novices to viewing sensorial presentations on a web page."[21]

The interest in providing a sensorial impression on the Internet is greater for so-called "material" products, for which the tactile properties are important when used, e.g., sweaters, duvet covers, shampoos, paper handkerchiefs, etc. It is less so for "geometric" products, for which these properties are not particularly important when used, e.g., DVDs, mugs or cups, batteries, packets of biscuits, etc. On certain major online trading sites such as Amazon, the "material" products have fewer tactile descriptions, however the company partially compensates for this by a generous policy of returns, reimbursement or exchange.

Experts have suggested other methods to compensate for the difficulty in providing sensorial impressions online:

- Use customers' comments relating to the products' sensorial effect in order to improve their shopping experience.
- Get users' testimonials regarding their tactile impressions when manipulating the products shown on the website.
- Link the online sale to the possibility of seeing, testing or trying the product in a store.

  The Monsieur Lacenaire brand of sweaters for men and many others use this system. The customer who buys online can go to a store. The sales staff are given information on him beforehand and can welcome him as if he were a "very important person" (VIP) during his visit.

- Allow the customer to try out the product for a limited time with the option of returning it. This policy enables the "endowment effect" felt by the consumer to be intensified.

  With its QuietComfort helmet, the company Bose guarantees return and full reimbursement if the customer is not satisfied. The reimbursement is of course subject to the helmet being returned in perfect condition. The sales formula with the principle "satisfied or reimbursed" is a practice increasingly seen online.

- Send "boxes." For several years now, the concept of sending boxes has been growing fast online. The interested consumer receives a selection of products that he can touch and compare. He keeps the ones he finds suitable and sends the others back. This online sales formula is developing in an increasing number of sectors such as textiles, cosmetics, holidays, gastronomy, etc.

  The company Smartbox proposes about 60 boxes on several themes: gastronomy, well-being, unusual holidays, premium gifts, bargains, etc. The products are available online and in points of sale. The cost of the box varies from about €50 to €400. For €13 per month, the company Birchbox offers a box containing a cosmetic surprise with four to six "beauty" miniatures to try out. Trois Fois Vin offers a subscription whose price varies from €19.90 to €39.90. It gives the right every month to two bottles of wine to sample that have been selected depending on the cost of the subscription. They are accompanied by videos, sheets with tasting advice, preferential prices for buying the bottles as part of the offer and so on.

The recent development of "haptic" technologies to give tactile impressions to screens is likely to provide interesting solutions to respond to certain sensory gaps on the Internet.

Make it easy to read websites by making them neuro-compatible

Only a few milliseconds are needed for the Internet user's brain to see whether a website is of interest to them. Once his impression is formed, it is difficult to make him change his opinion. In his book *Emotional Design: Why We Love or Hate Everyday Things*,[22] Professor Emeritus of cognitive sciences at the University of California, Donald Norman, shows that consumers who find the design of a website attractive, also find it easier to use. So, to make a good impression with Internet users, many companies employ professional web designers to create their websites as they know the rules for harmony, particularly those linked to the "golden number" ratio and the Vitruvian principles mentioned in Chapter 21. Once designed, it should be tested by a group of people corresponding to the target customers of the company. The number of openings and the bounce rate (Internet users going on to a second web page after consulting the first one) are good indicators of its attractiveness to users.

A website is all the more interesting to the neuro-consumer's brain if he has a greater impression of being in contact with a person rather than to the site's computer. Personalization using a sympathetic face with an air of interest adds to this impression.

According to neuroscientists, "self-centeredness" is one of the characteristics of the reptilian brain, which takes decisions rapidly. It seems preferable therefore to design the website with a focus on the consumer's expectations rather than the company's offers and characteristics.

Certain ideas enabling the design and particularly the packaging (see Chapter 17) or the messages (see Chapter 21) to be improved could help in creating attractive websites, for example:

- *Reward the visitor* so that, in return, he consults further pages on the website.
- *Exploit the sense of rarity* by alerts informing users that there is only a limited number of products available, that the stock will shortly run out, etc. Amazon frequently warns customers for certain products that there is a limited number still available in its stock.
- *Create websites that are easy to consult.* Less text, colors, illustrations or distractions, makes the website more appealing and easier to use.
- *Make the visitor's imagination work* with the use of plots or questions.
- *Avoid placing the logo or an important message* in the place on the web page that Dan Hill,[23] consultant and CEO of SensoryLogic, calls "the blind spot," i.e., the lower-right-hand part of the page. Hill suggests that the best place for a relevant message seems to be the bottom of the middle of the page. The criteria used to make a shop display eye-catching (see Chapter 17) can also be applied to a website.
- Faced with the large number of choices available on websites, the brain may become stressed and give up searching. To avoid this, the e-commerce sites help the customer to choose. Sometimes it is necessary to offer a guide that helps consumers to find their bearings in the hyper-choice world.

  Amazon tries to respond to this problem by orienting its customers. The company keeps track of their earlier searches and buying behavior on its site. On the basis of this data, it identifies interest "clusters" that it attaches to the customers to guide their choice towards new products linked to their expectations. For instance, to sell books, the company keeps the personal list of what consumers have previously shown an interest in without buying them immediately. Their attention is drawn to these books during later visits or by an advertisement sent by e-mail. Apple uses a system of collective intelligence called "Genius," which works with iTunes. It suggests music that customers are likely to appreciate by taking account of purchases already made or by comparing them with other persons with a similar consumer profile. By reducing the effort the brain makes to choose, the system increases users' enjoyment of the online experience and plays a part in increasing sales.

Notes

1. Jacques Lendrevie and Julien Levy, *Mercator*, Dunod, 2012.
2. Alan Turing, "Computing Machinery and Intelligence," *Mind*, vol. 59 (October 1950), pp. 433–460; John von Neumann, *The Computer and the Brain*, Yale University Press, 2012.
3. Gordon Earl Moore, "Cramming More Components onto Integrated Circuits," *Electronics*, vol. 38, no. 8 (1965), pp. 114–117.
4. Jan Zimmerman, *Web Marketing for Dummies* (3rd ed.), John Wiley & Sons, 2012.
5. Sehar Shoukat, "Cell Phone Addiction and Psychological and Physiological Health in Adolescents," *Experimental and Clinical Sciences Journal*, vol. 18 (February 2019), pp. 47–50; Tayana Panova and Xavier Carbonell, "Is Smartphone Addiction Really an Addiction?" *Journal of Behavioral Addictions*, vol. 7, no. 2 (June 2018), pp. 252–259.
6. Google, *Our Mobile Planet: France – Understanding the Mobile Consumer*, Ipsos MediaCT, May 2013, pp. 1–39.
7. Marie-Claude Vergara, *Le neuromarketing et le marketing sensoriel pour les points de vente: augmenter leur utilité dans l'amélioration du chiffre d'affaires*, MSc thesis, HEC Paris, May 2015.
8. François Cazals, *Stratégie digitale: La méthode des 6 C*, De Boeck Université, coll. HEC Paris, 2015.
9. http://lexpansion.lexpress.fr/entreprises/patrick-le-lay-president-directeur-general-de-tf1_1428488.html.
10. Seth Godin, *Permission Marketing: Turning Strangers into Friends and Friends into Customers*, Simon & Schuster, 2007.
11. Ibid.
12. Seth Godin, *The Purple Cow: Transform Your Business by Being Remarkable*, Portofolio Edition, 2009.
13. Brian Halligan and Darmesh Shah, *Inbound Marketing Revised and Updated: Attract, Engage and Delight Customers Online*, Wiley, 2014.
14. Gabriel Szapiro, *L'Inbound marketing selon la stratégie du Sherpa*, Jacques-Marie Laffont, 2015.
15. Don Peppers and Martha Rogers, *Enterprise One-to-One: Tools for Competing in the Interactive Age*, Bantam Doubleday, 1997.
16. Gitte Lindgaard, Cathy Dubeck and John Brown, "Attention Web Designers: You Have 50 Milliseconds to Make a Good First Impression," *Behavior and Information Technology*, vol. 25, no. 2 (March–April 2006), pp. 115–126.
17. Several illustrations of its offers are available at www.exhalia.com.
18. Daniel Kahneman, Jack L. Knetsch and Richard H. Thaler, "Anomalies: The Endowment Effect, Loss Aversion, and Status Quo Bias," *Journal of Economic Perspectives*, vol. 5, no. 1 (1999), pp. 193–206.
19. Margaux Limoges, *Dans quelles mesures le e-commerce souffre-t-il du déficit d'expérience tactile, et comment ce manque peut-il être compensé?* MIM dissertation, HEC Paris, March 2013.
20. Ibid.
21. Ibid.
22. Donald Norman, *Emotional Design: Why We Love (or Hate) Everyday Things*, Basic Books, 2005.
23. Dan Hill, *About Face: The Secrets of Emotionally Effective Advertising*, Kogan Page, 2010.

CHAPTER 25

The brain and the emergence of social networks

Gradually, and often without noticing, by regularly referring to the opinions of the online community when making purchasing decisions, the neuro-consumer is replacing his individual conscience with a collective conscience. This change is consolidated when it becomes a regular habit for his brain. The advent of social networks and online communities and the interactivity of objects with each other and with people is profoundly modifying the perceptions and behavior of the neuro-consumer's brain.

Digital communities today can form in record time. They call for rapid adaptation of companies' marketing and communication methods. They impose new standards on brands in order to meet their needs. Companies that fail to adapt run the risk of being excluded by a large number of customers influenced by social networks, which can prove detrimental to their development. Good understanding of community-based behavior and interactivity with consumers and the networks to which they belong are conditions for success. To achieve this, marketing and communication departments can call on professionals in the field. Their role is to recommend how to place digital and viral strategy at the heart of the company. The strategy must win over customers for the company and encourage their loyalty by including the increasing influence of the networks in its digital policy.

The advent of social networks and the interactivity of objects

For centuries, men have gathered together in communities. There are also communities in the animal kingdom and humans are attracted to some of them – for example wolves, monkeys, bees and ants – by the quality of their organization to the extent that they sometimes serve as models. Originally, grouping into communities enable the members to cope with a hostile environment. Over time, these gatherings have diversified. Currently, they cover many fields likely to draw people together: geography, ethnic background, belief and culture as well as cultural, scientific, artistic and sexual orientation.

On the basis of experiments carried out using magnetic resonance imaging (MRI), psychologist and psychiatrist Matthew Libermann, professor at the University of California, states that "our brains are built to think about the social world and our place within it."[1] According to Libermann, we are biologically orientated towards others. He argues that the brain is social by default and that our socialization does not come from cultural learning related to upbringing or social relationships but from a physiological relationship present at birth. Experiments carried out in the United States by several researchers including Wei Gaoa, Hongtu Zhub and their colleagues confirm this statement.[2] The behavior of the neuro-consumer's brain is subconsciously influenced by the norms of the social group to which it belongs.

Most types of individual behavior in the various communities obey common laws or imperatives: consideration for, and following of, "leader(s)," listening to experts in a given field, pressure from the group and how its members view the individual, the feeling of reciprocity and equity among them, requesting advice from those who have lived through certain experiences, etc. In certain tribes from different parts of the world, ideas propounded by the elders are respected and listened to because they come from individuals who are experienced due to what they have lived through. This is the case in many countries in Africa and Asia. Ancient tribes mostly get together in communities based on blood relationships. The new tribal world is organized to a greater extent around common projects and passions. Moreover, today an individual can belong to several "tribes."

Online social and community networks have appeared in recent years and the importance of their influence on neuro-consumers' behavior is spreading like wildfire across the globe. Companies trying to adapt to the digital revolution are faced with a new challenge conditioned by the influence of social networks and must adapt their traditional marketing strategies in order to meet their needs. The first difficulty for marketing is to understand the changes in the behavior of neuro-consumers caused by the emergence of these new types of community relationships. A second concern is to adapt marketing and communication strategies to respond to their effects on the neuro-consumer's brain. Many books about online social and community networks have been published, which enable readers who are interested to go into this subject in greater depth.[3]

Web 2.0: the emergence of social networks in the digital world

Once the Internet had become established, it was February 2004 before Marc Zuckerberg created the first community-based social network,

Facebook, in the United States. Since its launch, the rise of Facebook has been meteoric, as shown in David Fincher's film *The Social Network* (2010). If we compare Facebook to a country whose inhabitants are all subscribers, it is already bigger than China, the most populous country in the world. Facebook uses nearly 70 languages. It is present in most countries, on all the continents, and is the leading social network in the great majority of them.

Other networks have emerged since the creation of Facebook and while they also have many subscribers, they do not equal its power. Some are international, like those run by Google (YouTube and Google+), while others are more local, like Qzone (China), VK (Vkontakte, Russia) and LINE (Japan). Their objective and design show specific differences: Facebook is a network based on relationships and sharing; Twitter is a newsfeed; LinkedIn a professional network; WhatsApp (2014) a mobile communication platform; YouTube a video-sharing network; Pinterest an image-sharing network; Instagram (acquired by Facebook in 2012) a mobile application and a site for sharing photos; Jelly.co (2014) an image-sharing network. Some networks are specialized, for example dating networks like Meetic, eDarling and Ashley Madison. Others provide access to the collaborative economy between private individuals, such as Uber, Airbnb and BlaBlaCar. Their common characteristic is the extent to which they profoundly modify their subscribers' traditional behavior.

The creation of online community networks

In addition to subscribing to social networks, "internauts" create their own communities within which they interact using different digital media and, increasingly, mobile phones and tablets. Beyond the Facebook page, there are increasing numbers of interventions, such as "tweets" and "snaps" on forums, podcasts and contributions to "wikis," etc. The personal blog is the core of the interactive relationship with the community of subscribers that comes to visit it. For many years now, consumers have commented upon and recommended their purchases. Traditionally, they communicate with a limited group of friends and relations. With personal blogs or writing reviews on company websites about their products, their opinion is expressed and disseminated to a considerably bigger national and international audience.

An increasing number of amateur blogs assess or talk about companies and their brands, products, communication, etc. Companies work to create communities and maintain an interactive relationship with them. Beyond websites, they ask their digital marketing specialists to produce and manage a relational blog and sometimes a club based on regular use of the Internet. There are already many community-based company blogs and clubs.

> *Some are already well known and have many active members, for example Nespresso, Danone, British Airways, Tesco, Hilton, Lancôme and Harley Davidson. In the United Kingdom, Tesco has referenced more than 10 million members in its various specific clubs. The Harley Davidson club has nearly 800,000 subscribers.*

Companies benefit by developing permanent relational marketing with a group of customers who are attached to the brand.

Social customer relationship management (CRM) has the advantage of developing customer loyalty using well-adapted tools. It makes the company quickly aware of quality problems that might upset customers and enables it to respond without delay. It helps to enrich the value of brands by creating emotional attachment with the members of their community.

Really good blogs and clubs develop a relationship that goes beyond information and remuneration. They seek the affection of their subscribers.

> *The Harley Davidson club or HOG (Harley Owners Group) is a community of customers.[4] The company has created the biggest bikers club in the world with many active subscribers. It focuses on creating and maintaining a cognitive and emotional relationship with them. More than a mythical product, Harley Davidson is a community of enthusiasts grouped around the brand, its customers and dealerships. With the motto: "Ride to live, live to ride," members get together at weekends and share their passion. The company and its distributors work continuously to keep the community active. The HOG organizes many events and cavalcades through its 1,400 local clubs. The emotional aspect of the relational program is considered inseparable from the product. Beyond the machine, the motorcycle manufacturer offers an experience, a lifestyle, a relationship with a group that shares the same enthusiasm for the brand. Constantly updated communication informs club members of bike-related events.*

As well as creating their own community-based blogs and clubs, companies also contact and develop relationships with professional or semi-professional bloggers in order to interest them in their brand, their products or services or their communication.

> *McDonald's is interested in "mom bloggers" because they talk about the brand and sometimes have big audiences.[5] Some companies are considering how to create a "relational identity" for their brand by reorienting their strategies to interest the community-based and social networks.*

Web 3.0: connected objects

After the arrival of online communities and social networks, a new development is now under way: the interactivity of objects. Already known as "Web 3.0," it has been made possible thanks to the use of the IPV6 protocol. This offers quasi-infinite possibilities for addressing objects (340 billion billion addresses are available). Companies are starting to become interested in what some professionals already consider to be a new Eldorado for the Internet.

> *Applications have appeared or are at an advanced stage of development: "Google Glass" advanced reality glasses, Orange's Airbox Auto, connected watches, driverless automobiles, drones, robots, urban technologies for households, telemedicine services, e-health, etc.*

Forecasts speak of 80 billion connections by 2020.[6] The Web 3.0 interface enables multiple connections of objects between each other but also with human beings.

> *The new systems use developments of artificial intelligence technologies. They claim to make people's lives easier at no cost, to answer all their questions, to permit free local or international communication.*
>
> *Siri (Apple) and Cortona (Microsoft) respond to personal requests. Facebook wants to include an assistant in Messenger – Facebook M – which will respond to written requests such as booking a flight, a hotel room or a table in a restaurant. If necessary, its artificial intelligence system will be able to receive assistance from a team of humans: the "M Trainers."*
>
> *Google Now can manage an individual's daily timetable without any outside help. It anticipates the weather, suggests the optimal route to get to a meeting, calculates the journey time and enables the user to arrive on time under good conditions.*

In not too distant future, the Internet is expected to completely replace the director's secretary. The director who receives a simple invitation to the theater for two people will see his digital system confirm the booking, send an e-mail to his partner or friend to suggest they accompany him, book a babysitter for the children, book a restaurant for after the show, suggest postponing less important meetings until later dates available in his diary, etc., without any intervention whatsoever on his part.

With the huge increase in interconnections on the Internet and through online communities and social networks, we are seeing exponential growth in the collection, management, processing and communication of data.

According to François Cazals, "90% of the data ever produced by humanity have been generated over the last two years. The volume will continue to grow considerably to deal with the programmed interaction of objects with each other and with human beings."[7] At a conference, Eric Emerson Schmidt, executive chairman of Google, stated: "Every two days now we create as much information as we did from the dawn of civilization up until 2003."[8] We are entering the world of "big data." It presents new challenges that companies will have to deal with quickly.

All the phenomena produced by the advent of Web 2.0 and Web 3.0 have led to significant transformations in the behavior of the neuro-consumer-actor's brain.

Purchasing decisions: from individual to collective

Neuro-consumers' increasing interest in online communities and social networks has led to a change in the behavior of their brain once they join. Companies' communication needs to adapt quickly to deal with this change. They need to put into place specific viral e-marketing, call in experts with good knowledge of how the networks function and develop strategies to encircle customers using the whole range of tools available on the Internet.

Changes in the behavior of the neuro-consumer's brain when he joins a social network

At the beginning of the year 2000, a group of American Internet enthusiasts and creators of start-ups – Rick Levine, Christopher Locke, Doc Searls and David Weinberg – published a seminal work on social networks, *The Cluetrain Manifesto: The End of Business as Usual.*[9] As the authors put it: "Conversations are markets." This development calls for a profound change in the way companies behave. To succeed in this context, it is no longer enough to segment their clientele and then address each segment or even individual customers. Neuro-consumer-actors who use the social media require that communication aimed at them should be present where they talk, exchange ideas, voice their opinion, talk about brands and share their experience. Today, communication must reach the customer on his own ground. To be heard, communication cannot just reproduce its usual methods of approach but has to penetrate this new space, be accepted by the auditors and be part of their discussions. Good knowledge of how members of these networks behave and the types of relationship that they expect is essential. Many books have been written on this subject.[10]

- Members of communities and social networks no longer want communication to be thought of as a shop window, but as a fully integrated component of an exchange. They consider themselves as transmitters as much as receivers. They want communication to have several qualities:
    - *Interactivity and reactivity*: A dialog is above all an exchange. The internaut's comments must receive quick responses, particularly on mobile devices as when he communicates with friends.
    - *Esteem*: The internaut must have the impression of being listened to, respected and appreciated.
    - *Consistency*: Communication must not be limited to a time and a place chosen by the company. It must be continuous. The information transmitted must be updated depending on current events and the needs of the auditors.

Internauts do not want to be spoken to as customers, but rather as friends. The marketing specialist must take off his expert's hat and continually ask himself the question: "When you talk to a friend, what do you talk about?" The neuro-consumer-actor who belongs to a network is primarily interested in useful information. He wants to be entertained by a funny viral film. He likes to be amused by an original game. Means of communication are all the more appreciated when their interest or humor makes the internaut want to transmit them to the rest of the community.

To be able to please the internaut, the company must be prepared to deal with the influence of the community on individual purchasing decisions.

The company must also accept that it will lose control over its communication to a significant extent. At the risk of being severely criticized, its experts in relationships with the networks can only influence without transgressing the rules that govern relationships in the community. Products and brands are not only commented upon, but they can find themselves being condemned, hijacked or promoted. Publicity is sometimes modified and often parodied. The creativity of internauts, which is generally turned towards humor, is limitless. It can have a positive effect by helping to make a brand more well known, but also a negative effect by mocking or denigrating the company's messages.

The development of online communities and social networks is leading to a situation where collective conscience is gradually replacing individual conscience, where the opinion and recommendations of community members become more important than companies' communication and advertising. The decision about whether to buy a product or service is often influenced by their opinion. Nowadays, an increasing number of products and services are not bought if they receive negative opinions on social networks. The role of experts and leaders is becoming continually more important. Companies' communication policies are allocating an increasing amount of space to viral marketing. A new conception of marketing should enable companies to respond to this change.

Viral marketing

Viral marketing is dependent on the customer being completely satisfied. It is based on a policy of flawless quality and maximum customer loyalty.

Frederick Reichheld,[11] former professor at the University of Harvard and chairman of the Bain consultancy, suggests that among all the questions that could be put to customers, one is fundamental as its answer has the strongest correlation with their loyalty but also with positive word-of-mouth recommendation on their part. The question is: "What is the probability that you would recommend this product, company or brand to a friend or colleague?" From the information obtained, Reichheld produces an index called the "net promoter score." Those questioned must answer using a Likert scale ranging from 0 ("completely unlikely") to 10 ("completely likely"). Customers who give a score of 9 or 10 are called "promoters." "Detractors" score from 0 to 6. The evaluation is stringent because those who say they are "likely" get a score of 6 and are grouped with the "detractors." The final score is a percentage that corresponds to the number of "promoters" minus the number of "detractors." It is often negative. However, some companies, such as eBay and Amazon, have obtained a high "net promoter score" of 75% to 80% in many countries.

A good "net promoter score" is the essential starting point for developing a viral marketing strategy. Also called "buzz marketing" or "word of mouth," viral marketing is an essential tool for spreading, at very little cost, positive communication through online communities and social networks. Word of mouth has long been considered one of the most effective means of spreading an idea. Advertising specialist William Bernbach (1911–1982), cofounder of the DDB advertising agency stressed its importance: "You can't sell to someone who's not listening. Word of mouth is the best medium of all, and if blandness never sold a product, neither did irrelevant brilliance."[12]

The neuro-consumer-actor's brain, saturated by messages from online publicity and e-communication, tends to place less and less trust in those coming from traditional communication. He prefers to refer to the opinion of customers who have already tested brands, products and services. He seeks their evaluation by contacting them directly via online communities and social networks and on specialized blogs. He looks carefully at critiques from those who have already tried the product or service, e.g., at the number of stars given a product on Amazon. He is interested in Facebook "likes." Before selecting a tourist destination, he looks at the comments on TripAdvisor. To book a restaurant in North America, he looks at the evaluations on Yelp.com.

With the development of the Internet, blogs and online communities and social networks, recommendation is increasingly becoming more important than the actual communication and advertising of companies. Experts can become opinion leaders and sometimes transform themselves into "contaminators," veritable evangelists for a product, service or brand in their community. They have increasing influence on what the neuro-consumer-actor's brain perceives.

Companies' e-communication policy needs to adapt to the development of these new trends. Rather than trying to speak directly to customers, with the risk of seeing its messages lost in the hubbub of advertising, the company must get neuro-consumer-actors to talk among themselves about the company, the brand and the products. They trust the opinions of members of their network who have experience of the product or experts in the field more than messages sent out by companies.

To increase their effectiveness, companies introduce viral marketing strategies adapted to the expectations of customers who are part of the world of communities. The spread of the company's message is difficult, or even impossible, to control and therefore is essential to get it right. Errors in communication will be rapidly widely circulated through communities and can have a detrimental effect on the company if the messages do not meet members' expectations. The success of a viral marketing policy is largely dependent on the professionalism with which it is implemented.

To succeed, it is necessary to identify the communities of interest to the company, find their leaders, try to turn them into "contaminators" and place the message where it will be spread the most, in what the experts call "the hive." Its "humming" amplifies the communication, feeds the buzz, adapts the creativity of the company's e-communication media (sites, blogs, viral films, games, mobile applications, etc.) to the specific expectations of the listeners and the relational imperatives proper to each community. Seth Godin's book *Unleashing the Ideavirus: Stop Marketing at People! Turn Your Ideas into Epidemics by Helping Your Customers Do the Marketing for You*,[13] like that by Karim B. Stambouli and Eric Briones,[14] is full of recommendations. They help to avoid certain errors made by neophytes. They help to develop an effective word-of-mouth policy based on community relational marketing using techniques specific to viral communication.

Careful thinking about the new methods of segmentation proves to be useful when developing a viral marketing strategy. Some experts, like Gabriel Szapiro,[15] have identified ambassadors, early adopters, influencers, deciders and users. Others, such as market research company Forester,[16] have produced categories that they call inactive members, spectators, participants, collectors, critics, leaders and creators.

The need to use Internet and social network professionals

Beyond the themes mentioned above, communication with the communities and social networks is important in disseminating the value of a brand.

They form an effective "hive" to give it appealing content that will feed debates among members.

> *Converse sports shoes provide an example of how a brand can create a positive aura, widely spread across the social networks.*

Traditional marketing, mainly used to practice a form of unidirectional communication, finds itself helpless when it has to deal with social networks. To cope with this, companies have to call upon experts in the field, be they in-house or external consultants. As a result, new professions are emerging, such as webmaster, community manager, social network officer, etc., who can provide much-needed help.

The job of these new digital marketing experts is to organize a digital marketing strategy aimed at Internet-based customers and in collaboration with the communities and social networks. They must create a digital ecosystem that corresponds to the company's positioning and its values. They must integrate it throughout the company's value chain, remodeling structures and persuading the rest of the personnel of its value. Their roles include the following tasks:

- Collaborate in defining a digital strategy in relation with the networks and coordinated with the company's brand and communication strategy – define a relational orientation for the brand.
- Identify the communities that could interest the company.
- Choose the e-communication tools that will enable them to implement a strategy of encircling their customers.
- Create an ecosystem of company presence on the communities and social networks.
- Select the social media preferred by the consumers the company wishes to win over or whose loyalty it wants to increase.

> *Although Facebook, YouTube, Twitter, etc. are indispensable, other, more restricted networks may be better adapted to a specific clientele or market. LinkedIn and Viadeo enable the company to target professionals. Some local social networks have a good audience and a specific interest for internauts, e.g., Ozone, Sina Weibo and Pinduoduo in China; LINE in Japan; KakaoTalk in South Korea; VK (Vkontakte) in Russia; Xing in Germany; Les Copains d'Avant in France.*

- Select networks depending on how well they are adapted to the configuration of the message: short or long texts, images, photos, videos, audio, etc.
- Define an editorial policy for blogs and websites on the networks. A page is like a television channel. It needs to present a program whose content arouses the visitors' interest.

- Contribute to the production of written, audio, video, etc. content for blogs, viral films, games, etc.

  Dior has published a video on YouTube called "Secret Garden Versailles," which has received many visits. L'Occitane presents interesting reports related to the brand and to Provence. Isover produces videos that give a lot of advice to users.

- Choose national or international themes.

  Companies like BMW and L'Occitane are present on the national pages of the social networks in around 50 countries.

- Encourage, manage and coordinate the participation of the company's employees in the blogosphere and on the networks.

  Companies such as Microsoft and Orange encourage this type of collaboration.

- Solicit the participation of customers and the community. Journalist Jeff Hove calls this "crowdsourcing."[17] Many companies solicit the community to help them in different fields.

  Danette asked the community to help create the flavor for a biscuit. Hasbro, manufacturer of Monopoly, sought help with choosing a new French town to include in the game. French railway SNCF asked for help from communities in Europe to help improve facilities for mobile telephones in trains. Starbucks uses "Starbucks Ideas." Procter & Gamble takes a strong interest in ideas originating from the community for future developments with its "open innovation" concept.

Community-based sites are also solicited to help improve – at reduced cost – certain aspects of in-house communication, such as the creation or modification of a logo, design, graphic style, etc. Creads, a participative agency, mobilizes around 60,000 creative people. 99-Design and other sites also make creative experts from all over the world available to companies, at very low prices. Very small businesses and small-and-medium-sized enterprizes (SMEs) with limited means can benefit from these community-based experts in the following ways:

- choosing social media, booking, adapting, organizing and managing paid media: purchasing of links, referencing, placing of messages, etc.;
- collecting information and carrying out surveys of visitor profiles – how they use the different networks and what they expect – and spotting the appearance of crises or discontentment, tools such as Hubspot and Hootsuite help to categorize and analyze social network data;

- checking the effectiveness, relevance and profitability of digital communication policies and the messages sent out – such checks are all the more essential when these digital activities are remunerated.

The success of a digital communication strategy involving neuro-consumer-actors on the networks is highly dependent on the professionalism of those who implement it. This is absolutely essential in developing a policy of encircling neuro-consumers using the whole range of means available through the Internet.

Encircling the neuro-consumer with a coherent digital strategy

In order to be able to respond appropriately to all the expectations of customers who surf the Internet and subscribe to communities and social networks, the experts' main task is to develop an encircling policy based on the Internet ecosystem.

Policy for encircling a customer using a digital strategy

First, it is necessary to define an interactive strategy incumbent on e-communication, harmonized with the company's overall relational communication strategy (Figure 25.1).

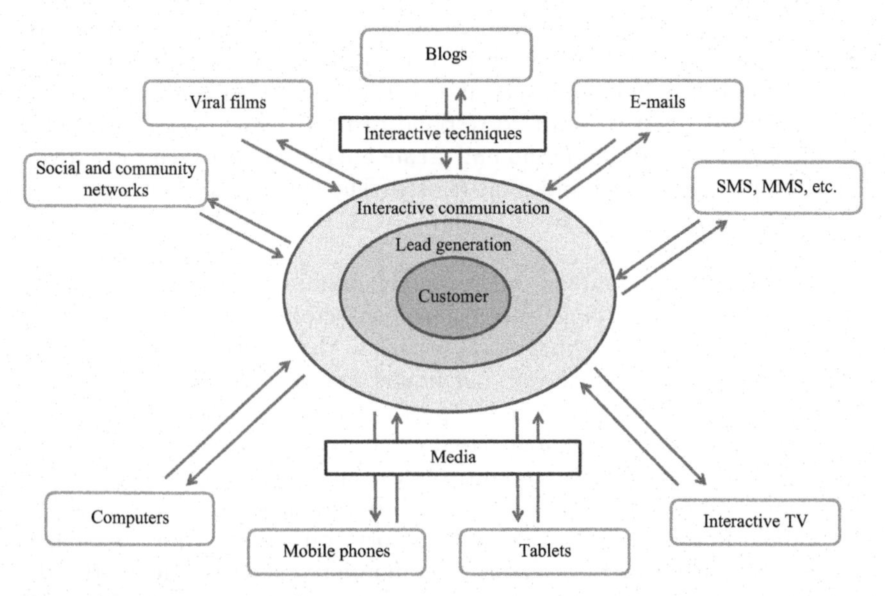

Figure 25.1 Internet encircling strategy

Information enabling the company to better know the neuro-consumers who show an interest in the brand or its products and services or who contact the company via the Internet generates leads that are stored and analyzed in the CRM system.

Because of its interactive nature, the company's blog is the hub of its relational and e-communication policy. It complements the website, which is more passive and often remains a catalogue of the products offered by the company. The blog becomes the relational soul of the company. It is destined to maintain a permanent relationship with the customer. To achieve this, it must attract the customer to visit it regularly by proposing themes, renewed periodically, matching his centers of interest. To do this, it can present information likely to please him: white papers, concrete case studies, guides to good practice, online lectures at set times announced by mail or "webinars," etc.

The blog can offer fun by creating games related to the company's products or brand characteristics and called "advertgames." A good, funny viral film or saga of such films attracts the attention of the neuro-consumer's brain and encourages the relationship with his community. They may be retransmitted or rewarded on online communities and social networks by the internaut and generate interest in meeting new visitors.

The company's presence in the appropriate communities and social networks provides an entry for the numerous innovations, propositions and relations presented in the blog. All the available interactive techniques (e-mail, SMS, MMS) that disseminate written messages, videos, images or photos are used. The messages are adapted so that they can be read on the various devices used by each customer: computer, tablet, interactive television, mobile phone, etc.

The strategy of using the Internet to encircle the neuro-consumer-actor enables the company to establish permanent interactivity with its customers. In order to avoid any dispersal, generating a risk of confusion for the internauts, the strategy must correspond to the orientation recommended by the positioning of the brand and create a harmony between the various digital networks, media and tools chosen. The concern for harmonization also calls for consistency to be achieved with the company's overall communication policy.

The strong attraction of the community-based networks leads us to consider what is leading neuro-consumer-actors to change gradually from the concept of an "individual brain" to that of a "collective brain." Neuroscientific research shows that brains can be synchronized around a sensory-based community. Music represents a good example of the emergence of auditory communities based on a common perception or feeling among different brains for certain sounds, musical themes, composers or musicians. Brains can also become synchronized around other senses, like taste, then leading to the creation of sensory communities. Just as companies are showing increasing interest in communities and social networks, the emergence of communities linked by the senses is an interesting subject of investigation for the neuromarketing and sensory marketing of tomorrow.

Notes

1. Matthew Libermann, *Social: Why Our Brains Are Wired to Connect*, Crown Publishing, 2013.
2. Wei Gao, Hongtu Zhu, Kelly S. Giovanello, J. Keith Smith, Dinggang Shen, John H. Gilmore and Weili Lin, "Evidence on the Emergence of the Brain's Default Network from 2-week-old to 2-year-old Healthy Pediatric Subjects," *PNAS*, vol. 106, no. 16 (April 2009), pp. 6790–6795.
3. Shama Hyder, *The Zen of Social Media Marketing: An Easier Way to Build Credibility, Generate Buzz, and Increase Revenue*, BenBella Books, 2016; David Meerman Scott, *The New Rules of Marketing & PR: How to Use Social Media, Online Video, Mobile Applications, Blogs, News Releases, and Viral Marketing to Reach Buyers Directly* (4th ed.), John Wiley & Sons, 2013; Romain Rissoan, *Réseaux Sociaux: Comprendre et maîtriser ces nouveaux outils de communication*, ENI Editions, coll. "Marketing Book," 2014; Mathieu Chartier, *Guide complet des réseaux sociaux*, 2013; Mélanie Hossier et Olivier Murat, *Faire du marketing sur les réseaux sociaux*, Eyrolles, 2014.
4. Jacques Lendrevie and Julien Levy, *Mercator*, Dunod, 2012.
5. Keith O'Brien, "How McDonald's Came Back Bigger Than Ever," *New York Time Magazine*, May 4, 2012.
6. François Cazals, *Stratégie digitale: La méthode des 6 C*, De Boeck Université, coll. HEC Paris, 2015.
7. Ibid.
8. Eric Emerson Schmidt, Techonomy Conference, Lake Tahoe, August 5, 2010.
9. Rick Levine, Christopher Locke, Doc Searls and David Weinberg, *The Cluetrain Manifesto: The End of Business as Usual*, Perseus Books, 2000.
10. Jacques Lendrevie and Julien Levy, "Le marketing des réseaux sociaux," in *Mercator*, pp, 571–635; Pierre Mercklé, *Sociologie des réseaux sociaux*, La Découverte, coll. "Repères," 2011.
11. Frederick Reichheld, "The One Number You Need to Grow," *Harvard Business Review*, December 2003 and *The Loyalty Effect: The Hidden Force Behind Growth, Profits, and Lasting Value*, Harvard Business Review Press, 1996.
12. Quoted in Karim B. Stambouli and Eric Briones, *Buzz Marketing: La Stratégie du bouche-à-oreille*, Éditions d'Organisation, 2002.
13. Seth Godin, *Unleashing the Ideavirus: Stop Marketing at People! Turn Your Ideas into Epidemics by Helping Your Customers Do the Marketing for You*, Dearborn Trade, 2001.
14. Stambouli and Briones, *Buzz Marketing*.
15. Gabriel Szapiro, *L'Inbound marketing selon la stratégie du Sherpa*, Jacques-Marie Laffont, 2015.
16. Lendrevie and Levy, *Mercator.*
17. Jeff Howe, "The Rise of Crowdsourcing," *Wired Magazine*, June 2006.

Points to remember

- The digital revolution, accompanied by the rapid development of mobile devices, is profoundly changing the behavior of the neuro-consumer's brain.
- The brain is gradually changing its individual conscience into a collective conscience.
- The Internet and mobile devices provide the neuro-consumer with access to a wide range of information and services and enable him to communicate with anyone anywhere on the planet instantaneously. These opportunities have led to a transformation in his behavior in many areas including education, entertainment, knowledge, purchases, easy access to a range of services, etc. However, this change is not without its dangers – dishonesty, fraud, addiction, mystification –his brain must adapt quickly.
- With interactivity, the former "neuro-consumer," is in the process of becoming a "neuro-consumer-actor."
- In order to cope with the new expectations of the neuro-consumer-actor's brain, companies are obliged to update their traditional marketing and communication policies. The emergence of e-marketing, m-marketing and e-communication has become a concrete reality. Companies are using new theories and methods that experts in the field have named "permission-based e-marketing," "desire marketing," "inbound marketing," etc.
- The establishment of a personalized relationship with each neuro-consumer has become a priority for many brands. The new technologies arising from "big data" and social CRM make this possible. The use of "one-to-one" marketing is developing in order to meet the relational needs and "self-centeredness" of the neuro-consumer-actor's brain.
- An individual spends a few milliseconds deciding whether or not to open a web page. Neuroscience techniques provide recommendations on how to facilitate this and encourage the consultation of sites. They are also contributing to reducing the bounce rate (failure to open a second page just after having consulted the first).
- One of the problems encountered by companies is the difficulty in giving a potential customer a sensory experience of the products presented online, other than through sight and hearing. Neuromarketing approaches are providing suggestions to compensate for this shortcoming, in order to appeal to all of the consumer's senses.
- Neuromarketing professionals are offering various solutions aimed at making the reading of websites "neuro-compatible" with the behavior of the neuro-consumer-actor's brain.

- The emergence of communities and social networks and the interactivity of objects with each other and with humans is transforming the perceptions of the neuro-consumer's brain.
- Since the appearance of the first social network, Facebook, created by Marc Zuckerberg in 2004, they have multiplied and grown exponentially all over the world.
- According to the seminal book, *The Cluetrain Manifesto*, "conversations have become markets." Companies' marketing and communication must adapt to the expectations of the brains of these new neuro-consumers who are linked to social networks. They must obey specific rules imposed by belonging to different communities as well as by the type of networks to which they subscribe.
- The use of viral marketing (or "buzz marketing") on the networks through online communities enables companies to significantly increase their sales. For this to work, it must be implemented professionally, in accordance with the rules of the particular communities and networks that the companies are targeting.
- In order to avoid significant errors due to a lack of understanding of this new behavioral ecosystem designed around online communities and social networks, it is essential that experts are consulted, be they in-house or external, which has resulted in the emergence of new professions, such as webmaster, community manager, social network officer, etc. Their role is to develop a digital ecosystem for the company, accompanied by strategies, actions, organization and in-house persuasion. All this must remain compatible with the expectations of the brains of neuro-consumer-actors grouped into networks.
- The establishment of a good digital and community-based policy implies the development of a coherent strategy for encircling internaut customers using all the means based on these techniques.
- Neuro-consumer-actors' preference for the social and collective is leading to the emergence of new types of communities related to the senses. This is already the case for music and taste. These new forms of association may be interesting research topics for neuromarketing and sensory marketing in the coming years.

CONCLUSION

A vision of the future

As the 21st century progresses, significant changes are taking place in consumer behavior. The brain is struggling to understand the vast and ever-increasing amount of information assailing it every day. It tries to identify what is essential among the huge number of messages reaching it from all over the globe. Submerged by the virtual, it has difficulty in separating reality from illusion. Inundated with evaluations made by the online communities and social networks to which it belongs, it struggles to form its own opinion. The neuro-consumer may sometimes feel that his brain is becoming increasingly collective, that recommendations by online communities take priority over his personal choice. His intelligence shows him that his rationality is sometimes more subject to the influence of his senses than to reason and that the subliminal effects from marketing, communication and commercial tricks used by companies can often condition his purchasing decisions.

For their part, companies, confronted with the profound changes in their customers' behavior and with the spectacular and rapid development of new information and communication technologies, are forced to reconsider the effectiveness and relevance of their marketing and communication methods. They must adapt to new types of marketing, e-communication and m-communication that addresses the brain directly through its interaction with the senses and adopt a new approach that includes collective thinking.

In this book, we have tried to anticipate these new concerns and challenges by looking at how the neuro-consumer's brain works, as well as its reactions to the subconscious influences to which it is permanently subjected. What will happen in the future?

The considerable amount of ongoing research into neuroscience and genetics can provide precious information about the behavior of the human brain and shed more light on the influence of features such as memory, emotion and desire on purchasing decisions as well as on the brain's perception of brands and communication. Greater knowledge of the role played by somatic markers and mirror neurons suggests that marketing should be adapted to take account of their existence and how they act on the brain. Knowledge of the brain's structure and of how it can differ with gender, age or personal circumstance, etc. is leading to new thinking on market segmentation in order to adapt to these specific characteristics.

Another major field of research relates to sensory experience. Whereas there is already much research into sight and hearing, more studies are being carried out to gain a better understanding of the influence of smell, touch and taste on purchasing. In order to create a customer experience to compete with the growth of online shopping, sales points are increasingly investing in what experts call "sensory marketing." This approach is also being used to improve the design of products and services and to try and make communication more relevant. With the online experience largely limited to appealing to the senses of sight and hearing, web developers must develop methods based on neuroscientific techniques to address this shortcoming, in order to appeal to all five of the senses.

Communication, e-communication and m-communication must acquire better knowledge of what pleases or displeases the brain of the consumer, who has, with the digital revolution become a "neuro-consumer-actor." The use of "nudges" will be increasingly based on evidence and thinking from neuroscientific analysis. At the same time, companies must understand that it is not enough to create individual relationships with neuro-consumers. A collective conscience is emerging, influenced by online communities and social networks, and they must adapt their communication and marketing strategies accordingly in order to survive.

Major reforms are under way in the methodology used for market research, the creation of messages used in communication and e-communication, sales techniques, etc. For several years most of the big multinationals such as Google, Facebook, Nike, Coca-Cola, Pepsi, McDonald's, Nespresso, Ikea and Accor have shown strong interest in the use of neuroscience and direct stimulation of the senses in their marketing, commercial, communication and brand policies. The Neuromarketing World Forum is becoming increasingly successful, attracting companies from all over the world in the majority of industries, such as Hershey's, Google, Unilever, Cartier, Fonterra, Johnson & Johnson, Tesco, Heineken, Nielsen and Ipsos.

In the future, neuroscientific methods and techniques will become essential tools for research in fields other than medicine, in particular psychology and sociology. As we have seen throughout this book, the number of applications in the disciplines of marketing, communication and sales is growing at an increasing rate. Analysis of the behavior of the neuro-consumer's brain is meeting with growing success. It complements traditional market research by looking at the attitudes and subconscious decision-making, often irrational, of human intelligence. When magnetic resonance imaging (MRI) and electroencephalography (EEG) are used to observe the effects on the brain of innovation, packaging, pricing and communication, significant improvements can be made.

The use of systems that address the senses directly are achieving noteworthy results. They make sales points more attractive, sales staff more effective and brands more relevant.

With the increase in online communities and social networks, understanding the gradual shift from an individual to a collective conscience has become increasingly essential for companies who want to attract generations Y and Z and should be the subject of much research over the next decade.

The speed with which the digital world is developing, its unprecedented influence on all generations, requires considerable thinking about how to anticipate the brain's reactions. In this environment of perpetual and rapid change, how will the neuro-consumer's brain behave? How will it react to the exponential growth in the quantity of information and the number of demands made upon it? The answers to these questions are as vital to the customers as to the companies that are trying to satisfy their needs with their products and services, their distribution and their communication.

These questions are also important for organizations and governments that must be able to understand how the brain acts and reacts in order to minimize the risk of their citizens becoming subject to subliminal manipulation.

In this book we have tried, on the basis of current scientific knowledge, to interpret the behavior of the consumer's brain. We have endeavored to debate the risk of manipulation in subliminal communication techniques that require the establishment of strong moral and ethical rules for companies that use them.

With a growing amount of research, our knowledge of the subconscious behavior of the brain and its fragility when faced with the risk of manipulation should considerably improve. It will be up to companies to take strong legal and ethical precautions when using such methods in order to improve their products, services and communication in order to satisfy their customers. More efficient than traditional tools, the new techniques derived from neuroscientific knowledge will not benefit companies unless they make a real contribution to improving consumer satisfaction and well-being, otherwise they run the risk of being denigrated by online communities and thereby detrimental to the company's brand image. Neuroscience can only be successfully used in marketing and communication if it is accompanied by a strong desire to satisfy customers' needs and expectations.

Our hope in writing this book is that it will inspire future research and lead to further improvement in our knowledge of the behavior of the marvelous tool that is the human brain, as well as the application of this knowledge to the fields of marketing, sales, communication, e-communication and m-communication. We hope that it will lead to that knowledge being developed, both for consumers and for companies, with the objective of further improving the close relationship that already exists between them.

SELECTED BIBLIOGRAPHY

This bibliography is intended as a source of further information for readers and is by no means a complete record of all works consulted.

Dan Ariely, *Predictably Irrational: The Hidden Forces That Shape Our Decisions*, Harper Perennial, 2010.

Richard Bandler and John Grinder, *Reframing: Neuro-Linguistic Programming and the Transformation of Meaning*, Real People Press, 1983.

Gregory Bateston, *Mind and Nature: A Necessary Unity*, Bantam Doubleday, 1988.

Darren Bridger, *Decoding the Irrational Consumer: How to Commission, Run and Generate Insights from Neuroscience Research*, Kogan Page, 2015.

Judith Butler, *Gender Trouble*, Routledge Kegan & Paul, 1990.

Jean-Pierre Changeux, *Neuronal Man*, Princeton University Press, 1997.

Patricia Churchland, *Braintrust: What Neuroscience Tells Us about Morality*, Princeton University Press, 2012.

Robert B. Cialdini, *Influence: The Psychology of Persuasion*, Harper Business, 2006.

Stanislas Dehaene, *Consciousness and the Brain: Deciphering How the Brain Codes Our Thoughts*, Penguin Random House, 2015.

Robert Dilts and Judith Delozier, *Encyclopedia of Systemic Programming and NLP New Coding*, NLP University Press, 2000.

Roger Dooley, *Brainfluence: 100 Ways to Persuade and Convince Consumers with Neuromarketing*, John Wiley & Sons, 2012.

David Eagleman, *Incognito: The Secret Lives of the Brain*, Vintage, 2012.

Robert East, Malcom Wright and Marc Vanhuele, *Consumer Behavior: Applications in Marketing*, Sage, 2008.

Paul Ekman, *The Nature of Emotion*, Oxford University Press, 2008.

Alberto Gallace and Charles Spence, *In Touch with the Future: The Sense of Touch from Cognitive Neuroscience to Virtual Reality*, Oxford University Press, 2014.

Patrick Georges, Anne-Sophie Bayle-Tourtoulou and Michel Badoc, *Neuromarketing in Action: How to Talk and Sell to the Brain*, Kogan Page, 2013.

Michael Gershon, *The Second Brain*, Harper Paperbacks, 1999.

Seth Godin, *Permission Marketing: Turning Strangers into Friends, and Friends into Customers* [1999], Simon & Schuster, 2007.

Seth Godin, *The Purple Cow: Transform Your Business by Being Remarkable*, Portofolio, 2009.

Daniel Goleman, *Emotional Intelligence*, Bantam Books, 1995.

Brian Halligan and Dharmesh Shah, *Inbound Marketing: Get Found Using Google, Social Media and Blogs*, John Wiley & Sons, 2009.

Brian Halligan and Dharmesh Shah, *Inbound Marketing Revised and Updated: Attract, Engage and Delight Customer Online*, John Wiley & Sons, 2014.

Dan Hill, *About Face: The Secrets of Emotionally Effective Advertising*, Kogan Page, 2010.

Bertil Hulten, Niklas Broweus and Marcus Van Dijk, *Sensory Marketing*, Palgrave Macmillan, 2009.

Jon Kabat-Zinn, *Wherever You Go, There You Are*, Hyperion, 1994.

Daniel Kahneman, *Thinking, Fast and Slow*, Penguin, 2012.

David Katz, *The World of Touch*, Lawrence Erlbaum Associates, 1989.

Chan Kim and Renée Mauborgne, *Blue Ocean Strategy*, Harvard Business School Press, 2005.

Bryan Kolb and Ian Q. Whishaw, *An Introduction to Brain and Behavior* (2nd ed.), Worth Publishing, 2006.

Aradhna Krishna, *Customer Sense: How the 5 Senses Influence Buying Behavior*, Palgrave Macmillan, 2013.

Aradhna Krishna (Ed.), *Sensory Marketing: Research on the Sensuality of Products*, Routledge, 2010.

Joseph Le Doux, *The Emotional Brain: The Mysterious Underpinnings of Emotional Life*, Touchstone, 1998.

Rick Levine, Christopher Locke, Doc Searls and David Weinberg, *The Cluetrain Manifesto: The End of Business as Usual*, Perseus Books, 2000.

Michael Lindstrom, *Brand Sense: Sensory Secrets Behind the Stuff We Buy*, Free Press 2010.

Martin Lindstrom, *Brandwashed: Tricks Companies Use to Manipulate Our Minds and Persuade Us to Buy*, Kogan Page, 2012.

Martin Lindstrom, *Buy.Ology: How Everything We Believe About What We Buy Is Wrong*, Random House Business, 2009.

Paul MacLean, *A Triune Concept of the Brain and Behavior*, Toronto University Press, 1974.

Read Montague, *Your Brian Is (Almost) Perfect: How We Make Decisions*, Plume Books, 2006.

Kate Newlin, *Passion Brands: Why Some Brands Are Just Gotta Have, Drive All Night For, Tell All Your Friends About*, Prometheus Books, 2009.

Donald Norman, *Emotional Design: Why We Love (or Hate) Everyday Things*, Basic Books, 2005.

David Ogilvy, *Confessions of an Advertising Man*, Southbank Publishing, 2011.

David Ogilvy, *Ogilvy on Advertising*, Crown Publishing, 1983.

Don Peppers and Martha Rogers, *Enterprise One-to-One: Tools for Competing in the Interactive Age*, Bantam Doubleday, 1997.

A.K. Pradeep, *The Buying Brain: Secrets for Selling to the Subconscious Mind*, Wiley, 2010.

Patrick Renvoisé and Christophe Morin, *Selling to the Old Brain*, SalesBrain, 2003.

Giacomo Rizzolatti and Corrado Sinigaglia, *Mirrors in the Brain: How Our Minds Share Actions, Emotions, and Experience*, Oxford University Press, 2008.

Eric Singler, *Nudge Management: Applying Behavioural Science to Boost Well-Being, Engagement and Performance at Work*, Pearson, 2018.

Michael R. Solomon, *Consumer Behavior: Buying, Having, and Being*, Pearson Education, 2014.

Richard H. Thaler and Cass R. Sunstein, *Nudge: Improving Decisions About Health, Wealth and Happiness*, Yale University Press, 2008.

John von Neumann, *The Computer and the Brain* [1958] (3rd ed.), Yale University Press, 2012.

Semir Zeki, *Inner Vision: An Exploration of Art and the Brain*, Oxford University Press, 1999.

Semir Zeki, *Splendors and Misery of the Brain: Love, Creativity and the Quest of Human Happiness*, Wiley Blackwell, 2008.

Leon Zurawicki, *Neuromarketing: Exploring the Brain of the Consumer*, Springer, 2010.

INDEX

Locators in *italics* refer to figures.

"4D's method" 206–210

Abercrombie & Fitch 97, 123, 128, 150, 190, 266
addiction 21, 275
adolescent brain 79–80
"advertainment" 234
"advertgames" 300
advertising: advent of the neuro-consumer 20–21; of desire 103–107, 108–110; of emotions 103–107, 108; hearing 126–128; inbound 108–110; post-testing 224–225; pre-testing 223–224; smell 136–141; touch in 150–151, 152; traditional studies and their limits 18–20; understanding the consumer confronted with advertising 225–228; use of neuroscientific research 1–2; visual 120–121; *see also* digital marketing; language in advertising; neuromarketing
aesthetics *see* esthetics; sight
affects (emotions) 12
age: adolescence 79–80; aging of the brain and its consequences among older people 80–81; digital environment 81–82; generational differences 81–84; young children 78–79
Air Berger 139
Alba, Joseph 123
Alcmaeon of Croton 23–24
Alzheimer, Alois 26
amateur blogs 290–291
amygdala 102, 105
ancient world 10–11, 23–24, 215–216
Aphrodite (Venus) 252
Apple 274
appropriation effect 152
architecture 235
Arfi, Stéphane 139–140
Ariely, Dan 182, 241
Aristotle 10, 23–24, 144, 216, 217, 234
art: digital images 153; "golden number" 235–236; sales outlet staging 191–194; subliminal communication 234–235
Asperger, Hans 26
ATAWAD (anytime, anywhere, any device) 273
attention: images of products 245; sense of sight 118
attention graph 208–209, *209*
auction websites 277
audio advertising 126–127; *see also* hearing
auditory system 125–126
augmented reality 292
autism 26
automatic reflexes, the brain 67–73
automotive industry 174

babies in advertising 89, 245
Babiloni, Fabio 33
"baby boomer" generation 83
Bach-y-Rita, George 61
Bach-y-Rita, Paul 61, 68
background music 128–129, 130, 197–198
Bacon, Francis 218
bacteria, digestive tract as a second brain 62–63
Badoc, Michel 88, 92, 242
bait: desire-based e-marketing 281; permission e-marketing 279–280
Bard, Arthur 93
Bard, Mitchell 93
Bardainne, Claire 153
Barthes, Roland 218, 220–221
Bateson, Gregory 27
Bayle-Tourtoulou 88, 92, 242
beauty: "golden number" 235–236; subliminal role of eroticism in the perception of messages 236–238
beginnings/ends 74
behavior *see* consumer behavior

behavioral economics 241
Berger, Hans 42
Bernard, Claude 75
Bernbach, William 295
Berne, Eric 41
Berthoz, Alain 103
big data 5: media revolution 271; personalized consumer relationships 281–282; Web 3.0 292–293
"bioesthetics" 172
biology *see* the brain
the "bipartite" brain 71–73
Blackwell, Roger D. 16
blind taste tests 29–30, 161
Bloch, Felix 43
blogs 290–291, 300
Boucicaut, Aristide 191–192
boxes, online retail 285
the brain 1; adapting the Internet to the brain's perception 278–282; addressing the subconscious brain 231–239; age 78–84; antiquity to the Middle Ages 23–24; automatic reflexes 67–73; cognitive ergonomics 4; computer or orchestral conductor? 61; desire 106–107; digestive tract as a second brain 61–63; digital revolution 5, 272–273; emotions 100–101, 103–106; as free or programmed? 64; gender 78, 84–89; instinctive intelligence 71–72, 73–76; knowledge about brain behavior 33–34; memory's role 92; Middle Ages to modern times 24–25; mirror neurons 95–98; modern times and the advent of neuroscience 25–27; scientific research 3; self-centeredness of the brain 207, 208, 209–210, 281, 286; shrewd intelligence 72–73, 74; social networks' influence on 293–299; somatic markers 94–95; structure and functions 57–58, *59–60*, 65–66; subliminal relationships 4; "triune" brain 64–66, *65*; understanding how it works 55–61; understanding the consumer confronted with advertising 225–228; a vision of the future 304–306; ways of subconsciously influencing the brain 241–246; weight of 57; *see also* neuroscientists; perception of the world
brain activity map (BAM) 33
brain manipulation 34, 36–37
brand extension 261–263
brand legitimacy 261–263
brand mythology 251–256
brand positioning 254–255
"brand-identity prism" 256–259
brands: and blind taste tests 29–30, 161; jingles 126–128; logos and designs 120; manipulation and ethics 36–37; packaging 177; pleasure 104; somatic markers 95, 264–267; subconscious decision-making 214; *see also* subliminal influence of brands
Breuer, Joseph 12, 67
Broca, Paul 25
Broca area 25
Burgelin, Olivier 221
Bush, George W. 232
Butler, Judith 85
Buyssens, Eric 221
buzz marketing 110, 295

Calvert, Gemma 31, 34
Carey, Benedict 183–184
Cato the Edler 256
celebrities 227
Central Intelligence Agency (CIA) 232
cerebral imaging 51
cerebral plasticity, and music 131
Chabris, Christopher 118
Changeux, Jean-Pierre 12–13, 85
character facet, "brand-identity prism" 257
characters in advertising 243
Charlier, Pascal 139
Chétochine, Georges 108
Childers, Terry 148, 152
children, reptilian brain 78–79; *see also* age
China, the Internet revolution 274–275
choices: cognitive dissonance 105–106; packaging presentation 178; *see also* purchasing decisions
Christianity 9–11, 24
Churchland, Patricia 9, 12
Cialdini, Robert 39, 123
Cicero 217, 218
Citrin, Varma 148
Classical philosophy 10–11
client segmentation 16
Coca-Cola taste test 29–30, 161
cognitive bias 19
cognitive dissonance 105–106, 246
cognitive ergonomics 4
cognitive sciences 13
cognitive specialists 49

Collins, Stephen M. 63
colors: goods and services sales outlets 122; influence on taste perception 123; packaging 176; sense of sight 115–116; sensory brands 265; sensory marketing 195–196; somatic markers 94–95; and taste 158–159, 174
communication: from the brain to the body 58–59; decoding communication signs 220–222; influence on the brain 213–214; research in communication and subconscious perceptions 222–225; sales training programs 208–209; social networks 293–299; subliminal relationships 4; use of neuroscientific research 1–2; a vision of the future 305; *see also* language in advertising
communities, online 290–291, 294, 300
competition: brand mythology 254–255; "brand-identity prism" 256–259
computer, brain as 61
computer-assisted axial tomography (CT) 42
congruence: background music 130; brands 266–267; sensory marketing 202–203; smell 200; videos 245–246
conscience 60
consumer behavior: advent of the neuro-consumer 20–21; client segmentation 16; emergence of neuromarketing 29–32; the Internet revolution 273–278; interview-based studies 14–15; market research 14–16; observation-based research 15; pursuit of happiness 9–11; sensory marketing 190; "tightwads" and "spendthrifts" 186–187; traditional studies and their limits 18–20; understanding the consumer confronted with advertising 225–228; a vision of the future 304–306; *see also* purchasing decisions
consumers 6; *see also* neuro-consumers
"contaminators" (online communication) 296
contrast 73, 207
cortex 59
creative context 228; *see also* innovation
cross-channel retail 277, 283, 299–300
crowdfunding 276–277
crowdsourcing 298
cultural facet, "brand-identity prism" 259
customer relationship management (CRM) 5; emotional frustration 108; observation-based research 15; permission e-marketing 280; personalized consumer relationships 281–282; policy for encircling a customer using a digital strategy 300; social CRM 5, 291
customer satisfaction 109, 188–189, 260–261
cutaneous system 145
cybernetics 27–28

Damasio, Antonio 9, 11–12, 25, 94, 99
Darwin, Charles 69–70
data: big data 5, 271; personalized consumer relationships 281–282; "semantic web" 273; Web 3.0 292–293
de Montaigne, Michel 11
de Saint- Exupéry, Antoine 101
de Saussure, Ferdinand 220
decision-making: advent of the neuro-consumer 20–21; background music 128–129; emotions and desires 103–107; experience-based and sensory marketing 167–168; influence of the Internet 276–278; instinctive vs. shrewd intelligence 71–73; nudge marketing 241; pain of purchase 182; rationality vs. irrationality 1–2; six stimuli of 73–74; social networks 288, 293–299; subliminal effects 213–214; touch 145–147, 150–151
"Deep Blue" computer 61, 273
Dehaene Stanislas 60
Delattre, Nicolas 48
Democritus 24
deontological rules 35–37
Descartes, Rene 11, 24
descriptions of products 152, 242–244
design: appealing to our sense of sight 119–120; innovation 172–174; sensory brands 265; traditional studies and their limits 20
desire: influence on the behavior of the brain 106–107; marketing of 108–110; philosophical theory 12
desire-based e-marketing 280–282
Desmet, Pieter 145
differentiation of company 207
digestive tract as a second brain 61–63
digital images 153
digital immigrants 81–82

digital marketing: defining 272; desire-based e-marketing 280–282; the Internet and social network professionals 296–299; permission e-marketing 278–280; policy for encircling a customer using a digital strategy 299–300; social networks' influence on purchasing decisions 293–299; viral marketing 295–296; a vision of the future 304–306; *see also* online retail
digital natives 81–82, 84
digital revolution 5, 271; adapting the Internet to the brain's perception 278–282; changes to the brain's perception 272–273; the Internet and the neuro-consumer's behavior 273–278; websites 282–286
Dionysus (Bacchus) 252–253
"divine ratio" 235
Dooley, Roger 56, 88, 183, 244, 245
double interaction of desire (DIAD) 108–109
Droulers, Olivier 170, 172, 224–225
Durand, Jacques 220

Eagleman, David 68, 69, 106–107, 116–117
ear, auditory system 125–126; *see also* hearing
education, neuroscientific courses 49–51
efficiency, advertising communications 227
egoism 73
Einstein, Albert 25, 57
Ekman, Paul 99
electoral campaigns 232, 239–240
electrocardiogram (ECG) 47
electrodermal activity (EDA) 47
electroencephalography (EEG): brand essence 262; language in advertising 225–226; as neuroscientific technique 42–43, 45; packaging 175–176
electromyography (EMG) 175–176, 225–226
electrophysiology, taste receptors 156–157
e-marketing *see* digital marketing
Emosense 139–140
emotional frustration 108
emotional intelligence 101–103
emotional memory 103–104
emotions: advertising communications 227–228; brands 259–260, 261, 262, 267; defining 99; gender and the brain 88; generational differences 81–82; influence on the behavior of the brain 100–101, 103–106; language in advertising 219–220, 224–225; marketing of 108; mirror neurons 96–97; philosophical theory 12; role of 99–101; sales training programs 208–209; six stimuli of decision-making 74
empathy 96–97
endowment effect 146–147, 184, 283, 284
ends/beginnings 74
energy efficiency, the brain 71, 72–73
Engel, James F. 16
Engen, Trygg 135
"engraving", the brain 71
enteric nervous system 62
entertainment in advertising 233–234
Epicurus 10–11
equilibrium (homeostasis) 9, 75–76
e-retail *see* online retail
ergonomics 202
eroticism: does sex sell? 238–239; perceptions 236–238
esthetics: "bioesthetics" 172; the brain 170–171; gustatory marketing 160; innovation 172–173; *see also* sight
ethical rules, neuroscientific research 35–37
ethos 218–219
European Association for Transactional Analysis (EATA) 42
experiential marketing: customers spending more time in sales outlets 188–189; purchasing decisions 167–168; vs. sensory marketing 189–190
exposure: mere-exposure effect 69; visual marketing 121
eye tracking: babies in advertising 245; gender differences 238; as neuroscientific technique 47–48; visual marketing 120–121

Facebook 254, 272, 289–290, 295, 297
Facebook M 292
facial electromyography (fEMG) 47
fear: marketing 101; role in advertising 105–106
female brain 85, 86–87, 88, 89
Field, Peter 246

Follett, Ken 107
food: blind taste tests 29–30, 161; menu descriptions 243–244; smell in sensory marketing 199; *see also* taste
fragrances *see* smell
Freud, Sigmund: automatic reflexes 67; social rules 101–102; subliminal role of eroticism in the perception of messages 237; the unconscious 12
frontal lobes 66
functional imagery techniques 28; *see also* electroencephalography; magnetic resonance imaging (MRI)
functional magnetic resonance imaging (fMRI): the brain's structure 59–60; brand essence 262; as neuroscientific technique 43–44

Gage, Phineas 100–101
Galen, Claudius 24
Gardner, Howard 101
gender: the brain 78, 84–89; erotic messages 238; online retail 284; the senses 86–87, 135; tactile marketing 151
Generation X 83–84
Generation Y 84
Generation Z 84
generational context: age and generational cohorts 81–84; the Internet revolution 276
geo-location 277
Georges, Patrick: intelligence 72, 74; mirror neurons 96, 97; neuromarketing 1; sales training programs 205; sensory policy 49; written communication 242–243
Gershon, Michael 62
Ghyka, Matila Costiesco 235
Ginger, Serge 85, 86–87
Giono, Jean 233
Go game 273
Godin, Seth 278–279
"golden number" 235–236
Goleman, Daniel 101
Google Glass 292
Google Now 292
gratuity economy 276
Greek mythology 251–253
Greek philosophy 10–11, 215–218
Greek science 23–24
Grinder, John 40, 41
grouping products 185–186
gustatory marketing 160–161

Halligan, Brian 280
"halo effect" 123
happiness, philosophical theory 9–11, 12
"happy hormone" 160
haptic properties 148–149, 152
haptic technology 152–153
Harley Davidson club 291
Harvey, Thomas Stoltz 57
hearing: in advertising 126–128; auditory system 125–126; background music 128–129; congruence of sound with the target and objective sought 130; gender 86; links with taste 131; sensory brands 265; sensory marketing 197–198; significance of sound 125; sound as an indicator of product quality 130–131; and taste 159–160
Hegel, Georges W.F. 234
Heidegger, Martin 234
Herbert, Wray 246
Hershey's 266
Herz, Rachel 135, 136, 138
Hippocrates 23–24, 220
historical context: the brain 23–25; philosophical theory 11–13; subconscious 67
Hoegg, Jo Andrea 123
Holley, André 13
homeostasis 9, 75–76
Honsfield, Godfrey 42
hormonal secretion: the brain 60–61; as neuroscientific technique 46–47
hormones: adolescence 79–80; gender and the brain 86, 87; motherhood 89; seratonin 160
Howe, Neil 82
human brain *see* the brain
Human Brain Project (HBP) 33, 61

IBM "Deep Blue" computer 61, 273
iconography 176
identity, "brand-identity prism" 256–259
illusion 117, 119
images of products: as compensation for being unable to touch 152; packaging 176; ways of subconsciously influencing the brain 244–245; *see also* sight
imaging *see* cerebral imaging; magnetic resonance imaging
implementation reaction time (IRT) program 225–226

implication 34, 36–37
impulsivity: adolescence 80; gender and the brain 88–89
inbound marketing 108–110
individual factors, need to touch 147–148
infants in advertising 89, 245
influence, research on 39
"informatik" 273
information: big data 5, 271; digital revolution 5, 271; influence of the Internet 278; permission e-marketing 278–280; policy for encircling a customer using a digital strategy 299–300
information processing, innovations in 51
innovation 167; the brain 170–172; design 172–174; packaging 174–178; product development and launch 169
instinctive intelligence 71–72, 73–76
insular cortex 170
intelligence: emotional 101–103; instinctive 71–72, 73–76; shrewd 72–73, 74; traps of 74–75
interactivity in communication 290–291, 294, 296
interactivity of objects 273, 292–293
Internet: adapting the Internet to the brain's perception 278–282; digital revolution's influence on the brain 272–273; generational differences 81–82; influence on the neuro-consumer's behavior 273–278; neuro-compatibility of websites 282–286
Internet of things (IOT) 273, 292–293
interpersonal relations, touch 149–150
interruption marketing 110
interview-based studies, consumer behavior 14–15
IQ 101
irrationality in purchasing decisions 2

Jacobson's organ 87
Jacquet, Florent 184–185
James, William 118
Jeannerod, Marc 13
jingles 126–128
Judeo-Christian religions 9–11
Jung, Carl Gustav 12

Kahneman, Daniel 72–73, 105
Kandinski, Vassily 234
Kant, Immanuel 10, 234
Kapferer, Jean-Noël 250
Kawabata, Hiro 170–171
Kim, Chan 110, 170
Kimura, Doreen 86
kinesthesis 144
Klatzky, Robert L. 145
Kleinschmidt, Andréas 61
Knutson, Brian 31, 182
Koestler, Arthur 66
Kolb, Bryan 55–56, 99
Krishna, Aradhna 136, 137–138, 149

Laborit, Henri 26
Laird, D. A. 137
language in advertising 215; philosophical theory 215–220; research in communication and subconscious perceptions 222–225; rhetoric 216–220; semiology and decoding communication signs 220–222; understanding the consumer confronted with advertising 225–228; ways of subconsciously influencing the brain 242–244
Largactil 26
last-minute purchase proposals 277
launch *see* product launch
Lauterbur, Paul 43
Lavazza 256, 257, 258, 259
layout of space: sensory marketing 194–195; superstores 191
Le Bihan, Denis 43, 44
Le Bon Marché 191–192
Le Doux, Joseph 102
Lecerf-Thomas, Bernagette 75
Leibniz, Gottfried Wilhelm 67
Lenoir, Frédéric 9
lesbian, gay, bisexual and transgender (LGBT) movements 85
Lewi, Georges 250–253
Libermann, Matthew 289
libido 237
lighting, sensory marketing 196
Likert scale 295
limbic brain 64–65, 79–80
limbic system 66
Limoges, Margaux 283–284
Lindstrom, Martin 18–19, 20–21, 31, 146, 239, 259–261
linguistics, neuro-linguistic programming 40
Loewenstein, Georges 182
Loftus, Elizabeth 34

logos (rhetoric) 218–219, 221
logos: appealing to our sense of sight 119–120; sensory brands 264
long-term memory 93
L'Oréal 127, 254, 266
luxury: brand perception 261, 262–263; price strategies 180, 184

MacLean, Paul D. 64–66, 69–70, 78
magnetic resonance imaging (MRI): automatic reflexes 70–71; ban on certain experiments 35; blind taste tests 30; brands 260–261; gender and the brain 85; as neuroscientific technique 42–45; social networks 289; ways of subconsciously influencing the brain 242
male brain 85, 86–87, 88
mammalian brain 64–65, 79–80
manipulation 34, 36–37
market research: consumer behavior 14–16; a vision of the future 305
marketers: aims of this book 6; understanding neuro-consumers 7–8
marketing: advent of the neuro-consumer 20–21; of desire 108–110; of emotions 108; emotions and desires 103–107; hearing 126–128; inbound 108–110; post-testing 224–225; pre-testing 223–224; smell 136–141; touch in 150–151, 152; traditional studies and their limits 18–20; understanding the consumer confronted with advertising 225–228; use of neuroscientific research 1–2; visual 120–121; *see also* digital marketing; language in advertising; neuromarketing
marketing surveys 224–225
materials, sensory marketing 196
Mauborgne, Renée 110
McCulloch, Warren 27
McDonald's 79, 127
memory: advertising communications 227–228; emotions 102; function of 92–94; neuroscientific research 33–34; philosophical theory 12; role in brain behavior 92; sight 116–117, 118; and smell 135–136, 137–138; somatic markers 94–95; subconscious influence 243; subliminal effects 246; types 93, *93*
mentalization facet, "brand-identity prism" 258
mere-exposure effect 69
microbiota 63
Microsoft 274, 279
Middle Ages 24–25
Mill, Johns Stuart 218
Milner, Peter 104
mindfulness 107
Miniard, Paul W. 16
mirror neurons 3, 95–98, 227
mistake-making, cognitive dissonance 105–106
misuse, ethical rules 35–36
mobile phones 276
Montague, Read 3, 8, 27, 29, 31, 54, 161, 260
"Moore's Law" 273
Morin, Christopher 30, 31, 66, 208
motherhood 89
Mouton, Jean 108–109
Müller, Johannes Peter 67
multi-channel retail 277, 283, 299–300
"multiprogrammed neuro-consumers" 276
music: background 128–129, 197–198; congruence 130; influence on cognitive and motor deficiencies 131; jingles 126–128; sensory marketing 197–198
mythology of brands 251–256

narratives in brands 255
"Need for Touch" (NfT) scale 147–148
negative messages 21
neocortex 66, 80, 246
neo-mammalian brain 66
nervous system: automatic reflexes 67; from the brain to the body 58–59; digestive tract as a second brain 62
Nespresso 266
net promoter scoring 295
neuro-consumers: advent of 7–8, 20–21; meaning of term 2; protection from neuromarketing 35–37
the neuro-consumer's brain *see* the brain
neuroendocrinology: the brain 60–61; as neuroscientific technique 46–47
"neuroesthetics" 172
NeuroFocus 36, 45, 173, 225–226
neuro-linguistic programming (NLP) 40–41
neurological iconic signature (NIS) t 173

neuro-marketing: consumer protection and misuse 35–37; emergence of 29–32; origins 3, 29
neuro-marketing companies, emergence of 31–32
Neuromarketing World Forum 305
neurons: communication from the brain to the body 58–59; digestive tract as a second brain 62; mirror neurons 95–98; number in the brain 58
neuro-philosophy 12
neuroscience, advent of 25–27; *see also* the brain
neuroscientific courses 49–51
neuroscientific diagnosis 49
neuroscientific research: the brain 3; deontological and ethical rules 35–37; knowledge about brain behavior 33–34; recent trends 1
neuroscientific techniques 7–8, 39; courses on 49–51; diagnosis and the emergence of new professions 49; electroencephalography 42–43, 45; hormonal secretion 46–47; magnetic resonance imaging 42–45; neuro-linguistic programming 40–41; peripheral tools 47–49; transactional analysis 41–42
neuroscientists: the brain 2, 3; historical perspective 2; knowledge about brain behavior 33–34; philosophical underpinnings 11–13; pursuit of happiness and consumer behavior 9–11
NeuroSense 31, 32, 173
NeuroSpin 33, 43
new information and communications technologies (NICTs) 274
newness 170; *see also* innovation
nuclear magnetic resonance (NMR) 43; *see also* magnetic resonance imaging (MRI)
nudge marketing 239–241
nudity 237–238

observation-based research 15
odor perception 134; *see also* smell
Ogilvy, David 18
older people: aging of the brain and its consequences among older people 80–81; digital immigrants 81–82; generational differences 82–83
Olds, James 104
olfactory cells 133
olfactory marketing 139–141, 198–200; *see also* smell
omni-channel retail 277, 283, 299–300
"one-to-one" customer relationships 282
online communities 290–291, 294, 300; *see also* social networks
online retail: appealing to touch/smell/taste 151–153, 282–285; influence of the Internet 276–278; neuro-compatibility of websites 282–286; personalized consumer relationships 281–282, 285; *see also* digital marketing
optical illusions 119
orchestral conductor, brain as 61

packaging 174–178
Packard, Vance 232–233
pain of purchase 182, 185, 207
paleo-mammalian brain 64–65
Palo Alto school 27–28
participatory economics of purchasing 276–277
Pascal, Blaise 67, 233
Pasteur, Louis 198
pathos 218–219
Peck, Joann 147–148, 152
Peninou, Georges 221–222
Pepsi taste test 29–30, 161
perceptions: adapting the Internet to the brain's perception 278–282; background music 128–129; the brain and objective reality 68–69; the brain as free or programmed 69–70; brand mythology 254–255; brands 259–263; digital revolution's influence on the brain 272–273; eroticism 236–238; objective reality or simple construction by the brain? 68–69; price 181–182; sight 116–117, 119, 121–123; smell 134–135, 138; traps of intelligence 74
Perelman, Chaïm 218
perfume: history of 198; price strategies 180
perfume diffusers 140
perfume in products 87, 137–138
peripheral vision 119
permission e-marketing 278–280
Perrier 257, 258, 259
personality facet, "brand-identity prism" 257

personalized consumer relationships 281–282, 285
personnel *see* staff
persuasion, and rhetoric 218–220
Pfaffman, Carl 156
philosophical theory 9; basis in neuroscience 11–13; language in advertising 215–220
photographs of people 245
photographs of products 152
photoreceptors 115
physical facet, "brand-identity prism" 257
Piaget, Jean 78
Pieters, Rick 120–121
Plassmann, Hilke 29, 183
Plato 10, 12, 216, 217
play, young children 79
pleasure: advertainment 234; ancient culture 10–11, 12; the brain's pleasure center 60, 104; brands 260; desire 12, 46; Freudian perspectives 237; influence of desire 106–107; purchasing a product 182, 185; taste 161; touching products 284
political advertising 232, 239–240
Ponzo, Mario 119
Poseidon (Neptune) 252
positioning, sensory marketing 197, 202
positron emission tomography (PET) 42–43, 227–228
possession effect 283
postmodernity 216
post-testing 224
Pradeep, A. K.: aging of the brain and its consequences among older people 81; gender and the brain 88; images of products 245; innovation 171–172; language in advertising 225–226; manipulation and ethics 36; mirror neurons 96; neuroscientific research 19; neuroscientific techniques 43, 45
Préel, Bernard 82
prefrontal cortex 30, 34, 59–60, 80, 92
Prensky, Marc 81–82
preprogramming of the brain 70
pre-testing 223–224
price: importance of 180–181; the neuro-consumer confronted with pricing policies 185–187; neuroscience's conception of 182–184; psychological price perceived by the customer 181–182; subconscious mental strategies 184–185; "tightwads" and "spendthrifts" 186–187
price comparisons 276
Pringle, Hamish 246
product development 169
product launch 169, 263
product quality: need to touch 148; sight as an influence on perception 121–123; sound as an indicator of 130–131
professionals: aims of this book 6; the emergence of new neuroscience professions 49, 51; the Internet and social networks 296–299; "spin doctors" 48, 49
programming: of the brain 69–70; neuro-linguistic programming 40
proprioception 144
Protagoras 216
Proust, Marcel 12, 94, 135, 137
Purcell, Edward Mills 43
purchasing decisions: advent of the neuro-consumer 20–21; background music 128–129; emotions and desires 103–107; experience-based and sensory marketing 167–168; influence of the Internet 276–278; instinctive vs. shrewd intelligence 71–73; nudge marketing 241; pain of purchase 182; rationality vs. irrationality 1–2; six stimuli of decision-making 73–74; social networks 288, 293–299; subliminal effects 213–214; touch 145–147, 150–151

qualifications: courses 49–51; emergence of new professionals 49
queer theory 84–85
Quintilian 217

radio, audio advertising 126–127
Ramachandran, Vilayanur 95–96
rationality: emotional intelligence 101; purchasing decisions 1–2
reaction time 71
reading websites 285–286
reality 68–69; *see also* perception of the world
recommendations: gender and the brain 88–89; word of mouth marketing 295

reflection facet, "brand-identity prism" 257–258
reflexes, the brain 67–73
Reichheld, Frederick 295
relationship facet, "brand-identity prism" 258–259; *see also* interpersonal relations, touch
religion 9–11
Renvoisé, Patrick 31, 66, 208
repetition: brand mythology 256; mere-exposure effect 69
reptilian brain 64–66; sales training programs 206–210; self-centeredness 286; young children 78–79
retail: background music 128–129; influence of the Internet 276–278; neuromarketing in 32; sensory and experience-based marketing 188–189, 194–203; sight as an influence on perception 121–123; smell 137; staging 191–194; themes 190–191; touch in purchasing decisions 145–147, 150–151; *see also* online retail
retention graph 208–209, *209*
retention of messages 246
rhetoric in modern times 216–220
risk taking, adolescence 80
Rizzolatti, Giacomo 95
Roman thought 24, 217, 256
Roullet, Bernard 170, 172, 224–225
Rousseau, Jean-Jacques 197–198
Roy, Charles Smart 42
Rubin's vase 117

sales departments: key factors 205; training programs 205–210
sales outlets *see* retail
"Sales Point" method 206
Scentair 139
Schifferstein, Hendrik 145
Schopenhauer, Arthur 11, 218
Schwartz, Barry 105–106
scientific research *see* neuroscientific research
secrets in brands 255
segmentation of market 16
Séguéla, Jacques 258
self-centeredness of the brain 207, 208, 209–210, 281, 286
Selling to the Old Brain method 206–210
"semantic web" 273
semiology and decoding communication signs 220–222
semiotics 220
"seniors" generation 82–83
the senses 3–4; and brands 264–267; congruence 162; gender 86–87, 135; marketing 162–163; perception of the world 68–69; power of 113–114; saturation 162; *see also* hearing; sight; smell; taste; touch
sensory brands 264–267
sensory marketing 162–163; customers spending more time in sales outlets 188–189; vs. experiential marketing 189–194; origins 188; purchasing decisions 167–168; in sales outlets 188–189, 194–203; a vision of the future 305; website context 282–285
sensory policy 49, *50*, 202, 266
seratonin 160
Servan-Schreiber, David 80–81
services sector: appealing to our sense of sight 121–123; staging 192–194
sex in marketing: does sex sell? 238–239; male brain 88; subliminal role of eroticism in the perception of messages 236–238
sexual desire: the brain 70; subconscious signals 107
Shah, Dharmesh 280
shape: innovation 171–172; packaging 177
Sherrington, Charles Scott, Sir 42
shops *see* retail
short-term memory 93
shrewd intelligence 72–73, 74
sight: brain's interpretations 116–119; eye-tracking research 120–121; goods and services sales outlets 121–123; importance of 115; logos and designs 119–120; as mechanism 115–116; packaging 174–176; sensory brands 264–265; sensory marketing 194–197; and taste 158–159, 160–161
signs in advertising 221–222
Simon-Kucher & Partners 180
Simons, Daniel 118
Singer, Gary 67
Singler, Eric 48, 175, 241
Sinigaglia, Corrado 95
Sixième Son 127
size of product/packaging 177
smart phones 276
Smartbox 285
smell: gender 86–87; importance of 133; link to the memory 135–136,

137–138; as mechanism 133–134; olfactory marketing and olfactory techniques 139–141; perception thresholds 134–135; research into marketing 136–138; sensory brands 265; sensory marketing 198–200; and taste 157–158; and touch 149; website context 282–285
Snowden, Edward 275
social conformism 19
social customer relationship management (CRM) 5, 291
social networks: emergence of 288–290; generational differences 81; interactivity of objects 292–293; online community networks 290–291; origins 272–273; policy for encircling a customer using a digital strategy 299–300; professionals 296–299; purchasing decisions 288, 293–299
social rules 101–102
Socrates 10, 215–216, 217
somatic markers: brands 95, 264–267; influence on the behavior of the brain 94–95; neuroscientists 3; philosophical theory 12
"sonic branding" 126
sophism 216–217
sound *see* hearing
"spendthrifts" 186–187
"spin doctors" 48, 49
Spinoza, Baruch (later known as Benedict de Spinoza) 11, 12, 107
Squire, Larry R. 171–172
staff: sales training programs 205–210; visual appearance 122–123
staging in retail 191–194
Steinbuch, Karl 273
Stern, William 101
"Story" method 206
storytelling in brands 255
Strauss, William 82
stress: homeostasis 75–76; instinctive intelligence 71–72; "triune" brain 64
subconscious: addressing the subconscious brain 231–239; automatic reflexes 67; the brain 56; communication 213; mastering desires 107; pain of purchase 207; price strategies 184–185; purchasing decisions 2; research in communication and subconscious perceptions 222–225; research trends 306; the senses 3–4; ways of subconsciously influencing the brain 241–246
subliminal communication 4; addressing the subconscious brain 231–239; art 234–235; ban on 232; defining 231; effectiveness 232–233; eroticism 236–238; measuring influence 213–214; nudges 239–241
subliminal influence of brands 4, 214, 250–251; "brand-identity prism" 256–259; the neuro-consumer and brand mythology 251–256; perception of brands 259–263; sensory brands 264–267
Sunstein, Cass R. 239–241
superstores 191
surprise effect 281
Szapiro, Gabriel 109–110, 280

tactile marketing 150–151, 283–284; *see also* touch
tangibility 74
taste: blind taste tests 29–30; and color 158–159, 174; gustatory marketing 160–161; and hearing 159–160; influence of color 123; links with sound 131; receptors on the tongue 156–157; sensory brands 265–266; sensory marketing 201; seratonin 160; and sight 158–159, 160–161; and smell 157–158; and touch 159; website context 282–285
technology: background music 198; digital revolution 5, 271; the Internet revolution 272–274; tactile information 152–153
Technology, Media and Telecommunications (TMT) Predictions 32
telemetry 48–49
testosterone 87
text on packaging 176–177
thalamus 135–136
Thaler, Richard 146–147, 239–241
themes, sales outlets 190–191
"tightwads" 186–187
tongue, taste receptors 156–157
"Top-Ten Sales" method 206
touch: appropriation effect 152; gender 87; how touch works and types of 144–145; importance of 144; individual differences in the need to touch 147–148; interaction

with other senses 149; interpersonal relations 149–150; product differences in the need to touch 148; purchasing decisions 145–147, 150–151; sensory brands 265; sensory marketing 201–202; tactile marketing 150–151; and taste 159; website context and being unable to touch 151–153, 282–285
touch-sensitive receptors 145
training programs, sales departments 205–210
transactional analysis (TA) 41–42
Trevisan, Enrico 184–185
"triune" brain 64–66, *65*
trust, language in advertising 219
Turing, Alan 61, 273
Tversky, Amos 105

Uberization 276–277
unique selling proposition (USP) 222, 257

value chains 297
value for money 183
van der Linden, Martial 33
verbalization 19
Verma, Ragini 85
Vésale, André 24
Vicary, James MacDonald 232
videos, ways of subconsciously influencing the brain 245–246
viral marketing 110, 295–296
vision 116–117; *see also* sight
visual marketing 120–121, 152
visualization, six stimuli of decision-making 74
"Vitruvian proportions" 235–236
von Clef, Julia 138
Von Goethe, Johann Wolfgang 11
von Helmholtz, Hermann Ludwig Ferdinand 68
von Neumann, John 27

Wansink, Brian 243–244
Web 1.0 272
Web 2.0 273, 289–290
Web 3.0 273, 292–293
Weber, Max 16
websites, neuro-compatibility 282–286
Wedel, Michel 120–121
Weihenmayer, Erik 68
Whishaw, Ian Q. 55–56, 99
Willis, Thomas 25
word of mouth marketing 295
working memory 93
World Wide Web (WWW) 272; *see also* Internet
written communication: descriptions of products 152; ways of subconsciously influencing the brain 242–244
written descriptions of products 152

Xenophon 216

young children, reptilian brain 78–79; *see also* age

Zajonc, Robert 256
Zeus (Jupiter) 252
zoning, sensory marketing 190
Zuckerberg, Marc 272, 289–290

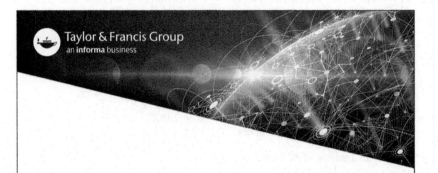